Springer Texts in Statistics

Advisors:
George Casella   Stephen Fienberg   Ingram Olkin

Springer
New York
Berlin
Heidelberg
Hong Kong
London
Milan
Paris
Tokyo

Springer Texts in Statistics

(continued after index)

Kenneth Lange

Applied Probability

Springer

Kenneth Lange
Department of Biomathematics
UCLA School of Medicine
Los Angeles, CA 90095-1766
USA
klange@ucla.edu

Library of Congress Cataloging-in-Publication Data
Lange, Kenneth.
    Applied probability / Kenneth Lange.
        p. cm. — (Springer texts in statistics)
    Includes bibliographical references and index.
    ISBN 0-387-00425-4 (alk. paper)
        1. Probabilities. 2. Stochastic processes. I. Title. II. Series.
QA273.L2684 2003
519.2—dc21   2003042436

ISBN 0-387-00425-4   Printed on acid-free paper.

Printed in the United States of America.

9 8 7 6 5 4 3 2 1   SPIN 10908198

Typesetting: Pages created by the author using a Springer T$_E$X macro package.

www.springer-ny.com

Springer-Verlag New York Berlin Heidelberg
A member of BertelsmannSpringer Science+Business Media GmbH

Preface

Despite the fears of university mathematics departments, mathematics education is growing rather than declining. But the truth of the matter is that the increases are occurring outside departments of mathematics. Engineers, computer scientists, physicists, chemists, economists, statisticians, biologists, and even philosophers teach and learn a great deal of mathematics. The teaching is not always terribly rigorous, but it tends to be better motivated and better adapted to the needs of students. In my own experience teaching students of biostatistics and mathematical biology, I attempt to convey both the beauty and utility of probability. This is a tall order, partially because probability theory has its own vocabulary and habits of thought. The axiomatic presentation of advanced probability typically proceeds via measure theory. This approach has the advantage of rigor, but it inevitably misses most of the interesting applications, and many applied scientists rebel against the onslaught of technicalities. In the current book, I endeavor to achieve a balance between theory and applications in a rather short compass. While the combination of brevity and balance sacrifices many of the proofs of a rigorous course, it is still consistent with supplying students with many of the relevant theoretical tools. In my opinion, it better to present the mathematical facts without proof rather than omit them altogether.

In the preface to his lovely recent textbook [153], David Williams writes, "Probability and Statistics used to be married; then they separated; then they got divorced; now they hardly see each other." Although this split is doubtless irreversible, at least we ought to be concerned with properly bringing up their children, applied probability and computational statistics. If we fail, then science as a whole will suffer. You see before you my attempt to give applied probability the attention it deserves. My other recent book [95] covers computational statistics and aspects of computational probability glossed over here.

This graduate-level textbook presupposes knowledge of multivariate calculus, linear algebra, and ordinary differential equations. In probability theory, students should be comfortable with elementary combinatorics, generating functions, probability densities and distributions, expectations, and conditioning arguments. My intended audience includes graduate students in applied mathematics, biostatistics, computational biology, computer science, physics, and statistics. Because of the diversity of needs, instructors are encouraged to exercise their own judgment in deciding what chapters and topics to cover.

Chapter 1 reviews elementary probability while striving to give a brief survey of relevant results from measure theory. Poorly prepared students should supplement this material with outside reading. Well-prepared students can skim Chapter 1 until they reach the less well-known material of the final two sections. Section 1.8 develops properties of the multivariate normal distribution of special interest to students in biostatistics and statistics. This material is applied to optimization theory in Section 3.3 and to diffusion processes in Chapter 11.

We get down to serious business in Chapter 2, which is an extended essay on calculating expectations. Students often complain that probability is nothing more than a bag of tricks. For better or worse, they are confronted here with some of those tricks. Readers may want to skip the final two sections of the chapter on surface area distributions on a first pass through the book.

Chapter 3 touches on advanced topics from convexity, inequalities, and optimization. Beside the obvious applications to computational statistics, part of the motivation for this material is its applicability in calculating bounds on probabilities and moments.

Combinatorics has the odd reputation of being difficult in spite of relying on elementary methods. Chapters 4 and 5 are my stab at making the subject accessible and interesting. There is no doubt in my mind of combinatorics' practical importance. More and more we live in a world dominated by discrete bits of information. The stress on algorithms in Chapter 5 is intended to appeal to computer scientists.

Chapters 6 through 11 cover core material on stochastic processes that I have taught to students in mathematical biology over a span of many years. If supplemented with appropriate sections from Chapters 1 and 2, there is sufficient material here for a traditional semester-long course in stochastic processes. Although my examples are weighted toward biology, particularly genetics, I have tried to achieve variety. The fortunes of this book doubtless will hinge on how compelling readers find these examples.

You can leaf through the Table of Contents to get a better idea of the topics covered in these chapters.

In the final two chapters on Poisson approximation and number theory, the applications of probability to other branches of mathematics come to the fore. These chapters are hardly in the mainstream of stochastic processes and are meant for independent reading as much as for classroom presentation.

All chapters come with exercises. These are not graded by difficulty, but hints are provided for some of the more difficult ones. My own practice is to require one problem for each hour and a half of lecture. Students are allowed to choose among the problems within each chapter and are graded on the best of the solutions they present. This strategy provides incentive for the students to attempt more than the minimum number of problems.

I would like to thank my former and current UCLA and University of Michigan students for their help in debugging this text. In retrospect, there were far more contributing students than I can possibly credit. At the risk of offending the many, let me single out Brian Dolan, Ruzong Fan, David Hunter, Wei-hsun Liao, Ben Redelings, Eric Schadt, Marc Suchard, Janet Sinsheimer, and Andy Ming-Ham Yip. I also thank John Kimmel of Springer–Verlag for his editorial assistance.

Finally, I dedicate this book to my mother, Alma Lange, on the occasion of her 80th birthday. Thanks, Mom, for your cheerfulness and generosity in raising me. You were, and always will be, an inspiration to the whole family.

Contents

1
Basic Notions of Probability Theory

1.1 Introduction

This initial chapter covers background material that every serious student of applied probability should master. In no sense is the chapter meant as a substitute for a previous course in applied probability or for a future course in measure-theoretic probability. Our comments are merely meant as reminders and as a bridge. Many mathematical facts will be stated without proof. This is unsatisfactory, but it is even more unsatisfactory to deny students the most powerful tools in the probabilist's toolkit. Quite apart from specific tools, the language and intellectual perspective of modern probability theory also furnish an intuitive setting for solving practical problems. Probability involves modes of thought that are unique within mathematics. As a brief illustration of the material reviewed, we derive properties of the multivariate normal distribution in the penultimate section of this chapter. Later chapters will build on the facts and vocabulary mentioned here and provide more elaborate applications. The concluding section of the current chapter mentions notions of convergence featured in more theoretical treatments of probability and statistics.

1.2 Probability and Expectation

The layman's definition of probability is the long-run frequency of success over a sequence of independent, identically constructed trials. Although this law of large numbers perspective is important, mathematicians have found it helpful to put probability theory on an axiomatic basis [19, 42, 49, 64, 126, 131, 152]. The modern theory begins with the notion of a sample space Ω and a collection $\mathcal{F}$ of subsets from Ω subject to the following conventions:

(1.2a) $\Omega \in \mathcal{F}$.

(1.2b)   If $A \in \mathcal{F}$, then its complement $A^c \in \mathcal{F}$.

(1.2c)   If $A_1, A_2, \ldots$ is a finite or countably infinite sequence of subsets from $\mathcal{F}$, then $\bigcup_i A_i \in \mathcal{F}$.

Any collection $\mathcal{F}$ satisfying these postulates is termed a σ-field or σ-algebra. Two immediate consequences of the definitions are that the empty set $\emptyset \in \mathcal{F}$ and that if $A_1, A_2, \ldots$ is a finite or countably infinite sequence of subsets from $\mathcal{F}$, then $\bigcap_i A_i = (\bigcup_i A_i^c)^c \in \mathcal{F}$. In probability theory, we usually substitute everyday language for set theory language. Table 1.1 provides a short dictionary for translating equivalent terms.

TABLE 1.1. A Brief Dictionary of Set Theory and Probability Terms

Set Theory	Probability	Set Theory	Probability
set	event	null set	impossible event
union	or	universal set	certain event
intersection	and	pairwise disjoint	mutually exclusive
complement	not	inclusion	implication

The axiomatic setting of probability theory is completed by introducing a probability measure or distribution Pr on the events in $\mathcal{F}$. This function should satisfy the properties:

(1.2d) $\Pr(\Omega) = 1$.

(1.2e) $\Pr(A) \geq 0$ for any $A \in \mathcal{F}$.

(1.2f) $\Pr(\bigcup_i A_i) = \sum_i \Pr(A_i)$ for any countably infinite sequence of mutually exclusive events $A_1, A_2, \ldots$ from $\mathcal{F}$.

A triple $(\Omega, \mathcal{F}, \Pr)$ constitutes a probability space. An event $A \in \mathcal{F}$ is said to be null when $\Pr(A) = 0$ and almost sure when $\Pr(A) = 1$.

Example 1.2.1 *Discrete Uniform Distribution*

One particularly simple sample space is the set $\Omega = \{1, \ldots, n\}$. Here the natural choice of $\mathcal{F}$ is the collection of all subsets of Ω. The uniform distribution (or normalized counting measure) attributes probability $\Pr(A) = \frac{|A|}{n}$ to a set A, where $|A|$ denotes the number of elements of A. ∎

Example 1.2.2 *Continuous Uniform Distribution*

A continuous analog of the discrete uniform distribution is furnished by Lebesgue measure on the unit interval $[0, 1]$. In this case, the best one can do is define $\mathcal{F}$ as the smallest σ-algebra containing all closed subintervals $[a, b]$ of $\Omega = [0, 1]$. The events in $\mathcal{F}$ are then said to be Borel sets. Henri Lebesgue was able to show how to extend the primitive identification $\Pr([a, b]) = b - a$ of the probability of an interval with its length to all Borel sets [131]. Invoking the axiom of choice from set theory, one can prove that it is impossible to attach a probability consistently to all subsets of $[0, 1]$. The existence of nonmeasurable sets makes the whole enterprise of measure-theoretic probability more delicate than mathematicians anticipated. Fortunately, one can ignore such subtleties in most practical problems.   ∎

The next example is designed to give readers a hint of the complexities involved in defining probability spaces.

Example 1.2.3 *Density in Number Theory*

Consider the natural numbers $\Omega = \{1, 2, \ldots\}$ equipped with the density function

$$\mathrm{den}(A) \quad = \quad \lim_{n \to \infty} \frac{|A \cap \{1, 2, \ldots, n\}|}{n}.$$

Clearly, $0 \le \mathrm{den}(A) \le 1$ whenever $\mathrm{den}(A)$ is defined. Some typical densities include $\mathrm{den}(\Omega) = 1$, $\mathrm{den}(\{j\}) = 0$, and $\mathrm{den}(\{j, 2j, 3j, 4j, \ldots\}) = 1/j$. Any σ-algebra $\mathcal{F}$ containing each of the positive integers $\{j\}$ fails the test of countable additivity stated in postulate (1.2f) above. Indeed,

$$\mathrm{den}(\Omega) \;\ne\; 0 \;=\; \sum_{j=1}^{\infty} \mathrm{den}(\{j\}).$$

Note that $\mathrm{den}(A)$ does satisfy the test of finite additivity. Of course, it is possible to define many legitimate probability distributions on the positive integers.   ∎

In practice, most questions in probability theory revolve around random variables rather than sample spaces. Readers will doubtless recall that a random variable X is a function from a sample space Ω into the real line R. This is almost correct. To construct a consistent theory of integration, one must insist that a random variable be measurable. This technical condition requires that for every constant c, the set $\{\omega \in \Omega : X(\omega) \le c\}$ be an event in the σ-algebra $\mathcal{F}$ attached to Ω. Measurability can also be defined in terms of the Borel sets $\mathcal{B}$ of R, which comprise the smallest σ-algebra containing all intervals $[a, b]$ of R. With this definition in mind, X is measurable if and only if the inverse image $X^{-1}(B)$ of every Borel set B is an event in $\mathcal{F}$. This is analogous to but weaker than defining continuity by requiring that the inverse image of every open set be open. Almost every conceivable function $X : \Omega \mapsto R$ qualifies as measurable. Formal verification of measurability usually invokes one or more of the many closure properties of measurable functions. For instance, measurability is preserved under the formation of finite sums, products, maxima, minima, and limits of measurable functions. For this reason, we seldom waste time checking measurability.

Measurable functions are candidates for integration. The simplest measurable function is the indicator 1_A of an event A. The integral or expectation $E(1_A)$ of 1_A is just the corresponding probability $Pr(A)$. Integration is first extended to simple functions $\sum_{i=1}^{n} c_i 1_{A_i}$ by the linearity device

$$E\left(\sum_{i=1}^{n} c_i 1_{A_i} \right) = \sum_{i=1}^{n} c_i E(1_{A_i})$$

$$= \sum_{i=1}^{n} c_i Pr(A_i)$$

and from there to the larger class of integrable functions by appropriate limit arguments. Although the rigorous development of integration is one of the intellectual triumphs of modern mathematics, we record here only some of the basic facts. The two most important are linearity and nonnegativity:

(1.2g) $E(aX + bY) = a E(X) + b E(Y)$.

(1.2h) $E(X) \geq 0$ for any $X \geq 0$.

From these basic properties, a host of simple results flow. For one example, the inequality $|E(X)| \leq E(|X|)$ holds whenever $E(|X|) < \infty$. For another example, taking expectations in the identity $1_{A \cup B} = 1_A + 1_B - 1_{A \cap B}$ produces the identity $Pr(A \cup B) = Pr(A) + Pr(B) - Pr(A \cap B)$. Of course, one can prove this and similar equalities without introducing expectations, but the application of the expectation operator often streamlines proofs.

One of the most impressive achievements of Lebesgue's theory of integration is that it identifies sufficient conditions for the interchange of limits and integrals. Fatou's lemma states that

$$E(\liminf_{n \to \infty} X_n) \leq \liminf_{n \to \infty} E(X_n)$$

for any sequence $X_1, X_2, \ldots$ of nonnegative random variables. Recall that $\liminf_{n \to \infty} a_n = \lim_{n \to \infty} \inf\{a_k\}_{k \geq n}$ for any sequence a_n. In the present case, each sample point ω defines a different sequence $a_n = X_n(\omega)$.

If the sequence of random variables X_n is increasing as well as nonnegative, then the monotone convergence theorem

$$\lim_{n \to \infty} E(X_n) = E(\lim_{n \to \infty} X_n) \tag{1.1}$$

holds, with the possibility $E(\lim_{n \to \infty} X_n) = \infty$ included. Again we need look no further than indicator functions to apply the monotone convergence theorem. Suppose $A_1 \subset A_2 \subset \cdots$ is an increasing sequence of events with limit $A_\infty = \cup_{n=1}^{\infty} A_n$. Then the continuity property

$$\lim_{n \to \infty} \Pr(A_n) \quad = \quad \Pr(A_\infty)$$

follows trivially from the monotone convergence theorem. Experts might rightfully object that this is circular reasoning because the continuity of probability is one of the ingredients that goes into constructing a rigorous theory of integration in the first place. However, this misses the psychological point that it is easier to remember and apply a general theorem than various special cases of it.

Example 1.2.4 *Borel-Cantelli Lemma*

Suppose a sequence of events $A_1, A_2, \ldots$ satisfies $\sum_{i=1}^{\infty} \Pr(A_i) < \infty$. The Borel-Cantelli lemma says only finitely many of the events occur. To prove this result, let 1_{A_i} be the indicator function of A_i, and let N be the infinite sum $\sum_{i=1}^{\infty} 1_{A_i}$. The monotone convergence theorem implies that

$$\mathrm{E}(N) \quad = \quad \sum_{i=1}^{\infty} \Pr(A_i).$$

If $\mathrm{E}(N) < \infty$ as assumed, then $N < \infty$ with probability 1. In other words, only finitely many of the A_i occur.   ∎

The dominated convergence theorem relaxes the assumptions that the sequence $X_1, X_2, \ldots$ is monotone and nonnegative but adds the requirement that all X_n satisfy $|X_n| \leq Y$ for some dominating random variable Y with finite expectation. Assuming that $\lim_{n \to \infty} X_n$ exists, the interchange (1.1) is again permissible. If the dominating random variable Y is constant, then most probabilists refer to the dominated convergence theorem as the bounded convergence theorem. Our next example illustrates the power of the dominated convergence theorem.

Example 1.2.5 *Differentiation Under an Expectation Sign*

Let X_t denote a family of random variables indexed by a real parameter t such that (a) $\frac{d}{dt} X_t(\omega)$ exists for all sample points ω and (b) $|\frac{d}{dt} X_t| \leq Y$ for some dominating random variable Y with finite expectation. We claim that $\frac{d}{dt} \mathrm{E}(X_t)$ exists and equals $\mathrm{E}(\frac{d}{dt} X_t)$. To prove this result, consider the difference quotient

$$\frac{\mathrm{E}(X_{t+\Delta t}) - \mathrm{E}(X_t)}{\Delta t} \quad = \quad \mathrm{E}\left(\frac{X_{t+\Delta t} - X_t}{\Delta t}\right).$$

For any sample point ω, the mean value theorem implies that

$$\left| \frac{X_{t+\Delta t}(\omega) - X_t(\omega)}{\Delta t} \right| \quad = \quad \left| \frac{d}{ds} X_s(\omega) \right|$$
$$\leq \quad Y(\omega)$$

for some s between t and $t + \Delta t$. Because the difference quotients converge to the derivative in a dominated fashion as Δt tends to 0, application of the dominated convergence theorem finishes the proof.

As a straightforward illustration, consider the problem of calculating the first moment of a random variable Z from its characteristic function $E(e^{itZ})$. Assuming that $E(|Z|)$ is finite, define the family of random variables $X_t = e^{itZ}$. It is then clear that the derivative $\frac{d}{dt} X_t(\omega) = iZ(\omega)e^{itZ(\omega)}$ exists for all sample points ω and that $Y = |iZe^{itZ}| = |Z|$ furnishes an appropriate dominating random variable. Hence, $E(Z) = -i\frac{d}{dt} E(e^{itZ})|_{t=0}$. ∎

1.3   Conditional Probability

Constructing a rigorous theory of conditional probability and conditional expectation is as much a chore as constructing a rigorous theory of integration. Fortunately, most of the theoretic results can be motivated starting with the simple case of conditioning on an event of positive probability. In this case, we define the conditional probability

$$\Pr(B \mid A) \;=\; \frac{\Pr(B \cap A)}{\Pr(A)}$$

of any event B relative to A. Because the conditional probability $\Pr(B \mid A)$ is a legitimate probability measure, it is possible to define the conditional expectation $E(Z \mid A)$ of any integrable random variable Z. Fortunately, this boils down to nothing more than

$$E(Z \mid A) \;=\; \frac{E(Z 1_A)}{\Pr(A)}. \tag{1.2}$$

Definition (1.2) has limited scope, and probabilists have generalized it by conditioning on a random variable rather than a single event. If X is a random variable taking only a finite number of values $x_1, \ldots, x_n$, then $E(Z \mid X)$ is the random variable defined by $E(Z \mid X = x_i)$ on the event $\{X = x_i\}$. Obviously, the conditional expectation operator inherits the properties of linearity and nonnegativity in Z from the ordinary expectation operator. In addition, there is the further connection

$$\begin{aligned}
E(Z) \;&=\; \sum_{i=1}^{n} E(Z \mid X = x_i)\Pr(X = x_i) \\
&=\; E[E(Z \mid X)] \tag{1.3}
\end{aligned}$$

between ordinary and conditional expectations. The final property worth highlighting,

$$E[f(X)Z] \;=\; E[f(X)\, E(Z \mid X)], \tag{1.4}$$

is a consequence of equation (1.3) and the obvious identity

$$E[f(X)Z \mid X] = f(X) E(Z \mid X).$$

Example 1.3.1 *The Hypergeometric Distribution*

Consider a finite sequence $X_1, \ldots, X_n$ of independent Bernoulli random variables with common success probability p. Here $\Pr(X_j = 1) = p$ and $\Pr(X_j = 0) = 1 - p$, and the sum $S_n = X_1 + \cdots + X_n$ follows a binomial distribution. The hypergeometric distribution can be recovered in this setting by conditioning. For $m < n$, define the shorter sum $S_m = X_1 + \cdots + X_m$ and calculate

$$
\begin{aligned}
\Pr(S_m = j \mid S_n = k) &= \frac{\binom{m}{j} p^j (1-p)^{m-j} \binom{n-m}{k-j} p^{k-j} (1-p)^{n-m+j-k}}{\binom{n}{k} p^k (1-p)^{n-k}} \\
&= \frac{\binom{m}{j} \binom{n-m}{k-j}}{\binom{n}{k}}.
\end{aligned}
$$

The mean of this hypergeometric distribution is just the conditional expectation $E(S_m \mid S_n = k)$. Using symmetry and the additivity of the conditional expectation operator, we find that

$$
\begin{aligned}
E(S_m \mid S_n = k) &= \sum_{i=1}^{m} E(X_i \mid S_n = k) \\
&= m \, E(X_1 \mid S_n = k) \\
&= \frac{m}{n} E(S_n \mid S_n = k) \\
&= \frac{mk}{n}.
\end{aligned}
$$

It is noteworthy that the identity $E(S_m \mid S_n) = \frac{m}{n} S_n$ does not require the X_j to be Bernoulli. ■

At the highest level of abstraction, we define conditional expectation $E(Z \mid \mathcal{G})$ relative to a sub σ-algebra $\mathcal{G}$ of the underlying σ-algebra $\mathcal{F}$. Here it is important to bear in mind that Z must be integrable and that in most cases $\mathcal{G}$ is the smallest σ-algebra making a random variable X or a random vector $(X_1, \ldots, X_n)$ measurable. The technical requirement that $E(Z \mid \mathcal{G})$ be measurable with respect to $\mathcal{G}$ then means that $E(Z \mid \mathcal{G})$ is a function of X or $(X_1, \ldots, X_n)$. Because $\mathcal{G}$ may have an infinity of events, we can no longer rely on defining $E(Z \mid \mathcal{G})$ by naively conditioning on events of positive probability. The usual mathematical trick of turning a theorem into a definition, however, comes to the rescue. Thus, $E(Z \mid \mathcal{G})$ is defined as the essentially unique integrable random variable that is measurable with respect to $\mathcal{G}$ and satisfies the analog

$$E[1_C Z] = E[1_C \, E(Z \mid \mathcal{G})] \tag{1.5}$$

of equation (1.4) for every event C in $\mathcal{G}$. Hidden in this definition is an appeal to the powerful Radon-Nikodym theorem of measure theory. The upshot of these indirect arguments is that the conditional expectation operator is perfectly respectable and continues to enjoy the basic properties mentioned earlier.

In our study of martingales in Chapter 10, we will encounter increasing σ-algebras. We write $\mathcal{F} \subset \mathcal{G}$ if every event of $\mathcal{F}$ is also an event $\mathcal{G}$. In other words, $\mathcal{F}$ is less informative than $\mathcal{G}$. The "tower property"

$$\mathrm{E}[\mathrm{E}(Z \mid \mathcal{G}) \mid \mathcal{F}] \;=\; \mathrm{E}(Z \mid \mathcal{F}) \tag{1.6}$$

holds in this case because equation (1.5) implies

$$
\begin{aligned}
\mathrm{E}[1_C \,\mathrm{E}(Z \mid \mathcal{G})] &= \mathrm{E}[\mathrm{E}(1_C Z \mid \mathcal{G})] \\
&= \mathrm{E}(1_C Z) \\
&= \mathrm{E}[1_C \,\mathrm{E}(Z \mid \mathcal{F})]
\end{aligned}
$$

for every C in $\mathcal{F}$.

1.4 Independence

Two events A and B are independent if and only if

$$\Pr(A \cap B) \;=\; \Pr(A)\Pr(B).$$

This definition is equivalent to $\Pr(B \mid A) = \Pr(B)$ when $\Pr(A) > 0$. A finite or countable sequence $A_1, A_2, \ldots$ of events is independent provided

$$\Pr\left(\bigcap_{j=1}^{n} A_{i_j}\right) \;=\; \prod_{j=1}^{n} \Pr(A_{i_j})$$

for all finite subsequences $A_{i_1}, \ldots, A_{i_n}$. Pairwise independence is insufficient to imply independence. A sequence of random variables $X_1, X_2, \ldots$ is independent whenever the sequence of events $A_i = \{X_i \leq c_i\}$ is independent for all possible choices of the constants c_i. In practice, one usually establishes the independence of two random variables U and V by exhibiting them as measurable functions $U = f(X)$ and $V = g(Y)$ of known independent random variables X and Y.

If X and Y are independent random variables with finite expectations, then Fubini's theorem implies $\mathrm{E}(XY) = \mathrm{E}(X)\,\mathrm{E}(Y)$. If X and Y are nonnegative, then equality continues to hold even when one or both random variables have infinite expectations. From equality (1.5), one can deduce that $\mathrm{E}(Y \mid X) = \mathrm{E}(Y)$ whenever Y is independent of X. This result extends to conditioning on a σ-algebra $\mathcal{G}$ provided Y is independent of every event in $\mathcal{G}$.

1.5   Distributions, Densities, and Moments

The distribution function $F(x)$ of a random variable X is defined by the formula $F(x) = \Pr(X \leq x)$. Readers will recall the familiar properties:

(1.5a) $0 \leq F(x) \leq 1$,

(1.5b) $F(x) \leq F(y)$ for $x \leq y$,

(1.5c) $\lim_{x \to y+} F(x) = F(y)$,

(1.5d) $\lim_{x \to -\infty} F(x) = 0$,

(1.5e) $\lim_{x \to \infty} F(x) = 1$,

(1.5f) $\Pr(a < X \leq b) = F(b) - F(a)$,

(1.5g) $\Pr(X = x) = F(x) - F(x-)$.

A random variable X is said to be discretely distributed if its possible values are limited to a sequence of points $x_1, x_2, \ldots$. In this case, its discrete density $f(x_i) = \Pr(X = x_i)$ satisfies:

(1.5h) $f(x_i) \geq 0$ for all i,

(1.5i) $\sum_i f(x_i) = 1$,

(1.5j) $F(x) = \sum_{x_i \leq x} f(x_i)$.

A continuously distributed random variable X has density $f(x)$ defined on the real line and satisfying:

(1.5k) $f(x) \geq 0$ for all x,

(1.5l) $\int_a^b f(x)\, dx = F(b) - F(a)$,

(1.5m) $\frac{d}{dx} F(x) = f(x)$ for almost all x.

One of the primary uses of distribution and density functions is in calculating expectations. If $h(x)$ is Borel measurable and the random variable $h(X)$ has finite expectation, then we can express its expectation as the integral

$$\mathrm{E}[h(X)] \;=\; \int h(x)\, dF(x).$$

One proves this result by transferring the probability measure on the underlying sample space to the real line via X. In this scheme, an interval $(a, b]$ is assigned probability $F(b) - F(a)$. In practice, we replace $dF(x)$ by counting measure or Lebesgue measure and evaluate $\mathrm{E}[h(X)]$ by $\sum_i h(x_i) f(x_i)$ in the discrete case and by $\int h(x) f(x)\, dx$ in the continuous case.

Counting measure and Lebesgue measure on the real line have infinite mass. Such infinite measures share many of the properties of probability measures. For instance, the dominated convergence theorem and Fubini's theorem continue to hold. We will use such properties without comment, relying on the student's training in advanced calculus to lend an air of respectability to our invocation of relevant facts.

The most commonly encountered expectations are moments. The nth moment of X is $\mu_n = \mathrm{E}(X^n)$. If we recenter X around its first moment (or mean) μ_1, then the nth central moment of X is $\mathrm{E}[(X - \mu_1)^n]$. As mentioned earlier, we can recover the mean of X from its characteristic function $\mathrm{E}(e^{itX})$ by differentiation. In general, if $\mathrm{E}(|X|^n)$ is finite, then

$$\mathrm{E}(X^n) \;=\; (-i)^n \frac{d^n}{dt^n} \, \mathrm{E}(e^{itX})|_{t=0}. \tag{1.7}$$

The characteristic function of a random variable always exists and uniquely determines the distribution function of the random variable [49, 92]. If X has density function $f(x)$, then $\mathrm{E}(e^{itX})$ coincides with the Fourier transform $\hat{f}(t)$ of $f(x)$ [43]. When $\hat{f}(t)$ is integrable, $f(x)$ is recoverable via the inversion formula

$$f(x) \;=\; \frac{1}{2\pi} \int_{-\infty}^{\infty} e^{-itx} \hat{f}(t) \, dt. \tag{1.8}$$

For a random variable X possessing moments of all orders, it is usually simpler to deal with the moment generating function $\mathrm{E}(e^{tX})$. For a nonnegative random variable X, we occasionally resort to the Laplace transform $\mathrm{E}(e^{-tX})$. (If X possesses a density function $f(x)$, then $\mathrm{E}(e^{-tX})$ is also the ordinary Laplace transform of $f(x)$ as defined in science and engineering courses.) When X is integer-valued as well as nonnegative, the probability generating function $\mathrm{E}(t^X)$ for $t \in [0, 1]$ also comes in handy. Each of these transforms $M(t)$ possesses the multiplicative property

$$M_{X_1 + \cdots + X_m}(t) \;=\; M_{X_1}(t) \cdots M_{X_m}(t)$$

for a sum of independent random variables $X_1, \ldots, X_m$. The simplest transforms involve constant random variables. For instance, the Laplace transform of the constant c is just e^{-ct}; the probability generating function of the positive integer n is t^n. Chapter 2 introduces more complicated examples.

The second central moment of a random variable is also called the variance of the random variable. Readers will doubtless recall the variance formula

$$\mathrm{Var}\left(\sum_{j=1}^{m} X_j\right) \;=\; \sum_{j=1}^{m} \mathrm{Var}(X_j) + \sum_{j=1}^{m}\sum_{k \neq j} \mathrm{Cov}(X_j, X_k) \tag{1.9}$$

for a sum of m random variables. Here

$$\mathrm{Cov}(X_j, X_k) \;=\; \mathrm{E}(X_j X_k) - \mathrm{E}(X_j)\,\mathrm{E}(X_k)$$

is the covariance between X_j and X_k. Independent random variables are uncorrelated and exhibit zero covariance, but independence is hardly necessary for two random variables to be uncorrelated. For random variables with zero means, the covariance function serves as an inner product and lends a geometric flavor to many probability arguments. For example, we can think of uncorrelated, zero-mean random variables as being orthogonal. Calculation of variances and covariances is often facilitated by the conditioning formulas

$$\begin{aligned} \mathrm{Var}(X) &= \mathrm{E}[\mathrm{Var}(X \mid Z)] + \mathrm{Var}[\mathrm{E}(X \mid Z)] \\ \mathrm{Cov}(X,Y) &= \mathrm{E}[\mathrm{Cov}(X,Y \mid Z)] + \mathrm{Cov}[\mathrm{E}(X \mid Z), \mathrm{E}(Y \mid Z)] \,(1.10) \end{aligned}$$

and by the simple formulas $\mathrm{Var}(cX) = c^2 \,\mathrm{Var}(X)$ and $\mathrm{Var}(X+c) = \mathrm{Var}(X)$ involving a constant c.

Example 1.5.1 *Best Predictor*

Consider a random variable X with finite variance. If Y is a second random variable defined on the same probability space, then it makes sense to inquire what function $f(Y)$ best predicts X. If we use mean square error as our criterion, then we must minimize $\mathrm{Var}[X - f(Y)]$. According to equation (1.10),

$$\begin{aligned} \mathrm{Var}[X - f(Y)] &= \mathrm{E}\{\mathrm{Var}[X - f(Y) \mid Y]\} + \mathrm{Var}\{\mathrm{E}[X - f(Y) \mid Y]\} \\ &= \mathrm{E}[\mathrm{Var}(X \mid Y)] + \mathrm{Var}[\mathrm{E}(X \mid Y) - f(Y)]. \end{aligned}$$

The term $\mathrm{E}[\mathrm{Var}(X \mid Y)]$ does not depend on the function $f(Y)$, and the term $\mathrm{Var}[\mathrm{E}(X \mid Y) - f(Y)]$ is minimized by taking $f(Y) = \mathrm{E}(X \mid Y)$. Thus, $\mathrm{E}(X \mid Y)$ is the best predictor. ∎

Table 1.2 lists the densities, means, and characteristic functions of some commonly occuring univariate distributions. Restrictions on the values and parameters of these distributions are not shown. Note that the beta distribution does not possess a simple characteristic function. The version of the geometric distribution given counts the number of Bernoulli trials until a success, not the number of failures.

In statistical applications, densities often depend on parameters. The parametric families displayed in Table 1.2 are typical. Viewed as a function of its parameters, a density $f(x)$, either discrete or continuous, is called a likelihood. For purposes of estimation, one can ignore any factor of $f(x)$ that depends only on the data x and not on the parameters. In maximum likelihood estimation, $f(x)$ is maximized with respect to its parameters. The parameters giving the maximum likelihood are the maximum likelihood estimates. These distinctions carry over to multidimensional densities.

TABLE 1.2. Common Distributions

Name	Density	Mean	Transform
Binomial	$\binom{n}{x}p^x(1-p)^{n-x}$	np	$(1-p+pe^{it})^n$
Poisson	$\frac{\lambda^x}{x!}e^{-\lambda}$	λ	$e^{\lambda(e^{it}-1)}$
Geometric	$(1-p)^{x-1}p$	$\frac{1}{p}$	$\frac{pe^{it}}{1-(1-p)e^{it}}$
Uniform	$\frac{1}{b-a}$	$\frac{a+b}{2}$	$\frac{e^{itb}-e^{ita}}{it(b-a)}$
Normal	$\frac{1}{\sqrt{2\pi\sigma^2}}e^{-(x-\mu)^2/2\sigma^2}$	μ	$e^{it\mu-\sigma^2t^2/2}$
Exponential	$\lambda e^{-\lambda x}$	$\frac{1}{\lambda}$	$\frac{\lambda}{\lambda-it}$
Beta	$\frac{\Gamma(\alpha+\beta)x^{\alpha-1}(1-x)^{\beta-1}}{\Gamma(\alpha)\Gamma(\beta)}$	$\frac{\alpha}{\alpha+\beta}$	
Gamma	$\frac{\lambda^\alpha x^{\alpha-1}}{\Gamma(\alpha)}e^{-\lambda x}$	$\frac{\alpha}{\lambda}$	$\left(\frac{\lambda}{\lambda-it}\right)^\alpha$

1.6   Convolution

If X and Y are independent random variables with distribution functions $F(x)$ and $G(y)$, then $F * G(z)$ denotes the distribution function of the sum $Z = X + Y$. Fubini's theorem permits us to write this convolution of distribution functions as

$$
\begin{aligned}
F * G(z) &= \int\int 1_{\{x+y\le z\}}\,dF(x)\,dG(y) \\
&= \int F(z-y)\,dG(y).
\end{aligned}
$$

If the random variable X possesses density $f(x)$, then executing the change of variables $w = x + y$ and interchanging the order of integration yield

$$
\begin{aligned}
F * G(z) &= \int\int_{-\infty}^{z-y} f(x)\,dx\,dG(y) \\
&= \int\int_{-\infty}^{z} f(w-y)\,dw\,dG(y) \\
&= \int_{-\infty}^{z}\int f(w-y)\,dG(y)\,dw.
\end{aligned}
$$

Thus, $Z = X + Y$ has density $\int f(z-y)\,dG(y)$. When Y has density $g(y)$, this simplifies to the convolution $f * g(z) = \int f(z-y)g(y)\,dy$ of the two density functions.

Other functions of X and Y produce similar results. For example, if we suppose that $Y > 0$, then the product $U = XY$ and ratio $V = X/Y$

have distribution functions $\int_0^\infty F(uy^{-1})\,dG(y)$ and $\int_0^\infty F(vy)\,dG(y)$, respectively. Differentiation of these by u and v lead to the corresponding densities $\int_0^\infty f(uy^{-1})y^{-1}\,dG(y)$ and $\int_0^\infty f(vy)y\,dG(y)$. Problem 11 asks the reader to verify these claims rigorously. Example 1.7.2 treats a ratio when the denominator possesses a density.

1.7   Random Vectors

Random vectors with dependent components arise in many problems. Tools for manipulating random vectors are therefore crucially important. For instance, we define the expectation of a random vector $X = (X_1, \ldots, X_n)^t$ componentwise by $\mathrm{E}(X) = [\mathrm{E}(X_1), \ldots, \mathrm{E}(X_n)]^t$. Linearity carries over from the scalar case in the sense that

$$
\begin{aligned}
\mathrm{E}(X+Y) &= \mathrm{E}(X)+\mathrm{E}(Y) \\
\mathrm{E}(AX) &= A\,\mathrm{E}(X)
\end{aligned}
$$

for a compatible random vector Y and a compatible constant matrix A. Similar definitions and results come into play for random matrices when we calculate the covariance matrix

$$
\begin{aligned}
\mathrm{Cov}(X,Y) &= \mathrm{E}\{[X-\mathrm{E}(X)][Y-\mathrm{E}(Y)]^t\} \\
&= \mathrm{E}(XY^t) - \mathrm{E}(X)\,\mathrm{E}(Y)^t
\end{aligned}
$$

of two compatible random vectors X and Y. The covariance operator is linear in each of its arguments and vanishes when these arguments are independent. Furthermore, one can readily check that

$$
\mathrm{Cov}(AX, BY) = A\,\mathrm{Cov}(X,Y)B^t
$$

for compatible constant matrices A and B. The variance matrix of X is expressible as $\mathrm{Var}(X) = \mathrm{Cov}(X,X)$ and is nonnegative definite.

We define the distribution function $F(x)$ of X via

$$
F(x) = \mathrm{Pr}\left(\cap_{i=1}^n \{X_i \le x_i\}\right).
$$

This function is increasing in each component x_i of x holding the other components fixed. The marginal distribution function of a subvector of X is recoverable by taking the limit of $F(x)$ as the corresponding components of x tend to ∞. The components of X are independent if and only if $F(x)$ factors as the product $\prod_{i=1}^n F_i(x_i)$ of the marginal distribution functions. In many practical problems, X possesses a density $f(x)$. We then have

$$
\mathrm{Pr}(X \in C) = \int_C f(x)\,dx
$$

for every Borel set C. Here the indicated integral is multidimensional. The distribution and density functions are related by

$$F(x) \;=\; \int_{-\infty}^{x_1} \cdots \int_{-\infty}^{x_n} f(y)\, dy_1 \cdots dy_n.$$

The marginal density of a subvector of X is recoverable by integrating over the corresponding components of x. Furthermore, the components of X are independent if and only if $f(x)$ factors as the product $\prod_{i=1}^{n} f_i(x_i)$ of the marginal densities. For discrete random vectors, similar results hold provided we interpret $f(x)$ as a discrete density and replace multiple integrals by multiple sums.

Conditional expectations are often conveniently calculated using conditional densities. Consider a bivariate random vector $X = (X_1, X_2)$ with density $f(x_1, x_2)$. The formula

$$f_{2|1}(x_2 \mid x_1) \;=\; \frac{f(x_1, x_2)}{f_1(x_1)}$$

determines the conditional density of X_2 given X_1. To compute the conditional expectation of a function $h(X)$ of X, we form

$$E[\, h(X) \mid X_1 = x_1] \;=\; \int h(x_1, x_2) f_{2|1}(x_2 \mid x_1)\, dx_2.$$

This works because

$$E[1_C(X_1) h(X)] \;=\; \int_C \int h(x_1, x_2) f_{2|1}(x_2 \mid x_1)\, dx_2 f_1(x_1)\, dx_1$$
$$= \; E\{1_C(X_1)\, E[\, h(X) \mid X_1]\}$$

mirrors equation (1.5).

Example 1.7.1 *Bayes' Rule*

In many statistical applications, it is common to know one conditional density $f_{2|1}(x_2 \mid x_1)$ but not the other $f_{1|2}(x_1 \mid x_2)$. Since

$$f_2(x_2) \;=\; \int f(x_1, x_2)\, dx_1$$
$$= \; \int f_1(x_1) f_{2|1}(x_2 \mid x_1)\, dx_1,$$

it follows that

$$f_{1|2}(x_1 \mid x_2) \;=\; \frac{f(x_1, x_2)}{f_2(x_2)}$$
$$= \; \frac{f_1(x_1) f_{2|1}(x_2 \mid x_1)}{f_2(x_2)}$$
$$= \; \frac{f_1(x_1) f_{2|1}(x_2 \mid x_1)}{\int f_1(x_1) f_{2|1}(x_2 \mid x_1)\, dx_1}.$$

Variants of this simple formula underlie all of Bayesian statistics.   ■

Often probability models involve transformations of one random vector into another. Let $T(x)$ be a continuously differentiable transformation of an open set U containing the range of X onto an open set V. The density $g(y)$ of the random vector $Y = T(X)$ is determined by the standard change of variables formula

$$
\begin{aligned}
\Pr(Y \in C) &= \int_{T^{-1}(C)} f(x)\, dx \\
&= \int_C f \circ T^{-1}(y)|\det dT^{-1}(y)|\, dy \qquad (1.11)
\end{aligned}
$$

from advanced calculus [77, 133]. Here we assume that $T(x)$ is invertible and that its differential (or Jacobian matrix)

$$
dT(x) = \left[\frac{\partial}{\partial x_j} T_i(x) \right]
$$

of partial derivatives is invertible at each point $x \in U$. Under these circumstances, the chain rule applied to $T^{-1}[T(x)] = x$ produces

$$
dT^{-1}(y)dT(x) = I
$$

for $y = T(x)$. This permits us to substitute $|\det T(x)|^{-1}$ for $|\det dT^{-1}(y)|$ in the change of variables formula.

Example 1.7.2 *Density of a Ratio*

Let X_1 and X_2 be independent random variables with densities $f_1(x_1)$ and $f_2(x_2)$. The transformation

$$
T \begin{pmatrix} x_1 \\ x_2 \end{pmatrix} = \begin{pmatrix} x_1/x_2 \\ x_2 \end{pmatrix} = \begin{pmatrix} y_1 \\ y_2 \end{pmatrix}
$$

has inverse

$$
T^{-1} \begin{pmatrix} y_1 \\ y_2 \end{pmatrix} = \begin{pmatrix} y_1 y_2 \\ y_2 \end{pmatrix} = \begin{pmatrix} x_1 \\ x_2 \end{pmatrix}
$$

and differential and Jacobian

$$
\begin{aligned}
dT(x) &= \begin{pmatrix} 1/x_2 & -x_1/x_2^2 \\ 0 & 1 \end{pmatrix} \\
\det dT(x) &= 1/x_2.
\end{aligned}
$$

The fact that $T(x)$ is undefined on $\{x : x_2 = 0\}$ is harmless since this closed set has probability 0. The change of variables formula (1.11) implies that $Y_1 = X_1/X_2$ and $Y_2 = X_2$ have joint density

$$
f_1(y_1 y_2) f_2(y_2)|y_2|.
$$

Integrating over y_2 gives the marginal density

$$\int_{-\infty}^{\infty} f_1(y_1 y_2) f_2(y_2)|y_2|\, dy_2$$

of Y_1.

As a concrete example, suppose that X_1 and X_2 have standard normal densities. Then the ratio $Y_1 = X_1/X_2$ has density

$$
\begin{aligned}
\frac{1}{2\pi} \int_{-\infty}^{\infty} e^{-(y_1 y_2)^2/2} e^{-y_2^2/2} |y_2|\, dy_2
&= \frac{1}{\pi} \int_0^{\infty} e^{-y_2^2(y_1^2+1)/2} y_2\, dy_2 \\
&= -\frac{1}{\pi(y_1^2+1)} e^{-y_2^2(y_1^2+1)/2} \Big|_0^{\infty} \\
&= \frac{1}{\pi(y_1^2+1)}.
\end{aligned}
$$

This is the density of a Cauchy random variable. Because X_1/X_2 and X_2/X_1 are identically distributed, the reciprocal of a Cauchy is Cauchy. ■

To recover the moments of a random vector X, we can differentiate its characteristic function $s \mapsto \mathrm{E}(e^{is^t X})$. In particular,

$$
\begin{aligned}
\mathrm{E}(X_j) &= -i\frac{\partial}{\partial s_j} \mathrm{E}(e^{is^t X})|_{s=0} \\
\mathrm{E}(X_j^2) &= -\frac{\partial^2}{\partial s_j^2} \mathrm{E}(e^{is^t X})|_{s=0} \\
\mathrm{E}(X_j X_k) &= -\frac{\partial^2}{\partial s_j \partial s_k} \mathrm{E}(e^{is^t X})|_{s=0}.
\end{aligned}
$$

The characteristic function of X uniquely determines its distribution.

1.8   Multivariate Normal Random Vectors

As an illustration of the material reviewed, we now consider the multivariate normal distribution. Among the many possible definitions, we adopt the one most widely used in stochastic simulation. Our point of departure will be random vectors with independent, standard normal components. If such a random vector X has n components, then its density is

$$\prod_{j=1}^{n} \frac{1}{\sqrt{2\pi}} e^{-x_j^2/2} = \left(\frac{1}{2\pi}\right)^{n/2} e^{-x^t x/2}.$$

As demonstrated in Chapter 2, the standard normal distribution has mean 0, variance 1, and characteristic function $e^{-s^2/2}$. It follows that X has mean vector $\mathbf{0}$, variance matrix I, and characteristic function

$$E(e^{is^t X}) \;=\; \prod_{j=1}^{n} e^{-s_j^2/2} \;=\; e^{-s^t s/2}.$$

We now define any affine transformation $Y = AX + \mu$ of X to be multivariate normal [124]. This definition has several practical consequences. First, it is clear that $E(Y) = \mu$ and $\mathrm{Var}(Y) = A\,\mathrm{Var}(X)A^t = AA^t = \Omega$. Second, any affine transformation $BY + \nu = BAX + B\mu + \nu$ of Y is also multivariate normal. Third, any subvector of Y is multivariate normal. Fourth, the characteristic function of Y is

$$E(e^{is^t Y}) \;=\; e^{is^t \mu}\, E(e^{is^t AX}) \;=\; e^{is^t \mu - s^t AA^t s/2} \;=\; e^{is^t \mu - s^t \Omega s/2}.$$

This enumeration omits two more subtle issues. One is whether Y possesses a density. Observe that Y lives in an affine subspace of dimension equal to or less than the rank of A. Thus, if Y has m components, then $n \geq m$ must hold in order for Y to possess a density. A second issue is the existence and nature of the conditional density of a set of components of Y given the remaining components. We can clarify both of these issues by making canonical choices of X and A based on the classical QR decomposition of a matrix, which follows directly from the Gram-Schmidt orthogonalization procedure. See Problem 15 or reference [31].

Assuming that $n \geq m$, we can write

$$A^t \;=\; Q\begin{pmatrix} R \\ \mathbf{0} \end{pmatrix}, \tag{1.12}$$

where Q is an $n \times n$ orthogonal matrix and R is an $m \times m$ upper triangular matrix with nonnegative diagonal entries. (If $n = m$, we omit the zero matrix in the QR decomposition.) It follows that

$$AX \;=\; (\,L \quad \mathbf{0}^t\,)\,Q^t X \;=\; (\,L \quad \mathbf{0}^t\,)\,Z.$$

In view of the change of variables formula (1.11) and the facts that the orthogonal matrix Q^t preserves inner products and has determinant ± 1, the random vector Z has n independent, standard normal components and serves as a substitute for X. Not only is this true, but we can dispense with the last $n - m$ components of Z because they are multiplied by the matrix $\mathbf{0}^t$. Thus, we can safely assume $n = m$ and calculate the density of $Y = LZ + \mu$ when L is invertible. In this situation, $\Omega = LL^t$ is termed the Cholesky decomposition, and the change of variables formula (1.11) shows that Y has density

$$
\begin{aligned}
f(y) \;&=\; \left(\frac{1}{2\pi}\right)^{n/2} |\det L^{-1}| e^{-(y-\mu)^t (L^{-1})^t L^{-1}(y-\mu)/2} \\
&=\; \left(\frac{1}{2\pi}\right)^{n/2} |\det \Omega|^{-1/2} e^{-(y-\mu)^t \Omega^{-1}(y-\mu)/2},
\end{aligned}
$$

where $\Omega = LL^t$ is the variance matrix of Y. In this formula for the density, the absolute value signs on $\det \Omega$ and $\det L^{-1}$ are redundant because these determinants are positive.

To address the issue of conditional densities, consider the compatibly partitioned vectors $Y^t = (Y_1^t, Y_2^t)$, $X^t = (X_1^t, X_2^t)$, $\mu^t = (\mu_1^t, \mu_2^t)$ and matrices

$$L = \begin{pmatrix} L_{11} & \mathbf{0} \\ L_{21} & L_{22} \end{pmatrix} \qquad \Omega = \begin{pmatrix} \Omega_{11} & \Omega_{12} \\ \Omega_{21} & \Omega_{22} \end{pmatrix}.$$

Now suppose that X is standard normal, that $Y = LX + \mu$, and that L_{11} has full rank. For $Y_1 = y_1$ fixed, the equation $y_1 = L_{11}X_1 + \mu_1$ shows that X_1 is fixed at the value $x_1 = L_{11}^{-1}(y_1 - \mu_1)$. Because no restrictions apply to X_2, we have

$$Y_2 = L_{22}X_2 + L_{21}L_{11}^{-1}(y_1 - \mu_1) + \mu_2.$$

Thus, Y_2 given Y_1 is normal with mean $L_{21}L_{11}^{-1}(y_1 - \mu_1) + \mu_2$ and variance $L_{22}L_{22}^t$. To express these in terms of the blocks of $\Omega = LL^t$, observe that

$$\begin{aligned} \Omega_{11} &= L_{11}L_{11}^t \\ \Omega_{21} &= L_{21}L_{11}^t \\ \Omega_{22} &= L_{21}L_{21}^t + L_{22}L_{22}^t. \end{aligned}$$

The first two of these equations imply that $L_{21}L_{11}^{-1} = \Omega_{21}\Omega_{11}^{-1}$. The last equation then gives

$$\begin{aligned} L_{22}L_{22}^t &= \Omega_{22} - L_{21}L_{21}^t \\ &= \Omega_{22} - \Omega_{21}(L_{11}^t)^{-1}L_{11}^{-1}\Omega_{12} \\ &= \Omega_{22} - \Omega_{21}\Omega_{11}^{-1}\Omega_{12}. \end{aligned}$$

These calculations do not require that Y_2 possess a density.

1.9 Convergence

Probabilists entertain many notions of convergence [19, 152]. The simplest of these is the pointwise convergence of a sequence of random variables X_n to a limit X. Because this form of convergence is allowed to fail on an event of probability 0, it is termed almost sure convergence. The usual calculus rules for dealing with limits apply to almost surely convergent sequences. The most celebrated almost sure convergence result is the strong law of large numbers. We refer readers to Example 10.3.1 for the statement and proof of one version of the strong law of large numbers.

More generally, a sequence X_n is said to converge to X in probability if for every $\epsilon > 0$

$$\lim_{n \to \infty} \Pr(|X_n - X| > \epsilon) = 0.$$

One can prove that X_n converges to X in probability if and only if every subsequence X_{n_m} of X_n possesses a subsubsequence $X_{n_{m_l}}$ converging to X almost surely. (See Problem 20.) This characterization makes it possible to generalize nearly all theorems involving almost surely convergent sequences to theorems involving sequences converging in probability. For instance, the dominated convergence theorem generalizes in this fashion. Convergence in probability can be deduced from the mean square convergence condition

$$\lim_{n \to \infty} \mathrm{E}(|X_n - X|^2) \;=\; 0$$

via Chebyshev's inequality, which is reviewed in Chapter 3.

Convergence in distribution is a very different matter. The underlying random variables need not even live on the same probability space. What is crucial is that the distribution functions $F_n(x)$ of the X_n converge to the distribution function $F(x)$ of X at every point of continuity x of $F(x)$. In other words, the way that the X_n attribute mass comes more and more to resemble the way X attributes mass. Besides this intuitive definition, there are other equivalent definitions useful in various contexts.

Proposition 1.9.1 *The following statements about the sequence X_n and its potential limit X are equivalent:*

(a) *The sequence of distribution functions $F_n(x)$ converges to $F(x)$ at every continuity point x of $F(x)$.*

(b) *The sequence of characteristic functions $\mathrm{E}(e^{isX_n})$ converges to the characteristic function $\mathrm{E}(e^{isX})$ for every real number s.*

(c) *The sequence of expectations $\mathrm{E}[f(X_n)]$ converges to $\mathrm{E}[f(X)]$ for every bounded continuous function $f(x)$ defined on the real line.*

(d) *There exist random variables Y_n and Y defined on a common probability space such that Y_n has distribution function $F_n(x)$, Y has distribution function $F(x)$, and Y_n converges to Y almost surely.*

The deep Skorokhod representation theorem mentioned in property (d) brings us full circle to almost sure convergence. It also enables us to prove some results with surprising ease. For instance, any continuous function $g(X_n)$ of a sequence converging in distribution also converges in distribution. This result follows trivially from applying $g(x)$ to the almost surely converging sequence Y_n. As another example, Fatou's lemma implies

$$
\begin{aligned}
-\mathrm{E}(|X|) \;&=\; -\mathrm{E}(|Y|) \\
&\le\; \liminf_{n \to \infty} \mathrm{E}(-|Y_n|) \\
&=\; \liminf_{n \to \infty} \mathrm{E}(-|X_n|),
\end{aligned}
$$

and this translates into $\mathrm{E}(|X|) \le \limsup_{n \to \infty} \mathrm{E}(|X_n|)$.

Doubtless the reader is already familiar with the central limit theorem. For the record, we now recall the simplest version. Let X_n be a sequence of independent, identically distributed (i.i.d.) random variables with common mean μ and common variance σ^2. Then the standardized random sums

$$\frac{S_n - n\mu}{\sqrt{n\sigma^2}} \quad = \quad \frac{\sum_{i=1}^{n}(X_i - \mu)}{\sqrt{n\sigma^2}}$$

converge in distribution to the standard normal distribution.

1.10   Problems

1. Let Ω be an infinite set. A subset $S \subset \Omega$ is said to be cofinite when its complement S^c is finite. Demonstrate that the family of subsets

$$\mathcal{F} \quad = \quad \{S \subset \Omega \colon S \text{ is finite or cofinite}\}$$

   is not a σ-algebra. What property fails? If we define $P(S) = 0$ for S finite and $P(S) = 1$ for S cofinite, then prove that $P(S)$ is finitely additive but not countably additive.

2. The symmetric difference $A \triangle B$ of two events A and B is defined as $(A \cap B^c) \cup (A^c \cap B)$. Show that $A \triangle B$ has indicator $|1_A - 1_B|$. Use this fact to prove the triangle inequality

$$\Pr(A \triangle C) \quad \leq \quad \Pr(A \triangle B) + \Pr(B \triangle C).$$

   It follows that if we ignore events of probability 0, then the collection of events forms a metric space.

3. Suppose A, B, and C are three events with $\Pr(A \cap B) > 0$. Show that A and C are conditionally independent given B if and only if the Markov property $\Pr(C \mid A \cap B) = \Pr(C \mid B)$ holds.

4. Prove the two conditioning formulas in equation (1.10) for calculating variances and covariances.

5. If X and Y are independent random variables with finite variances, then show that

$$\mathrm{Var}(XY) \quad = \quad \mathrm{Var}(X)\,\mathrm{Var}(Y) + \mathrm{E}(X)^2\,\mathrm{Var}(Y) + \mathrm{E}(Y)^2\,\mathrm{Var}(X).$$

6. Suppose X and Y are independent random variables with finite variances. Define Z to be either X or Y depending on the outcome of a coin toss. In other words, set $Z = X$ with probability p and $Z = Y$ with probability $q = 1-p$. Find the mean, variance, and characteristic function of Z.

7. Let $S_n = X_1 + \cdots + X_n$ be the sum of n independent random variables, each distributed uniformly over the set $\{1, 2, \ldots, m\}$. For example, imagine tossing a m-sided die n times and recording the total score. Calculate $E(S_n)$ and $\text{Var}(S_n)$.

8. Suppose Y has exponential density e^{-y} with unit mean. Given Y, let a point X be chosen uniformly from the interval $[0, Y]$. Show that X has density $E_1(x) = \int_x^\infty e^{-y} y^{-1} \, dy$ and distribution function $1 - e^{-x} + x E_1(x)$. Calculate $E(X)$ and $\text{Var}(X)$.

9. Let the random variable X have symmetric density $f(x) = f(-x)$. Prove that the corresponding distribution function $F(x)$ satisfies the identity $\int_{-a}^a F(x) \, dx = a$ for all $a \geq 0$ [135].

10. Suppose X has a continuous, strictly increasing distribution function $F(x)$ and $Y = -X$ has distribution function $G(y)$. Show that X is symmetrically distributed around some point μ if and only if the function $x \mapsto x - G^{-1}[F(x)]$ is constant, where $G^{-1}[G(y)] = y$ for all y.

11. Validate the formulas for the distribution and density functions of the product XY and the ratio X/Y of independent random variables X and $Y > 0$ given in Section 1.6. (Hint: Mimic the arguments used in establishing the convolution formulas.)

12. Let X_1 and X_2 be independent random variables with common exponential density $\lambda e^{-\lambda x}$ on $(0, \infty)$. Show that the random variables $Y_1 = X_1 + X_2$ and $Y_2 = X_1/X_2$ are independent, and find their densities.

13. Let $X_1, \ldots, X_n$ be a set of independent standard normal random variables. Prove that $\chi_n^2 = X_1^2 + \cdots + X_n^2$ has a gamma distribution. Calculate the mean and variance of χ_n^2.

14. Continuing Problem 13, let X be a multivariate normal random vector with mean vector μ and invertible variance matrix Ω. If X has n components, then show that the quadratic form $(X - \mu)^t \Omega^{-1} (X - \mu)$ has a χ_n^2 distribution.

15. For $n \geq m$, verify the QR decomposition (1.12). (Hints: Write

$$\begin{aligned} A^t &= (a_1, \ldots, a_m) \\ Q &= (q_1, \ldots, q_n) \\ R &= (r_1, \ldots, r_m). \end{aligned}$$

The Gram-Schmidt orthogonalization process applied to the columns of A^t yields orthonormal column vectors $q_1, \ldots, q_m$ satisfying

$$a_i = \sum_{j=1}^{i} q_j r_{ji}.$$

Complete this orthonormal basis by adding vectors $q_{m+1}, \ldots, q_n$.)

16. The Hadamard product $C = A \circ B$ of two matrices $A = (a_{ij})$ and $B = (b_{ij})$ of the same dimensions has entries $c_{ij} = a_{ij} b_{ij}$. If A and B are nonnegative definite matrices, then show that $A \circ B$ is nonnegative definite. If in addition A is positive definite, and B has positive diagonal entries, then show that $A \circ B$ is positive definite. (Hints: Let X and Y be multivariate normal random vectors with mean $\mathbf{0}$ and variance matrices A and B. Show that the vector Z with entries $Z_i = X_i Y_i$ has variance matrix $A \circ B$. To prove that $A \circ B$ is positive definite, demonstrate that $v^t Z$ has positive variance for $v \neq \mathbf{0}$. This can be done via the equality $\operatorname{Var}(v^t Z) = \operatorname{E}[(v \circ Y)^t A (v \circ Y)]$ based on formula (1.10).)

17. Consider a sequence of independent events $A_1, A_2, \ldots$ satisfying

$$\sum_{i=1}^{\infty} \Pr(A_i) = \infty.$$

As a partial converse to the Borel-Cantelli lemma, prove that infinitely many of the A_i occur. (Hints: Express the event that infinitely many of the events occur as $\cap_{n=1}^{\infty} \cup_{i=n}^{\infty} A_i$. Use the inequality $1 - x \leq e^{-x}$ to bound an infinite product by the exponential of an infinite sum.)

18. Use Problem 17 to prove that the pattern SFS of a success, failure, and success occurs infinitely many times in a sequence of Bernoulli trials. This result obviously generalizes to more complex patterns.

19. Consider a sequence $X_1, X_2, \ldots$ of independent random variables that are exponentially distributed with mean 1. Show that

$$1 = \limsup_{n \to \infty} \frac{X_n}{\ln n}$$

$$1 = \limsup_{n \to \infty} \frac{X_n - \ln n}{\ln \ln n}$$

$$1 = \limsup_{n \to \infty} \frac{X_n - \ln n - \ln \ln n}{\ln \ln \ln n}.$$

(Hints: Use the sums

$$\infty = \sum_{n=1}^{\infty} \frac{1}{n}$$

$$\infty = \sum_{n=1}^{\infty} \frac{1}{n \ln n}$$

$$\infty = \sum_{n=1}^{\infty} \frac{1}{n(\ln n)(\ln \ln n)}$$

from pages 54 and 55 of [133], and apply Problem 17.)

20. Suppose a sequence of random variables X_n converges to X in probability. Use the Borel-Cantelli lemma to show that some subsequence X_{n_m} converges to X almost surely. Now prove the full claim that X_n converges to X in probability if and only if every subsequence X_{n_m} of X_n possesses a subsubsequence $X_{n_{m_l}}$ converging to X almost surely.

21. Suppose X_n converges in probability to X and Y_n converges in probability to Y. Show that $X_n + Y_n$ converges in probability to $X + Y$, that $X_n Y_n$ converges in probability to XY, and that X_n/Y_n converges in probability to X/Y when $Y \neq 0$ almost surely.

22. Consider two sequences X_n and Y_n of random variables. Suppose that X_n converges in distribution to the random variable X and the difference $X_n - Y_n$ converges in probability to the constant 0. Prove that Y_n converges in distribution to X. (Hints: For one possible proof, invoke part (b) of Proposition 1.9.1. The inequality

$$|e^{isb} - e^{isa}| = \left| is \int_0^{b-a} e^{ist}\, dt \right|$$

$$\leq |s(b - a)|$$

will come in handy.)

2
Calculation of Expectations

2.1 Introduction

Many of the hardest problems in applied probability revolve around the calculation of expectations of one sort or another. On one level, this is merely a humble exercise in integration or summation. However, we should not be so quick to dismiss the intellectual challenges. Readers are doubtless already aware of the clever applications of characteristic and moment generating functions. This chapter is intended to review and extend some of the tools that probabilists routinely call on. Readers can consult the books [26, 48, 49, 62, 64, 126] for many additional examples of these tools in action.

2.2 Indicator Random Variables and Symmetry

Many counting random variables can be expressed as the sum of indicator random variables. If $S = \sum_{i=1}^{n} 1_{A_i}$ for events $A_1, \ldots, A_n$, then straightforward calculations and equation (1.9) give

$$
\begin{aligned}
\mathrm{E}(S) &= \sum_{i=1}^{n} \Pr(A_i) \\
\mathrm{Var}(S) &= \sum_{i=1}^{n} \Pr(A_i) + \sum_{i=1}^{n}\sum_{j \neq i} \Pr(A_i \cap A_j) - \mathrm{E}(S)^2.
\end{aligned} \qquad (2.1)
$$

Example 2.2.1 *Fixed Points of a Random Permutation*

There are $n!$ permutations π of the set $\{1,\ldots,n\}$. Under the uniform distribution, each of these permutations is equally likely. If A_i is the event that $\pi(i) = i$, then $S = \sum_{i=1}^{n} 1_{A_i}$ is the number of fixed points of π. By symmetry, $\Pr(A_i) = \frac{1}{n}$ and

$$
\begin{aligned}
\Pr(A_i \cap A_j) &= \Pr(A_j \mid A_i)\Pr(A_i) \\
&= \frac{1}{(n-1)n}.
\end{aligned}
$$

Hence, the formulas in (2.1) yield $E(S) = \frac{n}{n} = 1$ and

$$
\begin{aligned}
\mathrm{Var}(S) &= \frac{n}{n} + \sum_{i=1}^{n}\sum_{j \neq i} \frac{1}{(n-1)n} - 1^2 \\
&= 1.
\end{aligned}
$$

The equality $E(S) = \mathrm{Var}(S)$ suggests that S is approximately Poisson distributed. We will verify this conjecture in Example 4.2.1. ∎

Another simple way of evaluating moments is to exploit symmetry. Here is a example from sampling theory [32].

Example 2.2.2 *Sampling without Replacement*

Assume that m numbers $Y_1, \ldots, Y_m$ are drawn randomly without replacement from n numbers $x_1, \ldots, x_n$. It is of interest to calculate the mean and variance of the sample average $S = \frac{1}{m}\sum_{i=1}^{m} Y_i$. Clearly,

$$
E(S) = \frac{1}{m}\sum_{i=1}^{m} E(Y_i) = \bar{x},
$$

where $\bar{x}$ is the sample average of the x_i. To calculate the variance of S, let $s^2 = \frac{1}{n}\sum_{i=1}^{n}(x_i - \bar{x})^2$ denote the sample variance of the x_i. Now imagine filling out the sample to $Y_1, \ldots, Y_n$ so that all n values $x_1, \ldots, x_n$ are exhausted. Because the sum $Y_1 + \cdots + Y_n = n\bar{x}$ is constant, symmetry and equation (1.9) imply that

$$
\begin{aligned}
0 &= \mathrm{Var}(Y_1 + \cdots + Y_n) \\
&= ns^2 + n(n-1)\,\mathrm{Cov}(Y_1, Y_2).
\end{aligned}
$$

In verifying that $\mathrm{Cov}(Y_i, Y_j) = \mathrm{Cov}(Y_1, Y_2)$, it is helpful to think of the sampling being done simultaneously rather than sequentially. In any case, $\mathrm{Cov}(Y_1, Y_2) = -\frac{s^2}{n-1}$, and the formula

$$
\mathrm{Var}(S) = \frac{1}{m^2}\left[ms^2 + m(m-1)\,\mathrm{Cov}(Y_1, Y_2) \right]
$$

$$= \frac{1}{m^2}\left[ms^2 - \frac{m(m-1)s^2}{n-1}\right]$$
$$= \frac{(n-m)s^2}{m(n-1)}.$$

follows directly. ∎

The next problem, the first of a long line of problems in geometric probability, also yields to symmetry arguments [91].

Example 2.2.3 *Buffon Needle Problem*

Suppose we draw an infinite number of equally distant parallel lines on the plane R^2. If we drop a needle (or line segment) of fixed length randomly onto the plane, then the needle may or may not intersect one of the parallel lines. Figure 2.1 shows the needle intersecting a line. Buffon's problem is to calculate the probability of an intersection. Without loss of generality, we assume that the spacing between lines is 1 and the length of the needle is d. Let X_d be the random number of lines that the needle intersects. If $d < 1$, then X_d equals 0 or 1, and $\Pr(X_d = 1) = \mathrm{E}(X_d)$. Thus, Buffon's problem reduces to calculating an expectation for a short needle. Our task is to construct the function $f(d) = \mathrm{E}(X_d)$. This function is clearly nonnegative, increasing, and continuous in d.

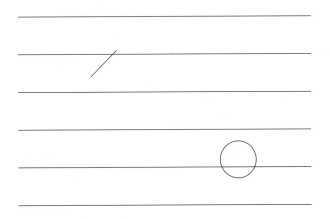

FIGURE 2.1. Diagram of the Buffon Needle Problem

Now imagine randomly dropping two needles simultaneously of length d_1 and d_2. The expected number of intersections of both needles obviously amounts to $\mathrm{E}(X_{d_1}) + \mathrm{E}(X_{d_2})$. This result holds whether we drop the two needles independently or dependently, as long as we drop them randomly. We can achieve total dependence by welding the end of one needle to the start of the other needle. If the weld is just right, then the two needles will form a single needle of length $d_1 + d_2$. This shows that

$$f(d_1 + d_2) \;=\; f(d_1) + f(d_2). \tag{2.2}$$

The only functions $f(d)$ that are nonnegative, increasing, and additive in d are the linear functions $f(d) = cd$ with $c \geq 0$. To find the proportionality constant c, we take the experiment of welding needles together to its logical extreme. Thus, a rigid wire of welded needles with perimeter p determines on average cp intersections. In the limit, we can replace the wire by any reasonable curve. The key to finding c is to take a circle of diameter 1. This particular curve has perimeter π and either is tangent to two lines or intersects the same line twice. Figure 2.1 depicts the latter case. The equation $2 = c\pi$ now determines $c = 2/\pi$ and $f(d) = 2d/\pi$.   ∎

2.3 Conditioning

A third way to calculate expectations is to condition. Two of the next three examples use conditioning to derive a recurrence relation. In the family planning model, the recurrence is difficult to solve exactly, but as with most recurrences, it is easy to implement by hand or computer.

Example 2.3.1 *Beta-Binomial Distribution*

Consider a random variable P with beta density

$$f_{\alpha\beta}(p) \;=\; \frac{\Gamma(\alpha+\beta)}{\Gamma(\alpha)\Gamma(\beta)} p^{\alpha-1}(1-p)^{\beta-1}$$

on the unit interval. In Section 2.9, we generalize the beta distribution to the Dirichlet distribution. In the meantime, the reader may recall the moment calculation

$$
\begin{aligned}
\mathrm{E}[P^i(1-P)^j] \;&=\; \int_0^1 p^i(1-p)^j f_{\alpha\beta}(p)\,dp \\
&=\; \frac{\Gamma(\alpha+\beta)}{\Gamma(\alpha)\Gamma(\beta)} \frac{\Gamma(\alpha+i)\Gamma(\beta+j)}{\Gamma(\alpha+\beta+i+j)} \int_0^1 f_{\alpha+i,\beta+j}(p)\,dp \\
&=\; \frac{(\alpha+i-1)\cdots\alpha(\beta+j-1)\cdots\beta}{(\alpha+\beta+i+j-1)\cdots(\alpha+\beta)}
\end{aligned}
$$

invoking the factorial property $\Gamma(x+1) = x\Gamma(x)$ of the gamma function. This gives, for example,

$$\mathrm{E}(P) \;=\; \frac{\alpha}{\alpha+\beta}$$

$$\mathrm{E}[P(1-P)] \;=\; \frac{\alpha\beta}{(\alpha+\beta)(\alpha+\beta+1)} \tag{2.3}$$

$$\mathrm{Var}(P) \quad = \quad \frac{\alpha\beta}{(\alpha+\beta)^2(\alpha+\beta+1)}.$$

Now suppose we carry out n Bernoulli trials with the same random success probability P pertaining to all n trials. The number of successes S_n follows a beta-binomial distribution. Application of equations (1.3) and (1.10) yields

$$\begin{aligned}
\mathrm{E}(S_n) &= n\,\mathrm{E}(P) \\
\mathrm{Var}(S_n) &= \mathrm{E}[\mathrm{Var}(S_n \mid P)] + \mathrm{Var}[\mathrm{E}(S_n \mid P)] \\
&= \mathrm{E}[nP(1-P)] + \mathrm{Var}(nP) \\
&= n\,\mathrm{E}[P(1-P)] + n^2\,\mathrm{Var}(P),
\end{aligned}$$

which can be explicitly evaluated using the moments in equation (2.3). Problem 1 provides the density of S_n. ∎

Example 2.3.2 *Repeated Uniform Sampling*

Suppose we construct a sequence of dependent random variables X_n by taking $X_0 = 1$ and sampling X_n uniformly from the interval $[0, X_{n-1}]$. To calculate the moments of X_n, we use the facts $\mathrm{E}(X_n^k) = \mathrm{E}[\mathrm{E}(X_n^k \mid X_{n-1})]$ and

$$\begin{aligned}
\mathrm{E}(X_n^k \mid X_{n-1}) &= \frac{1}{X_{n-1}} \int_0^{X_{n-1}} x^k dx \\
&= \frac{x^{k+1}}{X_{n-1}(k+1)} \Big|_0^{X_{n-1}} \\
&= \frac{1}{k+1} X_{n-1}^k.
\end{aligned}$$

Hence,

$$\mathrm{E}(X_n^k) \quad = \quad \frac{1}{k+1} \mathrm{E}(X_{n-1}^k) \quad = \quad \left(\frac{1}{k+1}\right)^n.$$

For example, $\mathrm{E}(X_n) = 2^{-n}$ and $\mathrm{Var}(X_n) = 3^{-n} - 2^{-2n}$. It is interesting that if we standardize by defining $Y_n = 2^n X_n$, then the mean $\mathrm{E}(Y_n) = 1$ is stable, but the variance $\mathrm{Var}(Y_n) = (\frac{4}{3})^n - 1$ tends to ∞.

The clouds of mystery lift a little when we rewrite X_n as the product

$$X_n \quad = \quad U_n X_{n-1} \quad = \quad \prod_{i=1}^n U_i$$

of independent uniform random variables $U_1, \ldots, U_n$ on $[0, 1]$. The product rule for expectations now gives $\mathrm{E}(X_n^k) = \mathrm{E}(U_1^k)^n = (k+1)^{-n}$. Although we cannot stabilize X_n, it is possible to stabilize $\ln X_n$. Indeed, Problem 2 notes that $\ln X_n = \sum_{i=1}^n \ln U_i$ follows a $-\frac{1}{2}\chi_{2n}^2$ distribution with mean $-n$ and variance n. Thus for large n, the central limit theorem implies that $(\ln X_n + n)/\sqrt{n}$ has an approximate standard normal distribution. ∎

Example 2.3.3 *Expected Family Size*

A married couple desires a family consisting of at least s sons and d daughters. At each birth the mother independently bears a son with probability p and a daughter with probability $q = 1 - p$. They will quit having children when their objective is reached. Let N_{sd} be the random number of children born to them. Suppose we wish to calculate the expected value $\mathrm{E}(N_{sd})$. Two cases are trivial. If either $s = 0$ or $d = 0$, then N_{sd} follows a negative binomial distribution. Therefore, $\mathrm{E}(N_{0d}) = d/q$ and $\mathrm{E}(N_{s0}) = s/p$. When both s and d are positive, the distribution of N_{sd} is not so obvious. Conditional on the sex of the first child, the random variable N_{sd} is either a probabilistic copy $N^*_{s-1,d}$ of $N_{s-1,d}$ or a probabilistic copy $N^*_{s,d-1}$ of $N_{s,d-1}$. Because in both cases the copy is independent of the sex of the first child, the recurrence relation

$$\begin{aligned}
\mathrm{E}(N_{sd}) &= p[1 + \mathrm{E}(N_{s-1,d})] + q[1 + \mathrm{E}(N_{s,d-1})] \\
&= 1 + p\,\mathrm{E}(N_{s-1,d}) + q\,\mathrm{E}(N_{s,d-1})
\end{aligned}$$

follows from conditioning on this outcome.

There are many variations on this idea. For instance, suppose we wish to compute the probability R_{sd} that they reach their quota of s sons before their quota of d daughters. Then the R_{sd} satisfy the boundary conditions $R_{0d} = 1$ for $d > 0$ and $R_{s0} = 0$ for $s > 0$. When s and d are both positive, we have the recurrence relation

$$R_{sd} = pR_{s-1,d} + qR_{s,d-1}.$$

■

2.4 Moment Transforms

Each of the moment transforms reviewed in Section 1.5 can be differentiated to capture the moments of a random variable. Equally important, these transforms often solve other theoretical problems with surprising ease. The next eight examples illustrate these two roles.

Example 2.4.1 *Characteristic Function of a Normal Density*

To find the characteristic function $\hat{\psi}(t) = \mathrm{E}(e^{itX})$ of a standard normal random variable X with density $\psi(x) = \frac{1}{\sqrt{2\pi}}e^{-x^2/2}$, we derive and solve a differential equation. Differentiation under the integral sign and integration by parts together imply that

$$\frac{d}{dt}\hat{\psi}(t) = \frac{1}{\sqrt{2\pi}}\int_{-\infty}^{\infty} e^{itx} ixe^{-\frac{x^2}{2}}\,dx$$

$$
\begin{aligned}
&= -\frac{i}{\sqrt{2\pi}} \int_{-\infty}^{\infty} e^{itx} \frac{d}{dx} e^{-\frac{x^2}{2}} dx \\
&= \frac{-i}{\sqrt{2\pi}} e^{itx} e^{-\frac{x^2}{2}} \Big|_{-\infty}^{\infty} - \frac{t}{\sqrt{2\pi}} \int_{-\infty}^{\infty} e^{itx} e^{-\frac{x^2}{2}} dx \\
&= -t\hat{\psi}(t).
\end{aligned}
$$

The unique solution to this differential equation with initial value $\hat{\psi}(0) = 1$ is $\hat{\psi}(t) = e^{-t^2/2}$.

If X is a standard normal random variable, then $\mu + \sigma X$ is a normal random variable with mean μ and variance σ^2. The general identity $\mathrm{E}[e^{it(\mu+\sigma X)}] = e^{it\mu}\,\mathrm{E}[e^{i(\sigma t)X}]$ permits us to express the characteristic function of the normal distribution with mean μ and variance σ^2 as

$$
\hat{\psi}_{\mu,\sigma^2}(t) \;=\; e^{it\mu}\hat{\psi}(\sigma t) \;=\; e^{it\mu - \frac{\sigma^2 t^2}{2}}.
$$

The first two derivatives

$$
\begin{aligned}
\frac{d}{dt}\hat{\psi}_{\mu,\sigma^2}(t) &= (i\mu - \sigma^2 t)e^{it\mu - \frac{\sigma^2 t^2}{2}} \\
\frac{d^2}{dt^2}\hat{\psi}_{\mu,\sigma^2}(t) &= -\sigma^2 e^{it\mu - \frac{\sigma^2 t^2}{2}} + (i\mu - \sigma^2 t)^2 e^{it\mu - \frac{\sigma^2 t^2}{2}}
\end{aligned}
$$

evaluated at $t = 0$ determine the mean μ and second moment $\sigma^2 + \mu^2$ as indicated in equation (1.7). ∎

Example 2.4.2 *Characteristic Function of a Gamma Density*

A random variable X with exponential density $\lambda e^{-\lambda x} 1_{\{x>0\}}$ has characteristic function

$$
\begin{aligned}
\int_0^{\infty} e^{itx} \lambda e^{-\lambda x} dx &= \frac{\lambda}{it - \lambda} e^{(it-\lambda)x} \Big|_0^{\infty} \\
&= \frac{\lambda}{\lambda - it}.
\end{aligned}
$$

An analogous calculation yields the Laplace transform $\lambda/(\lambda + t)$. Differentiation of either of these transforms produces

$$
\begin{aligned}
\mathrm{E}(X) &= \frac{1}{\lambda} \\
\mathrm{Var}(X) &= \frac{1}{\lambda^2}.
\end{aligned}
$$

The gamma density $\lambda^n x^{n-1} e^{-\lambda x} 1_{\{x>0\}}/\Gamma(n)$ is the convolution of n exponential densities with common intensity λ. The corresponding random variable X_n therefore has

$$
\mathrm{E}(X_n) \;=\; \frac{n}{\lambda}
$$

$$\mathrm{Var}(X_n) = \frac{n}{\lambda^2}$$

$$\mathrm{E}(e^{itX_n}) = \left(\frac{\lambda}{\lambda - it}\right)^n$$

$$\mathrm{E}(e^{-tX_n}) = \left(\frac{\lambda}{\lambda + t}\right)^n.$$

Problem 10 indicates that these results carry over to non-integer $n > 0$. ∎

Example 2.4.3 *Factorial Moments*

Let X be a nonnegative, integer-valued random variable. Repeated differentiation of its probability generating function $G(u) = \mathrm{E}(u^X)$ yields its factorial moments $\mathrm{E}[X(X-1)\cdots(X-j+1)] = \frac{d^j}{du^j}G(1)$. The first two central moments

$$\begin{aligned}
\mathrm{E}(X) &= G'(1) \\
\mathrm{Var}(X) &= \mathrm{E}[X(X-1)] + \mathrm{E}(X) - \mathrm{E}(X)^2 \\
&= G''(1) + G'(1) - G'(1)^2
\end{aligned}$$

are worth committing to memory. As an example, suppose X is Poisson distributed with mean λ. Then

$$G(u) = \sum_{k=0}^{\infty} \frac{\lambda^k}{k!} e^{-\lambda} u^k = e^{-\lambda(1-u)}.$$

Repeated differentiation yields $\frac{d^j}{du^j}G(1) = \lambda^j$. In particular, $\mathrm{E}(X) = \lambda$ and $\mathrm{Var}(X) = \lambda$. For another example, let X follow a binomial distribution with n trials and success probability p per trial. In this case $G(u) = (1-p+pu)^n$, $\mathrm{E}(X) = np$, and $\mathrm{Var}(X) = np(1-p)$. ∎

Example 2.4.4 *Random Sums*

Suppose $X_1, X_2, \ldots$ is a sequence of independent, identically distributed random variables. Consider the random sum $S_N = \sum_{i=1}^{N} X_i$, where the random number of terms N is independent of the X_i, and where we adopt the convention $S_0 = 0$. For example in an ecological study, the number of animal litters N in a plot of land might have a Poisson distribution to a good approximation. The random variable X_i then represents the number of offspring in litter i, and the compound Poisson random variable S_N counts the number of offspring over the whole plot.

If N has probability generating function $G(u) = \mathrm{E}(u^N)$, then the characteristic function of S_N is

$$\mathrm{E}(e^{itS_N}) = \sum_{n=0}^{\infty} \mathrm{E}(e^{itS_N} \mid N = n)\,\mathrm{Pr}(N = n)$$

$$= \sum_{n=0}^{\infty} E(e^{itX_1})^n \Pr(N = n)$$

$$= G[E(e^{itX_1})].$$

This composition rule carries over to other moment transforms. For instance, if the X_i are nonnegative and integer-valued with probability generating function $H(u)$, then a similar argument gives $E(u^{S_N}) = G[H(u)]$.

We can extract the moments of S_N by differentiation. Alternatively, conditioning on N produces

$$E(S_N) \;=\; E[N\,E(X_1)] \;=\; E(N)\,E(X_1)$$

and

$$\begin{aligned}
\mathrm{Var}(S_N) &= E[\mathrm{Var}(S_N \mid N)] + \mathrm{Var}[E(S_N \mid N)] \\
&= E[N\,\mathrm{Var}(X_1)] + \mathrm{Var}[N\,E(X_1)] \\
&= E(N)\,\mathrm{Var}(X_1) + \mathrm{Var}(N)\,E(X_1)^2.
\end{aligned}$$

For instance, if N has a Poisson distribution with mean λ and the X_i have a binomial distribution with parameters n and p, then S_N has

$$\begin{aligned}
E(u^{S_N}) &= e^{-\lambda[1-(1-p+pu)^n]} \\
E(S_N) &= \lambda n p \\
\mathrm{Var}(S_N) &= \lambda n p(1-p) + \lambda n^2 p^2
\end{aligned}$$

as its probability generating function, mean, and variance, respectively. ∎

Example 2.4.5 *Sum of Uniforms*

In Example 2.3.2, we considered the product of n independent random variables $U_1, \ldots, U_n$ uniformly distributed on $[0, 1]$. We now turn to the problem of finding the density of the sum $S_n = U_1 + \cdots + U_n$. Our strategy will be to calculate and invert the Laplace transform of the density of S_n, keeping in mind that the Laplace transform of a random variable coincides with the Laplace transform of its density. Because the Laplace transform of a single U_i is $\int_0^1 e^{-tx}\,dx = (1 - e^{-t})/t$, the Laplace transform of S_n is

$$\frac{(1 - e^{-t})^n}{t^n} \;=\; \frac{1}{t^n} \sum_{k=0}^{n} \binom{n}{k}(-1)^k e^{-kt}.$$

In view of the linearity of the Laplace transform, it therefore suffices to invert the term e^{-kt}/t^n. Since multiplication by e^{-kt} in the transform domain corresponds to an argument shift of k in the original domain, all we need to do is find the function with transform t^{-n} and shift it by k. We now make an inspired guess that the function x^{n-1} is relevant. Because the Laplace transform deals with functions defined on $[0, \infty)$, we exchange x^{n-1} for the function $(x)_+^{n-1}$, which equals 0 for $x \leq 0$ and x^{n-1} for $x > 0$. The change of variables $u = tx$ and the definition of the gamma function show that $(x)_+^{n-1}$ has transform

$$\int_0^\infty x^{n-1}e^{-tx}\,dx = \frac{1}{t^n}\int_0^\infty u^{n-1}e^{-u}\,du$$
$$= \frac{(n-1)!}{t^n}.$$

Up to a constant, this is just what we need. Hence, we conclude that S_n has density

$$f(x) = \frac{1}{(n-1)!}\sum_{k=0}^n \binom{n}{k}(-1)^k(x-k)_+^{n-1}.$$

The corresponding distribution function

$$F(x) = \frac{1}{n!}\sum_{k=0}^n \binom{n}{k}(-1)^k(x-k)_+^n$$

emerges after integration with respect to x.   ∎

Example 2.4.6 *A Nonexistence Problem*

Is it always possible to represent a random variable X as the difference $Y - Z$ of two independent, identically distributed random variables Y and Z? The answer is clearly no unless X is symmetrically distributed around 0. For a symmetrically distributed X, the question is more subtle. Suppose that Y and Z exist for such an X. Then the characteristic function of X reduces to

$$\mathrm{E}[e^{it(Y-Z)}] = \mathrm{E}(e^{itY})\,\mathrm{E}(e^{-itZ})$$
$$= \mathrm{E}(e^{itY})\,\mathrm{E}(e^{itY})^*$$
$$= |\mathrm{E}(e^{itY})|^2,$$

where the superscript $*$ denotes complex conjugation. (It is trivial to check that conjugation commutes with expectation for complex random variables possessing only a finite number of values, and this property persists in the limit for all complex random variables.) In any case, if the representation $X = Y - Z$ holds, then the characteristic function of X is nonnegative. Thus, to construct a counterexample, all we need to do is find a symmetrically distributed random variable whose characteristic functions fails the test of nonnegativity. For instance, if we take X to be uniformly distributed on $[-\frac{1}{2}, \frac{1}{2}]$, then its characteristic function

$$\int_{-\frac{1}{2}}^{\frac{1}{2}} e^{itx}\,dx = \frac{e^{itx}}{it}\Big|_{-\frac{1}{2}}^{\frac{1}{2}} = \frac{\sin(\frac{t}{2})}{\frac{t}{2}}$$

oscillates in sign.   ∎

Example 2.4.7 *Characterization of the Standard Normal Distribution*

Consider a random variable X with mean 0, variance 1, and characteristic function $\hat{\psi}(t)$. If X is standard normal and Y is an independent copy of X, then for all a and b the sum $aX + bY$ has the same distribution as $\sqrt{a^2 + b^2}X$. This distributional identity implies the characteristic function identity

$$\hat{\psi}(at)\hat{\psi}(bt) \;=\; \hat{\psi}\left(\sqrt{a^2 + b^2}\,t\right) \qquad (2.4)$$

for t.

Conversely, suppose the functional equation (2.4) holds for a random variable X with mean 0 and variance 1. Let us show that X possesses a standard normal distribution. The special case $a = -1$ and $b = 0$ of equation (2.4) amounts to $\hat{\psi}(-t) = \hat{\psi}(t)$, from which it immediately follows that $\hat{\psi}(t)^* = \hat{\psi}(-t) = \hat{\psi}(t)$. Thus, $\hat{\psi}(t)$ is real and even. It is also differentiable because $\mathrm{E}(|X|) < \infty$. Now define $g(t^2) = \hat{\psi}(t)$ for $t > 0$. The functional equation (2.4) entails the corresponding functional equation

$$g(a)g(b) \;=\; g(a + b) \qquad (2.5)$$

at $t = 1$. Taking $f(t) = \ln g(t)$ lands us right back at equation (2.2), except that it is no longer clear that $f(t)$ is monotone. Rather than rely on our previous hand-waving solution, we can differentiate equation (2.5), first with respect to a and then with respect to b. This yields

$$g'(a)g(b) \;=\; g'(a + b) \;=\; g(a)g'(b). \qquad (2.6)$$

If we take $b > 0$ sufficiently small, then $g(b) > 0$, owing to the continuity of $g(t)$ and the initial condition $g(0) = 1$. Dividing equation (2.6) by $g(b)$ and defining $\lambda = -g'(b)/g(b)$ leads to the differential equation $g'(a) = -\lambda g(a)$ with solution $g(a) = e^{-\lambda a}$. To determine λ, note that the equality

$$g''(t^2)4t^2 + g'(t^2)2 \;=\; \hat{\psi}''(t)$$

yields $-2\lambda = -1$ in the limit as t approaches 0. Thus, $\hat{\psi}(t) = e^{-t^2/2}$ as required.   ∎

Example 2.4.8 *Stirling's Formula*

The Fourier inversion formula (1.8) provides a quick derivation of Stirling's formula [22]. Our indirect attack uses arguments familiar from the proof of the central limit theorem. Consider a gamma random variable X_n with density $x^{n-1}e^{-x}/\Gamma(n)$. A brief calculation shows that the standardized random variable $S_n = (X_n - n)/\sqrt{n}$ has density

$$f_n(x) \;=\; \frac{\sqrt{n}(\sqrt{n}x + n)^{n-1}}{\Gamma(n)}e^{-\sqrt{n}x - n}.$$

Because X_n has characteristic function $1/(1-it)^n$ and

$$\ln(1-y) = -y - \frac{1}{2}y^2 + O(|y|^3)$$

for y near 0, we find that S_n has characteristic function

$$\hat{f}_n(t) = e^{-i\sqrt{n}t}\left(1 - \frac{it}{\sqrt{n}}\right)^{-n}$$

$$= e^{-i\sqrt{n}t - n\ln(1-it/\sqrt{n})}$$

$$= e^{-t^2/2 + O(1/\sqrt{n})}.$$

At the point $x = 0$, the Fourier inversion formula (1.8) gives

$$\frac{n^{n-1/2}e^{-n}}{\Gamma(n)} = f_n(0) = \frac{1}{2\pi}\int_{-\infty}^{\infty}\hat{f}_n(t)\,dt.$$

Taking limits in this equality therefore produces Stirling's formula

$$\lim_{n\to\infty}\frac{n^{n-1/2}e^{-n}}{\Gamma(n)} = \frac{1}{2\pi}\int_{-\infty}^{\infty}e^{-t^2/2}\,dt = \frac{1}{\sqrt{2\pi}},$$

provided we can justify the interchange of the limit and integral. Here the dominated convergence theorem comes into play with dominating function $(1 + t^2/2)^{-1}$. To prove that

$$\left|e^{-i\sqrt{n}t}\left(1 - \frac{it}{\sqrt{n}}\right)^{-n}\right| \le \left(1 + \frac{t^2}{2}\right)^{-1},$$

it suffices to show that

$$1 + \frac{t^2}{2} \le \left|\left(1 - \frac{it}{\sqrt{n}}\right)^n\right| = \left(1 + \frac{t^2}{n}\right)^{n/2}.$$

Equivalently, we must show that $g(s) = \frac{n}{2}\ln(1 + s/n) - \ln(1 + s/2)$ is nonnegative on the interval $[0, \infty)$ for $n \ge 2$. This simple calculus problem we leave to the reader. ∎

2.5 Tail Probability Methods

Consider a nonnegative random variable X with distribution function $F(x)$. The right-tail probability $\Pr(X > t) = 1 - F(t)$ turns out to helpful in calculating certain expectations relative to X. Let $h(t)$ be an integrable function on each finite interval $[0, x]$. If we define $H(x) = H(0) + \int_0^x h(t)\,dt$

and suppose that $\int_0^\infty |h(t)|[1-F(t)]\,dt < \infty$, then Fubini's theorem justifies the calculation

$$
\begin{aligned}
\mathrm{E}[H(X)] &= H(0) + \mathrm{E}\left[\int_0^X h(t)\,dt\right] \\
&= H(0) + \int_0^\infty \int_0^x h(t)\,dt\,dF(x) \\
&= H(0) + \int_0^\infty \int_t^\infty dF(x)h(t)\,dt \qquad (2.7) \\
&= H(0) + \int_0^\infty h(t)[1 - F(t)]\,dt.
\end{aligned}
$$

If X is concentrated on the integers $\{0, 1, 2, \ldots\}$, the right-tail probability $1 - F(t)$ is constant except for jumps at these integers. Equation (2.7) therefore reduces to

$$
\mathrm{E}[H(X)] = H(0) + \sum_{k=0}^\infty [H(k+1) - H(k)][1 - F(k)]. \qquad (2.8)
$$

Example 2.5.1 *Moments from Right-Tail Probabilities*

The choices $h(t) = nt^{n-1}$ and $H(0) = 0$ yield $H(x) = x^n$. Hence, equations (2.7) and (2.8) become

$$
\mathrm{E}[X^n] = n \int_0^\infty t^{n-1}[1 - F(t)]\,dt
$$

and

$$
\mathrm{E}[X^n] = \sum_{k=0}^\infty [(k+1)^n - k^n][1 - F(k)],
$$

respectively. For instance, if X is exponentially distributed, then the right-tail probability $1 - F(t) = e^{-\lambda t}$ and $E(X) = \int_0^\infty e^{-\lambda t}dt = \lambda^{-1}$. If X is geometrically distributed with failure probability q, then $1 - F(k) = q^k$ and $E(X) = \sum_{k=0}^\infty q^k = (1 - q)^{-1}$. ∎

Example 2.5.2 *Laplace Transforms*

Equation (2.7) also determines the relationship between the Laplace transform $\mathrm{E}(e^{-sX})$ of a nonnegative random variable X and the ordinary Laplace transform $\tilde{F}(s)$ of its distribution function $F(x)$. For this purpose, we choose $h(t) = -se^{-st}$ and $H(0) = 1$. The resulting integral $H(x) = e^{-sx}$ and equation (2.7) together yield the formula

$$
\mathrm{E}(e^{-sX}) = 1 - s \int_0^\infty e^{-st}[1 - F(t)]\,dt
$$

$$= s \int_0^\infty e^{-st} F(t)\, dt$$

$$= s\tilde{F}(s).$$

For example, if X is exponentially distributed with intensity λ, then the Laplace transform $\mathrm{E}(e^{-sX}) = \lambda/(s+\lambda)$ mentioned in Example 2.4.2 leads to $\tilde{F}(s) = \lambda/[s(s+\lambda)]$. ∎

2.6   Moments of Reciprocals and Ratios

Ordinarily we differentiate Laplace transforms to recover moments. However, to recover an inverse moment, we need to integrate [34]. Suppose X is a positive random variable with Laplace transform $L(t)$. If $n > 0$, then Fubini's theorem and the change of variables $s = tX$ shows that

$$\int_0^\infty t^{n-1} L(t)\, dt = \mathrm{E}\left(\int_0^\infty t^{n-1} e^{-tX}\, dt\right)$$

$$= \mathrm{E}\left(X^{-n} \int_0^\infty s^{n-1} e^{-s}\, ds\right)$$

$$= \mathrm{E}\left(X^{-n}\right) \Gamma(n).$$

The formula

$$\mathrm{E}\left(X^{-n}\right) = \frac{1}{\Gamma(n)} \int_0^\infty t^{n-1} L(t)\, dt \tag{2.9}$$

can be evaluated exactly in some cases. In other cases, for instance when n fails to be an integer, the formula can be evaluated numerically.

Example 2.6.1 *Mean and Variance of an Inverse Gamma*

Because a gamma random variable X with intensity λ and shape parameter β has Laplace transform $L(t) = [\lambda/(t+\lambda)]^\beta$, formula (2.9) gives

$$\mathrm{E}\left(X^{-1}\right) = \int_0^\infty \left(\frac{\lambda}{t+\lambda}\right)^\beta dt$$

$$= \frac{\lambda}{\beta - 1}$$

for $\beta > 1$ and

$$\mathrm{E}\left(X^{-2}\right) = \int_0^\infty t\left(\frac{\lambda}{t+\lambda}\right)^\beta dt$$

$$= \int_0^\infty \lambda\left(\frac{\lambda}{t+\lambda}\right)^{\beta-1} dt - \int_0^\infty \lambda\left(\frac{\lambda}{t+\lambda}\right)^\beta dt$$

$$= \frac{\lambda^2}{\beta - 2} - \frac{\lambda^2}{\beta - 1}$$

for $\beta > 2$. It follows that $\mathrm{Var}(X^{-1}) = \lambda^2/[(\beta-1)^2(\beta-2)]$ for $\beta > 2$. ∎

To calculate the expectation of a ratio X^m/Y^n for a positive random variable Y and an arbitrary random variable X, we consider the mixed characteristic function and Laplace transform $M(s,t) = \mathrm{E}(e^{isX-tY})$. Assuming that $\mathrm{E}(|X|^m) < \infty$ for some positive integer m, we can write

$$\frac{\partial^m}{\partial s^m} M(s,t) = \mathrm{E}\left[(iX)^m e^{isX-tY}\right]$$

by virtue of Example 1.2.5 with dominating random variable $|X|^k e^{-tY}$ for the kth partial derivative. For $n > 0$ and $\mathrm{E}(|X|^m Y^{-n}) < \infty$, we now invoke Fubini's theorem and calculate

$$
\begin{aligned}
\int_0^\infty t^{n-1} \frac{\partial^m}{\partial s^m} M(0,t)\, dt &= \int_0^\infty t^{n-1} \mathrm{E}\left[(iX)^m e^{-tY}\right] dt \\
&= \mathrm{E}\left[\int_0^\infty t^{n-1} (iX)^m e^{-tY}\, dt\right] \\
&= \mathrm{E}\left[\frac{(iX)^m}{Y^n} \int_0^\infty r^{n-1} e^{-r}\, dr\right] \\
&= \mathrm{E}\left[\frac{(iX)^m}{Y^n}\right] \Gamma(n).
\end{aligned}
$$

Rearranging this yields

$$\mathrm{E}\left(\frac{X^m}{Y^n}\right) = \frac{1}{i^m \Gamma(n)} \int_0^\infty t^{n-1} \frac{\partial^m}{\partial s^m} M(0,t)\, dt. \tag{2.10}$$

Example 2.6.2 *Mean of a Beta Random Variable*

If U and V are independent gamma random variables with common intensity λ and shape parameters α and β, then the ratio $U/(U+V)$ has a beta distribution with parameters α and β. The reader is asked to prove this fact in Problem 20. It follows that the mixed characteristic function and Laplace transform

$$
\begin{aligned}
M_{U,U+V}(s,t) &= \mathrm{E}\left[e^{isU-t(U+V)}\right] \\
&= \mathrm{E}\left[e^{(is-t)U}\right] \mathrm{E}\left[e^{-tV}\right] \\
&= \left(\frac{\lambda}{\lambda - is + t}\right)^\alpha \left(\frac{\lambda}{\lambda + t}\right)^\beta.
\end{aligned}
$$

Equation (2.10) consequently gives the mean of the beta distribution as

$$
\begin{aligned}
\mathrm{E}\left(\frac{U}{U+V}\right) &= \frac{1}{i} \int_0^\infty \alpha i \frac{\lambda^\alpha}{(\lambda+t)^{\alpha+1}} \left(\frac{\lambda}{\lambda+t}\right)^\beta dt \\
&= -\frac{\alpha \lambda^{\alpha+\beta}}{(\alpha+\beta)(\lambda+t)^{\alpha+\beta}} \Big|_0^\infty \\
&= \frac{\alpha}{\alpha+\beta}.
\end{aligned}
$$

2.7 Reduction of Degree

The method of reduction of degree also involves recurrence relations. However, instead of creating these by conditioning, we now employ integration by parts and simple algebraic transformations. The Stein and Chen lemmas given below find their most important applications in the approximation theories featured in the books [16, 141].

Example 2.7.1 *Stein's Lemma*

Suppose X is normally distributed with mean μ and variance σ^2 and $g(x)$ is a differentiable function such that $|g(X)(X - \mu)|$ and $|g'(X)|$ have finite expectations. Stein's lemma [141] asserts that

$$E[g(X)(X - \mu)] \;=\; \sigma^2 E[g'(X)].$$

To prove this formula, we note that integration by parts produces

$$\frac{1}{\sqrt{2\pi\sigma^2}} \int_{-\infty}^{\infty} g(x)(x - \mu) e^{-\frac{(x-\mu)^2}{2\sigma^2}} \, dx$$

$$= \lim_{a_n \to -\infty} \lim_{b_n \to +\infty} \left[\frac{-\sigma^2 g(x) e^{-\frac{(x-\mu)^2}{2\sigma^2}}}{\sqrt{2\pi\sigma^2}} \Bigg|_{a_n}^{b_n} + \frac{\sigma^2}{\sqrt{2\pi\sigma^2}} \int_{a_n}^{b_n} g'(x) e^{-\frac{(x-\mu)^2}{2\sigma^2}} \, dx \right]$$

$$= \frac{\sigma^2}{\sqrt{2\pi\sigma^2}} \int_{-\infty}^{\infty} g'(x) e^{-\frac{(x-\mu)^2}{2\sigma^2}} \, dx.$$

The boundary terms vanish for carefully chosen sequences a_n and b_n tending to $\pm\infty$ because the integrable function $|g(x)(x - \mu)| \exp[-\frac{(x-\mu)^2}{2\sigma^2}]$ cannot be bounded away from 0 as $|x|$ tends to ∞. To illustrate the repeated application of Stein's lemma, take $g(x) = (x - \mu)^{2n-1}$. Then the important moment identity

$$\begin{aligned} E[(X - \mu)^{2n}] &= \sigma^2 (2n - 1) E[(X - \mu)^{2n-2}] \\ &= \sigma^{2n}(2n - 1)(2n - 3) \cdots 1 \end{aligned}$$

follows immediately ∎

Example 2.7.2 *Reduction of Degree for the Gamma*

A random variable X with gamma density $\lambda^\alpha x^{\alpha-1} e^{-\lambda x}/\Gamma(\alpha)$ on $(0, \infty)$ satisfies the analogous reduction of degree formula

$$E[g(X)X] \;=\; \frac{1}{\lambda} E[g'(X)X] + \frac{\alpha}{\lambda} E[g(X)].$$

Provided the required moments exist and $\lim_{x \to 0} g(x)x^\alpha = 0$, the integration by parts calculation

$$\frac{\lambda^\alpha}{\Gamma(\alpha)} \int_0^\infty g(x)x x^{\alpha-1} e^{-\lambda x} dx$$

$$= \frac{\lambda^{\alpha}}{\Gamma(\alpha)\lambda} \left[-g(x)xx^{\alpha-1}e^{-\lambda x} \Big|_0^{\infty} + \int_0^{\infty} g'(x)xx^{\alpha-1}e^{-\lambda x}dx \right.$$

$$\left. + \int_0^{\infty} g(x)\alpha x^{\alpha-1}e^{-\lambda x}dx \right]$$

is valid. The special case $g(x) = x^{n-1}$ yields the recurrence relation

$$\mathrm{E}(X^n) = \frac{(n-1+\alpha)}{\lambda}\mathrm{E}(X^{n-1})$$

for the moments of X.   ■

Example 2.7.3 *Chen's Lemma*

Chen [28] pursues the formula $\mathrm{E}[Zg(Z)] = \lambda\,\mathrm{E}[g(Z+1)]$ for a Poisson random variable Z with mean λ as a kind of discrete analog to Stein's lemma. The proof of Chen's result

$$\sum_{j=0}^{\infty} jg(j)\frac{\lambda^j e^{-\lambda}}{j!} = \lambda\sum_{j=1}^{\infty} g(j)\frac{\lambda^{j-1} e^{-\lambda}}{(j-1)!}$$

$$= \lambda\sum_{k=0}^{\infty} g(k+1)\frac{\lambda^k e^{-\lambda}}{k!}$$

is almost trivial. The choice $g(z) = z^{n-1}$ gives the recurrence relation

$$\mathrm{E}(Z^n) = \lambda\,\mathrm{E}[(Z+1)^{n-1}]$$

$$= \lambda\sum_{k=0}^{n-1} \binom{n-1}{k}\mathrm{E}(Z^k)$$

for the moments of Z.   ■

2.8   Spherical Surface Measure

In this section and the next, we explore probability measures on surfaces. Surface measures are usually treated using differential forms and manifolds [139]. With enough symmetry, one can dispense with these complicated mathematical objects and fall back on integration in R^n. This concrete approach has the added benefit of facilitating the calculation of certain expectations.

Let $g(\|x\|)$ be any probability density such as $e^{-\pi\|x\|^2}$ on R^n that depends only on the Euclidean distance $\|x\|$ of a point x from the origin. Given a choice of $g(\|x\|)$, one can define the integral of a continuous, real-valued function $f(s)$ on the unit sphere $S_{n-1} = \{x \in \mathrm{R}^n : \|x\| = 1\}$ by

$$\int_{S_{n-1}} f(s)\, d\omega_{n-1}(s) \;=\; a_{n-1} \int f\!\left(\frac{x}{\|x\|}\right) g(\|x\|)\, dx \qquad (2.11)$$

for a positive constant a_{n-1} to be specified [14]. It is trivial to show that this yields an invariant integral in the sense that

$$\int_{S_{n-1}} f(Ts)\, d\omega_{n-1}(s) \;=\; \int_{S_{n-1}} f(s)\, d\omega_{n-1}(s)$$

for any orthogonal transformation T. In this regard note that $|\det(T)| = 1$ and $\|Tx\| = \|x\|$. Taking $f(s) = 1$ produces a total mass of a_{n-1} for the surface measure ω_{n-1}.

Of course, the constant a_{n-1} is hardly arbitrary. We can pin it down by proving the product measure formula

$$\int h(x)\, dx \;=\; \int_0^\infty \int_{S_{n-1}} h(rs)\, d\omega_{n-1}(s)\, r^{n-1}\, dr \qquad (2.12)$$

for any integrable function $h(x)$ on R^n. Formula (2.12) says that we can integrate over R^n by cumulating the surface integrals over successive spherical shells. To prove (2.12), we interchange orders of integration as needed and execute the successive changes of variables $t = r\|x\|^{-1}$, $z = tx$, and $t = \|z\|r^{-1}$. These maneuvers turn the right-hand side of formula (2.12) into

$$\int_0^\infty \int_{S_{n-1}} h(rs)\, d\omega_{n-1}(s)\, r^{n-1}\, dr$$

$$= a_{n-1} \int_0^\infty \int h(rx/\|x\|) g(\|x\|)\, dx\, r^{n-1}\, dr$$

$$= a_{n-1} \int \int_0^\infty h(rx/\|x\|)\, r^{n-1}\, dr\, g(\|x\|)\, dx$$

$$= a_{n-1} \int \int_0^\infty h(tx)(t\|x\|)^{n-1}\|x\|\, dt\, g(\|x\|)\, dx$$

$$= a_{n-1} \int_0^\infty \int h(tx)\|x\|^n g(\|x\|)\, dx\, t^{n-1}\, dt$$

$$= a_{n-1} \int_0^\infty \int h(z)(\|z\|/t)^n g(\|z\|/t)\, t^{-n}\, dz\, t^{n-1}\, dt$$

$$= a_{n-1} \int \int_0^\infty (\|z\|/t)^{n-1} g(\|z\|/t)\|z\|t^{-2}\, dt\, h(z)\, dz$$

$$= a_{n-1} \int_0^\infty r^{n-1} g(r)\, dr \int h(z)\, dz.$$

This establishes (2.12) provided we take $a_{n-1} \int_0^\infty r^{n-1} g(r)\, dr = 1$. For the choice $g(r) = e^{-\pi r^2}$, we calculate

$$\int_0^\infty r^{n-1} e^{-\pi r^2}\, dr = \int_0^\infty \left(\frac{t}{\pi}\right)^{(n-2)/2} e^{-t} \frac{1}{2\pi}\, dt = \frac{\Gamma(\frac{n}{2})}{2\pi^{n/2}}. \qquad (2.13)$$

Thus, the surface area a_{n-1} of S_{n-1} reduces to $2\pi^{n/2}/\Gamma(\frac{n}{2})$. Omitting the constant a_{n-1} in the definition (2.11) yields the uniform probability distribution on S_{n-1}.

Besides offering a method of evaluating integrals, formula (2.12) demonstrates that the definition of surface measure does not depend on the choice of the function $g(\|x\|)$. In fact, consider the extension

$$h(x) = f\left(\frac{x}{\|x\|}\right) 1_{\{1 \le \|x\| \le d\}}$$

of a function $f(x)$ on S_{n-1}. If we take $d = \sqrt[n]{n+1}$, then $\int_1^d r^{n-1}\, dr = 1$, and formula (2.12) amounts to

$$\int_{S_{n-1}} f(s)\, d\omega_{n-1}(s) = \frac{\int h(x)\, dx}{\int_1^d r^{n-1} dr} = \int h(x)\, dx,$$

which, as Baker notes [14], affords a definition of the surface integral that does not depend on the choice of the probability density $g(\|x\|)$. As a byproduct of this result, it follows that the surface area a_{n-1} of S_{n-1} also does not depend on $g(\|x\|)$.

Example 2.8.1 *Moments of $\|x\|$ Relative to $e^{-\pi \|x\|^2}$*

Formula (2.12) gives

$$\int \|x\|^k e^{-\pi \|x\|^2}\, dx = \int_0^\infty \int_{S_{n-1}} r^k e^{-\pi r^2}\, d\omega_{n-1}(s) r^{n-1} dr$$

$$= a_{n-1} \int_0^\infty r^{n+k-1} e^{-\pi r^2}\, dr$$

$$= \frac{a_{n-1}}{a_{n+k-1}}.$$

Negative as well as positive values of $k > -n$ are permitted.   ∎

Example 2.8.2 *Integral of a Polynomial*

The function $f(x) = x_1^{k_1} \cdots x_n^{k_n}$ is a monomial when $k_1, \ldots, k_n$ are nonnegative integers. A linear combination of monomials is a polynomial. To find the integral of $f(x)$ on S_{n-1}, it is convenient to put $k = \sum_{j=1}^n k_j$ and use the probability density $g(\|x\|) = a_{n+k-1}\|x\|^k e^{-\pi\|x\|^2}/a_{n-1}$. With these choices,

$$
\begin{aligned}
\int_{S_{n-1}} f(s)\,d\omega_{n-1}(s) &= a_{n-1}\frac{a_{n+k-1}}{a_{n-1}}\int \frac{x_1^{k_1}\cdots x_n^{k_n}}{\|x\|^k}\|x\|^k e^{-\pi\|x\|^2}\,dx \\
&= a_{n+k-1}\int x_1^{k_1}\cdots x_n^{k_n} e^{-\pi\|x\|^2}\,dx \\
&= a_{n+k-1}\prod_{j=1}^{n}\int_{-\infty}^{\infty} x_j^{k_j} e^{-\pi x_j^2}\,dx_j.
\end{aligned}
$$

If any k_j is odd, then the corresponding one-dimensional integral in the last product vanishes. Hence, the surface integral of the monomial vanishes as well. If all k_j are even, then the same reasoning that produced equation (2.13) leads to

$$
\begin{aligned}
\int_{-\infty}^{\infty} x_j^{k_j} e^{-\pi x_j^2}\,dx_j &= 2\int_0^{\infty} x_j^{k_j} e^{-\pi x_j^2}\,dx_j \\
&= \frac{\Gamma(\frac{k_j+1}{2})}{\pi^{(k_j+1)/2}}.
\end{aligned}
$$

It follows that

$$
\begin{aligned}
\int_{S_{n-1}} x_1^{k_1}\cdots x_n^{k_n}\,d\omega_{n-1}(s) &= \frac{2\pi^{(n+k)/2}}{\Gamma(\frac{n+k}{2})}\prod_{j=1}^{n}\frac{\Gamma(\frac{k_j+1}{2})}{\pi^{(k_j+1)/2}} \\
&= \frac{2\prod_{j=1}^{n}\Gamma(\frac{k_j+1}{2})}{\Gamma(\frac{n+k}{2})}
\end{aligned}
$$

when all k_j are even.   ∎

2.9   Dirichlet Distribution

The Dirichlet distribution generalizes the beta distribution. As such, it lives on the unit simplex $T_{n-1} = \{x \in R_+^n : \|x\|_1 = 1\}$, where $\|x\|_1 = \sum_{j=1}^{n}|x_j|$ and

$$
R_+^n = \{x \in R^n : x_j > 0,\ j = 1,\ldots,n\}.
$$

By analogy with our definition (2.11) of spherical surface measure, one can define the simplex surface measure μ_{n-1} on T_{n-1} through the equation

$$
\int_{T_{n-1}} f(s)\,d\mu_{n-1}(s) = b_{n-1}\int_{R_+^n} f\left(\frac{x}{\|x\|_1}\right)g(\|x\|_1)\,dx \qquad (2.14)
$$

for any continuous function $f(s)$ on T_{n-1}. In this setting, $g(\|x\|_1)$ is a probability density on R_+^n that depends only on the distance $\|x\|_1$ of x from the origin.

One can easily show that this definition of surface measure is invariant under permutation of the coordinates. One can also prove the product measure formula

$$\int_{\mathsf{R}_+^n} h(x)\, dx \;=\; \frac{1}{\sqrt{n}} \int_0^\infty \int_{T_{n-1}} h(rs)\, d\mu_{n-1}(s)\, r^{n-1}\, dr. \qquad (2.15)$$

The appearance of the factor $1/\sqrt{n}$ here can be explained by appealing to geometric intuition. In formula (2.15) we integrate $h(x)$ by summing its integrals over successive slabs multiplied by the thicknesses of the slabs. Now the thickness of a slab amounts to nothing more than the distance between two slices $(r+dr)T_{n-1}$ and rT_{n-1}. Given that the corresponding centers of mass are $(r+dr)n^{-1}\mathbf{1}$ and $rn^{-1}\mathbf{1}$, the slab thickness is $dr/\sqrt{n}$.

The proof of formula (2.15) is virtually identical to the proof of formula (2.12). In the final stage of the proof, we must set

$$\frac{b_{n-1}}{\sqrt{n}} \int_0^\infty r^{n-1} g(r)\, dr \;=\; 1.$$

The choice $g(r) = e^{-r}$ immediately gives $\int_0^\infty r^{n-1}e^{-r}dr = \Gamma(n)$. It follows that the surface area b_{n-1} of T_{n-1} is $\sqrt{n}/\Gamma(n)$. Omitting the constant b_{n-1} in the definition (2.14) yields the uniform probability distribution on T_{n-1}.

As before we evaluate the moment

$$\begin{aligned}
\int_{\mathsf{R}_+^n} \|x\|_1^k e^{-\|x\|_1}\, dx &= \frac{1}{\sqrt{n}} \int_0^\infty \int_{T_{n-1}} r^k e^{-r}\, d\mu_{n-1}(s) r^{n-1} dr \\
&= \frac{b_{n-1}}{\sqrt{n}} \int_0^\infty r^{n+k-1} e^{-r} dr \\
&= \frac{\Gamma(n+k)}{\Gamma(n)}.
\end{aligned}$$

For the multinomial $f(x) = x_1^{k_1} \cdots x_n^{k_n}$ with $k = \sum_{j=1}^n k_j$, we then evaluate

$$\begin{aligned}
\int_{T_{n-1}} f(s)\, d\mu_{n-1}(s) &= b_{n-1} \frac{\Gamma(n)}{\Gamma(n+k)} \int_{\mathsf{R}_+^n} \frac{x_1^{k_1} \cdots x_n^{k_n}}{\|x\|_1^k} \|x\|_1^k e^{-\|x\|_1}\, dx \\
&= b_{n-1} \frac{\Gamma(n)}{\Gamma(n+k)} \int_{\mathsf{R}_+^n} x_1^{k_1} \cdots x_n^{k_n} e^{-\|x\|_1}\, dx \\
&= b_{n-1} \frac{\Gamma(n)}{\Gamma(n+k)} \prod_{j=1}^n \Gamma(k_j + 1)
\end{aligned}$$

using the probability density

$$g(\|x\|_1) \;=\; \frac{\Gamma(n)}{\Gamma(n+k)} \|x\|_1^k e^{-\|x\|_1}$$

on R_+^n. This calculation identifies the Dirichlet distribution

$$\frac{\Gamma(k)}{b_{n-1}\Gamma(n)\prod_{j=1}^n \Gamma(k_j)}\prod_{j=1}^n s_j^{k_j-1}$$

as a probability density on T_{n-1} relative to μ_{n-1} with moments

$$\mathrm{E}\left(s_1^{l_1}\cdots s_n^{l_n}\right) = \frac{\Gamma(k)\prod_{j=1}^n \Gamma(k_j + l_j)}{\Gamma(k+l)\prod_{j=1}^n \Gamma(k_j)},$$

where $l = \sum_{j=1}^n l_j$. Note that $k_j > 0$ need not be an integer.

2.10   Problems

1. Show that the beta-binomial distribution of Example 2.3.1 has discrete density

   $$\Pr(S_n = k) = \binom{n}{k}\frac{\Gamma(\alpha + \beta)\Gamma(\alpha + k)\Gamma(\beta + n - k)}{\Gamma(\alpha)\Gamma(\beta)\Gamma(\alpha + \beta + n)}.$$

2. Prove that the $\ln X_n = \sum_{i=1}^n \ln U_i$ random variable of Example 2.3.2 follows a $-\frac{1}{2}\chi_{2n}^2$ distribution.

3. A noncentral chi-square random variable X has a χ_{n+2Y}^2 distribution conditional on a Poisson random variable Y with mean λ. Show that $\mathrm{E}(X) = n + 2\lambda$ and $\mathrm{Var}(X) = 2n + 8\lambda$.

4. Consider an urn with $b \geq 1$ black balls and $w \geq 0$ white balls. Balls are extracted from the urn without replacement until a black ball is encountered. Show that the number of balls N_{bw} extracted has mean $\mathrm{E}(N_{bw}) = (b+w+1)/(b+1)$. (Hint: Derive a recurrence relation and boundary conditions for $\mathrm{E}(N_{bw})$ and solve.)

5. Give a recursive method for computing the second moments $\mathrm{E}(N_{sd}^2)$ in the family planning model.

6. In the family planning model, suppose the couple has an upper limit m on the number of children they can afford. Hence, they stop whenever they reach their goal of s sons and d daughters or m total children, whichever comes first. Let N_{sdm} now be their random number of children. Give a recursive method for computing $\mathrm{E}(N_{sdm})$.

7. Let X be a nonnegative integer-valued random variable with probability generating function $Q(s)$. Find the probability generating functions of $X + k$ and kX in terms of $Q(s)$ for any nonnegative integer k.

8. Let $S_n = X_1 + \cdots + X_n$ be the sum of n independent random variables, each distributed uniformly over the set $\{1, 2, \ldots, m\}$. Find the probability generating function of S_n, and use it to calculate $E(S_n)$ and $\mathrm{Var}(S_n)$.

9. Consider a sequence $X_1, X_2, \ldots$ of independent, integer-valued random variables with common logarithmic distribution

$$\Pr(X_i = k) = -\frac{q^k}{k \ln(1 - q)}$$

for $k \geq 1$. Let N be a Poisson random variable with mean λ that is independent of the X_i. Show that the random sum $S_N = \sum_{i=1}^{N} X_i$ has a negative binomial distribution. Note that the required "number of successes" in the negative binomial need not be an integer [48].

10. Suppose X has gamma density $\lambda^\beta x^{\beta-1} e^{-\lambda x} / \Gamma(\beta)$ on $(0, \infty)$, where β is not necessarily an integer. Show that X has characteristic function $(\frac{\lambda}{\lambda-it})^\beta$ and Laplace transform $(\frac{\lambda}{\lambda+t})^\beta$. Use either of these to calculate the mean and variance of X. (Hint: For the characteristic function, derive and solve a differential equation. Alternatively, calculate the Laplace transform directly by integration and show that it can be extended to an analytic function in a certain region of the complex plane.)

11. Let X have the gamma density defined in Problem 10. Conditional on X, let Y have a Poisson distribution with mean X. Prove that Y has probability generating function

$$E(s^Y) = \left(\frac{\lambda}{\lambda + 1 - s}\right)^\beta.$$

12. Show that the bilateral exponential density $\frac{1}{2} e^{-|x|}$ has characteristic function $1/(1 + t^2)$. Use this fact to calculate its mean and variance.

13. Calculate the Laplace transform of the probability density

$$\frac{1 + a^2}{a^2} e^{-x} [1 - \cos(ax)] 1_{\{x \geq 0\}}.$$

14. Prove the elementary inequalities

$$\ln(n - 1)! \ \leq \ \int_1^n \ln t \, dt \ = \ n \ln n - n + 1 \ \leq \ \ln n!$$

that point the way to Stirling's formula.

15. Suppose the right-tail probability of a nonnegative random variable X satisfies $|1 - F(x)| \leq cx^{-n-\epsilon}$ for all sufficiently large x, where n is a positive integer, and ϵ and c are positive real numbers. Show that $E(X^n)$ is finite.

16. Let the positive random variable X have Laplace transform $L(t)$. Prove that $E[(aX + b)^{-1}] = \int_0^\infty e^{-bt} L(at)\, dt$ for $a \geq 0$ and $b > 0$.

17. Suppose X has a binomial distribution with success probability p over n trials. Show that

$$E\left(\frac{1}{X+1}\right) = \frac{1 - (1-p)^{n+1}}{(n+1)p}.$$

18. Let χ_n^2 and χ_{n+2}^2 be chi-square random variables with n and $n+2$ degrees of freedom, respectively. Demonstrate that

$$E[f(\chi_n^2)] = nE\left[\frac{f(\chi_{n+2}^2)}{\chi_{n+2}^2}\right]$$

for any well-behaved function $f(x)$ for which the two expectations exist. Use this identity to calculate the mean and variance of χ_n^2 [26].

19. Consider a negative binomial random variable X with density

$$\Pr(X = k) = \binom{k-1}{n-1} p^n q^{k-n}$$

for $q = 1 - p$ and $k \geq n$. Prove that for any function $f(x)$

$$E[qf(X)] = E\left[\frac{(X-n)f(X-1)}{X-1}\right].$$

Use this identity to calculate the mean of X [78].

20. One can generate the Dirichlet distribution by a different mechanism than the one developed in the text [89]. Take n independent gamma random variables $X_1, \ldots, X_n$ of unit scale and form the ratios

$$Y_i = \frac{X_i}{\sum_{j=1}^n X_j}.$$

Here X_i has density $x_i^{k_i-1} e^{-x_i}/\Gamma(k_i)$ on $(0, \infty)$ for some $k_i > 0$. Clearly, each $Y_i \geq 0$ and $\sum_{i=1}^n Y_i = 1$. Show that $(Y_1, \ldots, Y_n)^t$ follows a Dirichlet distribution on T_{n-1}.

21. Continuing Problem 20, calculate $E(Y_i)$, $\mathrm{Var}(Y_i)$, and $\mathrm{Cov}(Y_i, Y_j)$ for $i \neq j$. Also show that $(Y_1 + Y_2, Y_3, \ldots, Y_n)^t$ has a Dirichlet distribution.

3
Convexity, Optimization, and Inequalities

3.1 Introduction

Convexity is one of the key concepts of mathematical analysis and has interesting consequences for optimization theory, statistical estimation, inequalities, and applied probability. Despite this fact, students seldom see convexity presented in a coherent fashion. It always seems to take a backseat to more pressing topics. The current chapter is intended as a partial remedy to this pedagogical gap.

Our emphasis will be on convex functions rather than convex sets. It is helpful to have a variety of tests to recognize such functions. We present such tests and discuss the important class of log-convex functions. A strictly convex function has at most one minimum point. This property tremendously simplifies optimization. For a few functions, we are fortunate enough to be able to find their optima explicitly. For other functions, we must iterate. Section 3.4 introduces a class of optimization algorithms that exploit convexity. These algorithms are ideal for high-dimensional problems in statistics.

The concluding section of this chapter rigorously treats several inequalities. Our inclusion of Bernstein's proof of Weierstrass's approximation theorem provides a surprising application of Chebyshev's inequality and illustrates the role of probability theory in illuminating problems outside its usual sphere of influence. The less familiar inequalities of Jensen, Schlömilch, and Hölder find numerous applications in optimization theory and functional analysis.

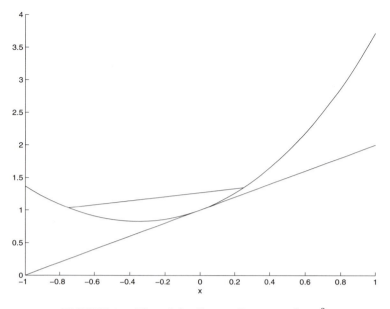

FIGURE 3.1. Plot of the Convex Function $e^x + x^2$

3.2   Convex Functions

A set $S \subset \mathsf{R}^m$ is said to be convex if the line segment between any two points x and y of S lies entirely within S. Formally, this means that whenever $x, y \in S$ and $\alpha \in [0, 1]$, the point $z = \alpha x + (1 - \alpha)y \in S$. In general, any convex combination $\sum_{i=1}^{n} \alpha_i x_i$ of points $x_1, \ldots, x_n$ in S must also reside in S. Here, the coefficients α_i must be nonnegative and sum to 1.

Convex functions are defined on convex sets. A real-valued function $f(x)$ defined on a convex set S is convex provided

$$f[\alpha x + (1 - \alpha)y] \quad \leq \quad \alpha f(x) + (1 - \alpha)f(y) \tag{3.1}$$

for all $x, y \in S$ and $\alpha \in [0, 1]$. Figure 3.1 depicts how in one dimension definition (3.1) requires the chord connecting two points on the curve $x \mapsto f(x)$ to lie above the curve. If strict inequality holds in (3.1) for every $x \neq y$ and $\alpha \in (0, 1)$, then $f(x)$ is said to strictly convex. One can prove by induction that inequality (3.1) extends to

$$f\left(\sum_{i=1}^{n} \alpha_i x_i\right) \quad \leq \quad \sum_{i=1}^{n} \alpha_i f(x_i)$$

for any convex combination of points from S.

Figure 3.1 also illustrates how a tangent line to the curve lies below the curve. This property characterizes convex functions. In stating and proving the property, we will deal with differentiable functions. Recall that $f(x)$ is differentiable at x if there exists a row vector $df(x)$ satisfying

$$f(y) \;=\; f(x) + df(x)(y - x) + r(y - x)$$

for all y near x and a remainder $r(y - x)$ that is $o(\|y - x\|)$ [31, 77]. Here $o(t)$ denotes a quantity such that $\lim_{t \to 0_+} o(t)t^{-1} = 0$. If the differential $df(x)$ exists, then the first partial derivatives of $f(x)$ exist at x as well, and we can identify $df(x)$ with the row vector of these partials.

Proposition 3.2.1 *Let $f(x)$ be a differentiable function on the open, convex set $S \subset \mathsf{R}^m$. Then $f(x)$ is convex if and only if*

$$f(y) \;\geq\; f(x) + df(x)(y - x) \tag{3.2}$$

for all $x, y \in S$. Furthermore, $f(x)$ is strictly convex if and only if strict inequality holds in inequality (3.2) when $y \neq x$.

Proof: If $f(x)$ is convex, then we can rearrange the inequality (3.1) to give

$$\frac{f[\alpha x + (1 - \alpha)y] - f(x)}{1 - \alpha} \;\leq\; f(y) - f(x).$$

Letting α tend to 1 proves inequality (3.2). To demonstrate the converse, let $z = \alpha x + (1 - \alpha)y$. Then with obvious notational changes, inequality (3.2) implies

$$\begin{aligned}
f(x) &\geq f(z) + df(z)(x - z) \\
f(y) &\geq f(z) + df(z)(y - z).
\end{aligned}$$

Multiplying the first of these inequalities by α and the second by $1 - \alpha$ and adding the results produce

$$\alpha f(x) + (1 - \alpha)f(y) \;\geq\; f(z) + df(z)(z - z) \;=\; f(z),$$

which is just inequality (3.1). The claims about strict convexity are left to the reader. ∎

Example 3.2.1 *Linear Functions*

For a linear function $f(x) = a^t x + b$, both of the inequalities (3.1) and (3.2) are equalities. Thus, a linear function is convex. ∎

Example 3.2.2 *Euclidean Norm*

The Euclidean norm $f(x) = \|x\| = \sqrt{\sum_{i=1}^m x_i^2}$ satisfies the standard triangle inequality and the homogeneity condition $\|cx\| = |c|\,\|x\|$. Thus,

$$\|\alpha x + (1 - \alpha)y\| \;\leq\; \|\alpha x\| + \|(1 - \alpha)y\| \;\leq\; \alpha\|x\| + (1 - \alpha)\|y\|$$

for any $\alpha \in [0, 1]$. It is noteworthy that $\|x\| = |x|$ is not differentiable at $x = 0$ when $m = 1$. However, closer examination of the proof of Proposition 3.2.1 makes it clear that the condition

$$f(y) \geq f(x) + c^t(y - x)$$

suffices for some vector c. This "supporting hyperplane" condition is certainly true at $x = 0$ for $f(x) = |x|$ provided we choose $|c| \leq 1$. ∎

It is useful to have simpler tests for convexity than inequality (3.1) or (3.2). One such test involves the second differential $d^2 f(x)$ of a function $f(x)$. We can view $d^2 f(x)$ as the Hessian matrix of second partial derivatives of $f(x)$.

Proposition 3.2.2 *Consider a twice continuously differentiable function $f(x)$ on the open, convex set $S \subset \mathbb{R}^m$. If its second differential $d^2 f(x)$ is positive semidefinite, then $f(x)$ is convex. If $d^2 f(x)$ is positive definite, then $f(x)$ is strictly convex.*

Proof: Note that $f(z)$ is a twice continuously differentiable function of the real variable α along the line $z = \alpha x + (1 - \alpha)y$. Executing a second-order Taylor expansion around $\alpha = 1$ therefore gives

$$f(y) = f(x) + df(x)(y - x) + \frac{1}{2}(y - x)^t d^2 f(z)(y - x)$$

for some z on the line between x and y. The claim follows directly from this equality and Proposition 3.2.1. ∎

Example 3.2.3 *Positive Definite Quadratic Functions*

Proposition 3.2.2 implies that the quadratic function $f(x) = \frac{1}{2}x^t A x + b^t x + c$ is strictly convex for any positive definite matrix A. ∎

Even Proposition 3.2.2 can be difficult to apply. The next proposition helps us to recognize convex functions by their closure properties.

Proposition 3.2.3 *Convex functions satisfy the following:*

(a) *If $f(x)$ is convex and $g(x)$ is convex and increasing, then the functional composition $g \circ f(x)$ is convex.*

(b) *If $f(x)$ is convex, then the functional composition $f(Ax + b)$ of $f(x)$ with an affine function $Ax + b$ is convex.*

(c) *If $f(x)$ and $g(x)$ are convex and α and β are nonnegative constants, then $\alpha f(x) + \beta g(x)$ is convex.*

(d) *If $f(x)$ and $g(x)$ are convex, then $\max\{f(x), g(x)\}$ is convex.*

(e) *If $f_n(x)$ is a sequence of convex functions, then $\lim_{n \to \infty} f_n(x)$ is convex whenever it exists.*

Proof: To prove assertion (a), we calculate

$$\begin{aligned}
g \circ f[\alpha x + (1 - \alpha)y] &\leq g[\alpha f(x) + (1 - \alpha)f(y)] \\
&\leq \alpha g \circ f(x) + (1 - \alpha)g \circ f(y).
\end{aligned}$$

The remaining assertions are left to the reader.   ■

Part (a) of Proposition 3.2.3 implies that $e^{f(x)}$ is convex when $f(x)$ is convex and that $f(x)^{\alpha}$ is convex when $f(x)$ is nonnegative and convex and $\alpha > 1$. One case not covered the Proposition is products. The counterexample $x^3 = x^2 x$ shows that the product of two convex functions is not necessarily convex.

Example 3.2.4 *Differences of Convex Functions*

Although the class of convex functions is rather narrow, most well-behaved functions can be expressed as the difference of two convex functions. For example, consider a polynomial $p(x) = \sum_{m=0}^{n} p_m x^m$. The second derivative test shows that x^m is convex whenever m is even. If m is odd, then x^m is convex on $[0, \infty)$, and $-x^m$ is convex on $(-\infty, 0)$. Therefore,

$$x^m = \max\{x^m, 0\} - \max\{-x^m, 0\}$$

is the difference of two convex functions. Because the class of differences of convex functions is closed under the formation of linear combinations, it follows that $p(x)$ belongs to this larger class.   ■

A positive function $f(x)$ is said to be log-convex if and only if $\ln f(x)$ is convex. Log-convex functions have excellent closure properties as documented by the next proposition.

Proposition 3.2.4 *Log-convex functions satisfy the following:*

(a) *If $f(x)$ is log-convex, then $f(x)$ is convex.*

(b) *If $f(x)$ is convex and $g(x)$ is log-convex and increasing, then the functional composition $g \circ f(x)$ is log-convex.*

(c) *If $f(x)$ is log-convex, then the functional composition $f(Ax + b)$ of $f(x)$ with an affine function $Ax + b$ is log-convex.*

(d) *If $f(x)$ is log-convex, then $f(x)^{\alpha}$ and $\alpha f(x)$ are log-convex for any $\alpha > 0$.*

(e) *If $f(x)$ and $g(x)$ are log-convex, then $f(x) + g(x)$, $f(x)g(x)$, and $\max\{f(x), g(x)\}$ are log-convex.*

(f) *If $f_n(x)$ is a sequence of log-convex functions, then $\lim_{n \to \infty} f_n(x)$ is log-convex whenever it exists and is positive.*

Proof: Assertion (a) follows from part (a) of Proposition 3.2.3 after composing the functions e^x and $\ln f(x)$. To prove that the sum of log-convex functions is log-convex, let $h(x) = f(x) + g(x)$. Then Hölder's inequality implies that

$$
\begin{aligned}
h[\alpha x + (1-\alpha)y] &= f[\alpha x + (1-\alpha)y] + g[\alpha x + (1-\alpha)y] \\
&\leq f(x)^\alpha f(y)^{1-\alpha} + g(x)^\alpha g(y)^{1-\alpha} \\
&\leq [f(x) + g(x)]^\alpha [f(y) + g(y)]^{1-\alpha} \\
&= h(x)^\alpha h(y)^{1-\alpha}.
\end{aligned}
$$

The remaining assertions are left to the reader. ∎

Example 3.2.5 *Gamma Function*

Gauss's representation of the gamma function

$$
\Gamma(z) = \lim_{n \to \infty} \frac{n! n^z}{z(z+1) \cdots (z+n)}
$$

shows that it is log-convex on $(0, \infty)$ [75]. Indeed, one can easily check that n^z and $(z+k)^{-1}$ are log-convex and then apply the closure of the set of log-convex functions under the formation of products and limits. Note that invoking convexity in this argument is insufficient because the set of convex functions is not closed under the formation of products. Alternatively, one can deduce log-convexity from Euler's definition

$$
\Gamma(z) = \int_0^\infty x^{z-1} e^{-x} dx
$$

by viewing the integral as the limit of Riemann sums, each of which is log-convex. ∎

3.3   Minimization of Convex Functions

Optimization theory is much simpler for convex functions than for ordinary functions [72, 111, 119]. For instance, we have the following:

Proposition 3.3.1 *Suppose that $f(x)$ is a convex function on the convex set $S \subset \mathrm{R}^m$. If z is a local minimum of $f(x)$, then it is also a global minimum, and the set $\{x : f(x) = f(z)\}$ is convex.*

Proof: If $f(x) \leq f(z)$ and $f(y) \leq f(z)$, then

$$
\begin{aligned}
f[\alpha x + (1-\alpha)y] &\leq \alpha f(x) + (1-\alpha)f(y) \\
&\leq f(z)
\end{aligned}
\tag{3.3}
$$

for any $\alpha \in [0,1]$. This shows that the set $\{x : f(x) \le f(z)\}$ is convex. Now suppose that $f(x) < f(z)$. Strict inequality then prevails between the extreme members of inequality (3.3) provided $\alpha > 0$. Taking $y = z$ and α close to 0 shows that z cannot serve as a local minimum. Thus, z must be a global minimum. ∎

Example 3.3.1 *Piecewise Linear Functions*

The function $f(x) = |x|$ on the real line is piecewise linear. It attains its minimum of 0 at the point $x = 0$. The convex function $f(x) = \max\{1, |x|\}$ is also piecewise linear, but it attains its minimum throughout the interval $[-1, 1]$. In both cases the set $\{y : f(y) = \min_x f(x)\}$ is convex. In higher dimensions, the convex function $f(x) = \max\{1, \|x\|\}$ attains its minimum of 1 throughout the closed ball $\|x\| \le 1$. ∎

Proposition 3.3.2 *Let $f(x)$ be a convex, differentiable function on the convex set $S \subset \mathrm{R}^m$. If the point $z \in S$ satisfies*

$$df(z)(x - z) \ge 0$$

for every point $x \in S$, then z is a global minimum of $f(x)$. In particular, any stationary point of $f(x)$ is a global minimum.

Proof: This assertion follows immediately from inequality (3.2) characterizing convex functions. ∎

Example 3.3.2 *Minimum of x on $[0, \infty)$.*

The convex function $f(x) = x$ has derivative $df(x) = 1$. On the convex set $[0, \infty)$, we have $df(0)(x - 0) = x \ge 0$ for any $x \in [0, \infty)$. Hence, 0 provides the minimum of x. ∎

Example 3.3.3 *Minimum of a Positive Definite Quadratic Function*

The quadratic function $f(x) = \frac{1}{2} x^t A x + b^t x + c$ has differential

$$df(x) = x^t A + b^t$$

for A symmetric. Assuming that A is also invertible, the sole stationary point of $f(x)$ is $-A^{-1} b$. This point furnishes the minimum of $f(x)$ when A is positive definite. ∎

Example 3.3.4 *Maximum Likelihood for the Multivariate Normal*

The sample mean and sample variance

$$\bar{y} = \frac{1}{k} \sum_{j=1}^{k} y_j$$

$$S = \frac{1}{k} \sum_{j=1}^{k} (y_j - \bar{y})^t (y_j - \bar{y})$$

are also the maximum likelihood estimates of the theoretical mean μ and theoretical variance Ω of a random sample $y_1, \ldots, y_k$ from a multivariate normal. To prove this fact, we first note that maximizing the loglikelihood function

$$-\frac{k}{2} \ln \det \Omega - \frac{1}{2} \sum_{j=1}^{k} (y_j - \mu)^t \Omega^{-1} (y_j - \mu)$$

$$= -\frac{k}{2} \ln \det \Omega - \frac{k}{2} \mu^* \Omega^{-1} \mu + \left(\sum_{j=1}^{k} y_j \right)^* \Omega^{-1} \mu - \frac{1}{2} \sum_{j=1}^{k} y_j^* \Omega^{-1} y_j$$

$$= -\frac{k}{2} \ln \det \Omega - \frac{1}{2} \mathrm{tr} \left[\Omega^{-1} \sum_{j=1}^{k} (y_j - \mu)(y_j - \mu)^t \right]$$

constitutes a special case of the previous example with $A = k\Omega^{-1}$ and $b = -\Omega^{-1} \sum_{j=1}^{k} y_j$. This leads to the same estimate, $\hat{\mu} = \bar{y}$, regardless of the value of Ω.

To estimate Ω, we invoke the Cholesky decompositions $\Omega = LL^t$ and $S = MM^t$ under the assumption that both Ω and S are invertible. Given that $\Omega^{-1} = (L^{-1})^t L^{-1}$ and $\det \Omega = (\det L)^2$, the loglikelihood becomes

$$k \ln \det L^{-1} - \frac{k}{2} \mathrm{tr} \left[(L^{-1})^t L^{-1} MM^t \right]$$

$$= k \ln \det \left(L^{-1} M \right) - \frac{k}{2} \mathrm{tr} \left[(L^{-1}M)(L^{-1}M)^t \right] - k \ln | \det M |$$

using the cyclic permutation property of the matrix trace function. Because products and inverses of lower triangular matrices are lower triangular, the matrix $R = L^{-1}M$ ranges over the set of lower triangular matrices with positive diagonal entries as L ranges over the same set. This permits us to reparameterize and estimate $R = (r_{ij})$ instead of L. Up to an irrelevant constant, the loglikelihood reduces to

$$k \ln \det R - \frac{k}{2} \mathrm{tr}(RR^t) \quad = \quad k \sum_{i} \ln r_{ii} - \frac{k}{2} \sum_{i} \sum_{j=1}^{i} r_{ij}^2.$$

Clearly, this is maximized by taking $r_{ij} = 0$ for $j \neq i$. Differentiation of the concave function $k \ln r_{ii} - \frac{k}{2} r_{ii}^2$ shows that it is maximized by taking $r_{ii} = 1$. In other words, the maximum likelihood estimator $\hat{R}$ is the identity matrix I. This implies that $\hat{L} = M$ and consequently that $\hat{\Omega} = S$. ∎

3.4   The MM Algorithm

Most practical optimization problems defy exact solution. In this section we discuss a minimization method that relies heavily on convexity arguments and is particularly useful in high-dimensional problems such as image reconstruction [101]. We call this method the MM algorithm; the first M of this two-stage algorithm stands for majorize and the second M for minimize. When it is successful, the MM algorithm substitutes a simple optimization problem for a difficult optimization problem. Simplicity can be attained by (a) avoiding large matrix inversions, (b) linearizing an optimization problem, (c) separating the variables of an optimization problem, (d) dealing with equality and inequality constraints gracefully, and (e) turning a nondifferentiable problem into a smooth problem. The price we pay for simplifying the original problem is that we must iterate.

A function $g(x \mid x_n)$ is said majorize a function $f(x)$ at x_n provided

$$
\begin{aligned}
f(x_n) &= g(x_n \mid x_n) & \text{(3.4)}\\
f(x) &\leq g(x \mid x_n) & x \neq x_n.
\end{aligned}
$$

In other words, the surface $x \mapsto g(x \mid x_n)$ lies above the surface $x \mapsto f(x)$ and is tangent to it at the point $x = x_n$. Here x_n represents the current iterate in a search of the surface $x \mapsto f(x)$. In the MM algorithm, we minimize the surrogate function $g(x \mid x_n)$ rather than the actual function $f(x)$. If x_{n+1} denotes the minimum of $g(x \mid x_n)$, then we can show that the MM procedure forces $f(x)$ downhill. Indeed, the inequality

$$
\begin{aligned}
f(x_{n+1}) &= g(x_{n+1} \mid x_n) + f(x_{n+1}) - g(x_{n+1} \mid x_n)\\
&\leq g(x_n \mid x_n) + f(x_n) - g(x_n \mid x_n) & \text{(3.5)}\\
&= f(x_n)
\end{aligned}
$$

follows directly from the fact $g(x_{n+1} \mid x_n) \leq g(x_n \mid x_n)$ and definition (3.4). The descent property (3.5) lends the MM algorithm remarkable numerical stability. When $f(x)$ is strictly convex, one can show with a few additional mild hypotheses that the iterates x_n converge to the global minimum of $f(x)$ regardless of the initial point x_0.

With obvious changes, the MM algorithm applies to maximization rather than minimization. To maximize a function $f(x)$, we minorize it by a surrogate function $g(x \mid x^n)$ and maximize $g(x \mid x^n)$ to produce the next iterate x^{n+1}. In this case, the letters MM stand for minorize/maximize rather than majorize/minimize. Chapter 6 discusses an MM algorithm for maximum likelihood estimation in transmission tomography. Here is a simpler example relevant to sports.

Example 3.4.1 *Bradley-Terry Model of Ranking*

In the sports version of the Bradley and Terry model [23, 85], each team i in a league of teams is assigned a rank parameter $r_i > 0$. Assuming ties are impossible, team i beats team j with probability $r_i/(r_i + r_j)$. If this outcome occurs y_{ij} times during a season of play, then the probability of the whole season is

$$L(r) = \prod_{i,j} \left(\frac{r_i}{r_i + r_j}\right)^{y_{ij}},$$

assuming the games are independent. To rank the teams, we find the values $\hat{r}_i$ that maximize $f(r) = \ln L(r)$. The team with the largest $\hat{r}_i$ is considered best, the team with the smallest $\hat{r}_i$ is considered worst, and so forth. In view of the fact that $-\ln u$ is convex, inequality (3.2) implies

$$
\begin{aligned}
f(r) &= \sum_{i,j} y_{ij} \Big[\ln r_i - \ln(r_i + r_j) \Big] \\
&\geq \sum_{i,j} y_{ij} \Big[\ln r_i - \ln(r_i^n + r_j^n) + \frac{r_i + r_j - r_i^n - r_j^n}{r_i^n + r_j^n} \Big] \\
&= g(r \mid r^n),
\end{aligned}
$$

where the superscript n indicates iteration number. Equality occurs in this minorizing inequality when $r = r^n$. Differentiating $g(r \mid r^n)$ with respect to r_i and setting the result equal to 0 produces the next iterate

$$r_i^{n+1} = \frac{\sum_{j \neq i} y_{ij}}{\sum_{j \neq i}(y_{ij} + y_{ji})/(r_i^n + r_j^n)}.$$

Because $L(r) = L(cr)$ for any $c > 0$, we constrain $r_1 = 1$ and omit the update r_1^{n+1}. In this example, the MM algorithm separates parameters and allows us to minimize $g(r \mid r^n)$ parameter by parameter. The values $\hat{r}_i$ are referred to as maximum likelihood estimates.   ■

Example 3.4.2 *Least Absolute Deviation Regression*

Statisticians often estimate parameters by the method of least squares. This classical method suffers from the fact that it is strongly influenced by observations far removed from their predicted values. To review the situation, consider p independent experiments with outcomes $y_1, \ldots, y_p$. We wish to predict y_i from q covariates $x_{i1}, \ldots, x_{iq}$ known in advance. For instance, y_i might be the height of the ith child in a classroom of p children. Relevant covariates might be the heights x_{i1} and x_{i2} of i's mother and father and the sex of i coded as $x_{i3} = 1$ for a girl and $x_{i4} = 1$ for a boy. Here we take $q = 4$ and force $x_{i3}x_{i4} = 0$ so that only one sex is possible. If we use a linear predictor $\sum_{j=1}^{q} x_{ij}\theta_j$ of y_i, it is natural to estimate the regression coefficients θ_j by minimizing the sum of squares

$$f(\theta) = \sum_{i=1}^{p} \Big(y_i - \sum_{j=1}^{q} x_{ij}\theta_j \Big)^2.$$

Differentiating $f(\theta)$ with respect to θ_j and setting the result equal to 0 produce

$$\sum_{i=1}^{p} x_{ij} y_i \;=\; \sum_{i=1}^{p} \sum_{k=1}^{q} x_{ij} x_{ik} \theta_k.$$

If we let y denote the column vector with entries y_i and X denote the matrix with entry x_{ij} in row i and column j, these q normal equations can be written in vector form as

$$X^t y \;=\; X^t X \theta$$

and solved as

$$\hat{\theta} \;=\; (X^t X)^{-1} X^t y.$$

In the method of least absolute deviation regression, we replace $f(\theta)$ by

$$h(\theta) \;=\; \sum_{i=1}^{p} \left| y_i - \sum_{j=1}^{q} x_{ij} \theta_j \right|.$$

Traditionally, one simplifies this expression by defining the residual

$$r_i(\theta) \;=\; y_i - \sum_{j=1}^{q} x_{ij} \theta_j.$$

We are now faced with minimizing a nondifferentiable function. Fortunately, the MM algorithm can be implemented by exploiting the convexity of the function $-\sqrt{u}$ in inequality (3.2). Because

$$-\sqrt{u} \;\geq\; -\sqrt{u^n} - \frac{u - u^n}{2\sqrt{u^n}},$$

we find that

$$
\begin{aligned}
h(\theta) \;&=\; \sum_{i=1}^{p} \sqrt{r_i(\theta)^2} \\
&\leq\; h(\theta^n) + \frac{1}{2} \sum_{i=1}^{p} \frac{r_i^2(\theta) - r_i^2(\theta^n)}{\sqrt{r_i^2(\theta^n)}} \\
&=\; g(\theta \mid \theta^n).
\end{aligned}
$$

Minimizing $g(\theta \mid \theta^n)$ is accomplished by minimizing the weighted sum of squares

$$\sum_{i=1}^{p} w_i(\theta^n) r_i(\theta)^2$$

with ith weight $w_i(\theta^n) = |r_i(\theta^n)|^{-1}$. A slight variation of the above argument for minimizing a sum of squares leads to

$$\theta^{n+1} = [X^t W(\theta^n) X]^{-1} W(\theta^n) X^t y,$$

where $W(\theta^n)$ is the diagonal matrix with ith diagonal entry $w_i(\theta^n)$. Unfortunately, the possibility that some $w_i(\theta^n) = \infty$ cannot be ruled out. Problem 13 suggests a simple remedy. ∎

3.5 Moment Inequalities

Inequalities give important information about the magnitude of probabilities and expectations without requiring their exact calculation. The Cauchy-Schwarz inequality $|E(XY)| \le E(X^2)^{1/2} E(Y^2)^{1/2}$ is one of the most useful of the classical inequalities. It is also one of the easiest to remember because it is equivalent to the fact that a correlation coefficient must lie on the interval $[-1, 1]$. Equality occurs in the Cauchy-Schwarz inequality if and only if X is a linear function of Y or vice versa.

Markov's inequality is another widely applied bound. Let $g(x)$ be a nonnegative, increasing function, and let X be a random variable such that $g(X)$ has finite expectation. Then Markov's inequality

$$\Pr(X \ge c) \le \frac{E[g(X)]}{g(c)}$$

holds for any constant c for which $g(c) > 0$. This result follows upon taking expectations in the inequality $g(c)1_{\{X \ge c\}} \le g(X)$. Chebyshev's inequality is the special case of Markov's inequality with $g(x) = x^2$ applied to the random variable $X - E(X)$. In large deviation theory, we take $g(x) = e^{tx}$ and $c > 0$ and choose $t > 0$ to minimize the right hand side of the inequality $\Pr(X \ge c) \le e^{-ct} E(e^{tX})$ involving the moment generating function of X. Problem 18 provides a typical example of this Chernoff bound. Our next example involves a nontrivial application of Chebyshev's inequality.

Example 3.5.1 *Weierstrass's Approximation Theorem*

Weierstrass showed that a continuous function $f(x)$ on $[0, 1]$ can be uniformly approximated to any desired degree of accuracy by a polynomial. Bernstein's lovely proof of this fact relies on applying Chebyshev's inequality to a binomial random variable S_n with n trials and success probability x per trial [49]. The corresponding candidate polynomial is defined by the expectation

$$E\left[f\left(\frac{S_n}{n}\right)\right] = \sum_{k=0}^{n} f\left(\frac{k}{n}\right)\binom{n}{k} x^k (1-x)^{n-k}.$$

Note that $E(S_n/n) = x$ and $\text{Var}(S_n/n) = x(1-x)/n \le 1/(4n)$. Now given an arbitrary $\epsilon > 0$, one can find by the uniform continuity of $f(x)$ a $\delta > 0$ such that $|f(u) - f(v)| < \epsilon$ whenever $|u - v| < \delta$. If $||f||_\infty = \sup |f(x)|$ on $[0, 1]$, then Chebyshev's inequality implies

$$
\begin{aligned}
& \left| E\left[f\left(\frac{S_n}{n} \right) \right] - f(x) \right| \\
\le\ & E\left[\left| f\left(\frac{S_n}{n} \right) - f(x) \right| \right] \\
\le\ & \epsilon \Pr\left(\left| \frac{S_n}{n} - x \right| < \delta \right) + 2||f||_\infty \Pr\left(\left| \frac{S_n}{n} - x \right| \ge \delta \right) \\
\le\ & \epsilon + \frac{2||f||_\infty x(1-x)}{n\delta^2} \\
\le\ & \epsilon + \frac{||f||_\infty}{2n\delta^2}.
\end{aligned}
$$

Taking $n \ge ||f||_\infty/(2\epsilon\delta^2)$ then gives $\left| E\left[f\left(\frac{S_n}{n} \right) \right] - f(x) \right| \le 2\epsilon$ regardless of the chosen $x \in [0, 1]$. ∎

Proposition 3.5.1 (Jensen's Inequality) *Let the values of the random variable W be confined to the possibly infinite interval (a, b). If $h(w)$ is convex on (a, b), then $E[h(W)] \ge h[E(W)]$, provided both expectations exist. For a strictly convex function $h(w)$, equality holds in Jensen's inequality if and only if $W = E(W)$ almost surely.*

Proof: For the sake of simplicity, suppose that $h(w)$ is differentiable. If we let $v = E(W)$, then Jensen's inequality follows from Proposition 3.2.1 after taking expectations in the inequality

$$
h(W) \ \ge \ h(v) + dh(v)(W - v). \tag{3.6}
$$

If $h(w)$ is strictly convex, and W is not constant, then inequality (3.6) is strict with positive probability. Hence, strict inequality prevails in Jensen's inequality. ∎

Jensen's inequality is the key to a host of other inequalities. Here are two nontrivial examples.

Example 3.5.2 *Schlömilch's Inequality for Weighted Means*

If X is a positive random variable, then we define the weighted mean function $M(p) = E(X^p)^{\frac{1}{p}}$. For the sake of argument, we assume that $M(p)$ exists and is finite for all real p. Typical values of $M(p)$ are $M(1) = E(X)$ and $M(-1) = 1/E(X^{-1})$. To make $M(p)$ continuous at $p = 0$, it turns out that we should set $M(0) = e^{E(\ln X)}$. The reader is asked to check this fact in Problem 23. Here we are more concerned with proving Schlömilch's assertion that $M(p)$ is an increasing function of p. To simplify our proof, let us assume that $a \leq X \leq b$ for positive constants a and b. This condition permits us to differentiate under the expectation sign in

$$
\begin{aligned}
\frac{d}{dp} \ln M(p) &= \frac{d}{dp} \left[\frac{1}{p} \ln E(e^{p \ln X}) \right] \\
&= -\frac{1}{p^2} \ln E(X^p) + \frac{1}{p^2 E(X^p)} p E(X^p \ln X) \\
&= \frac{E(X^p \ln X^p) - E(X^p) \ln E(X^p)}{p^2 E(X^p)}.
\end{aligned}
$$

Because $f(u) = u \ln u$ is a convex function, Jensen's inequality implies that $E(X^p \ln X^p) - E(X^p) \ln E(X^p) \geq 0$. Hence, $M(p)$ has a nonnegative derivative and is increasing both to the left and right of 0. Continuity of $M(p)$ at $p = 0$ then completes the proof that $M(p)$ is increasing. To relax the boundedness condition on X, we assume $p \leq q$, replace X by

$$
X_n = \frac{1}{n} 1_{\{X < \frac{1}{n}\}} + X 1_{\{\frac{1}{n} \leq X \leq n\}} + n 1_{\{X > n\}},
$$

and take limits in the valid inequality $E(X_n^p)^{\frac{1}{p}} \leq E(X_n^q)^{\frac{1}{q}}$. Note that the dominated convergence theorem with dominating functions $X_n^p \leq 1 + X^p$ and $X_n^q \leq 1 + X^q$ justifies this limit for $p, q \neq 0$. If either $p = 0$ or $q = 0$, then Schlömilch's claim follows from dividing the obvious Jensen's inequality $E(\ln X^r) \leq \ln E(X^r)$ by r and exponentiating the result.

When the random variable X is defined on the space $\{1, \ldots, n\}$ equipped with the uniform distribution, Schlömilch's inequalities for $p = -1$, 0, and 1 reduce to the classical inequalities

$$
\frac{1}{\frac{1}{n} \left(\frac{1}{x_1} + \cdots + \frac{1}{x_n} \right)} \leq \left(x_1 \cdots x_n \right)^{\frac{1}{n}} \leq \frac{1}{n} \left(x_1 + \cdots + x_n \right)
$$

relating the harmonic, geometric, and arithmetic means.   ■

Example 3.5.3 *Hölder's Inequality*

Consider two random variables X and Y and two numbers $p > 1$ and $q > 1$ such that $p^{-1} + q^{-1} = 1$. Then Hölder's inequality

$$
|E(XY)| \leq E(|X|^p)^{\frac{1}{p}} E(|Y|^q)^{\frac{1}{q}} \tag{3.7}
$$

generalizes the Cauchy-Schwarz inequality whenever the indicated expectations on its right exist. To prove (3.7), it clearly suffices to assume that X and Y are nonnegative. It also suffices to take $E(X^p) = E(Y^q) = 1$ once we divide the left-hand side of (3.7) by its right-hand side. Now set $r = p^{-1}$, and let Z be a random variable equal to $u \geq 0$ with probability r and equal to $v \geq 0$ with probability $1 - r$. Schlömilch's inequality $M(0) \leq M(1)$ for Z says

$$u^r v^{1-r} \leq ru + (1-r)v.$$

If we substitute X^p for u and Y^q for v in this inequality and take expectations, then we find that $E(XY) \leq r + 1 - r = 1$ as required. ∎

3.6 Problems

1. On which intervals are the following functions convex: e^x, e^{-x}, x^n, $|x|^p$ for $p \geq 1$, $\sqrt{1+x^2}$, $x \ln x$, and $\cosh x$? On these intervals, which functions are log-convex?

2. Show that Riemann's zeta function

$$\zeta(s) = \sum_{n=1}^{\infty} \frac{1}{n^s}$$

is log-convex for $s > 1$.

3. Demonstrate that the function $f(x) = x^n - na \ln x$ is convex on $(0, \infty)$ for $a > 0$. Where does its minimum occur?

4. Prove the strict convexity assertions of Proposition 3.2.1.

5. Prove parts (b), (c), and (d) of Proposition 3.2.3.

6. Prove the unproved assertions of Proposition 3.2.4.

7. Suppose that $f(x)$ is a convex function on the real line. If a and y are vectors in R^m, then show that $f(a^t y)$ is a convex function of y. For $m > 1$ show that $f(a^t y)$ is not strictly convex.

8. Let $f(x)$ be a continuous function on the real line satisfying

$$f\left[\frac{1}{2}(x+y)\right] \leq \frac{1}{2}f(x) + \frac{1}{2}f(y).$$

Prove that $f(x)$ is convex.

9. If $f(x)$ is a nondecreasing function on the interval $[a, b]$, then show that $g(x) = \int_a^x f(y)\, dy$ is a convex function on $[a, b]$.

10. Heron's classical formula for the area of a triangle with sides of length a, b, and c is $\sqrt{s(s-a)(s-b)(s-c)}$, where $s = (a+b+c)/2$ is the semiperimeter. Using inequality (3.7) show that the triangle of fixed perimeter with greatest area is equilateral.

11. Let $H_n = 1 + \frac{1}{2} + \cdots + \frac{1}{n}$. Using inequality (3.7), verify the inequality $n\sqrt[n]{n+1} \le n + H_n$ for any positive integer n (Putnam Competition 1975).

12. Show that the loglikelihood $L(r)$ in Example 3.4.1 is concave under the reparameterization $r_i = e^{\theta_i}$.

13. Suppose in Example 3.4.2 we minimize the function

$$h_\epsilon(\theta) = \sum_{i=1}^{p} \left\{ \left[y_i - \sum_{j=1}^{q} x_{ij}\theta_j \right]^2 + \epsilon \right\}^{1/2}$$

instead of $h(\theta)$ for a small, positive number ϵ. Show that the same MM algorithm applies with revised weights $w_i(\theta^n) = 1/\sqrt{r_i(\theta^n)^2 + \epsilon}$.

14. Suppose the random variables X and Y have densities $f(u)$ and $g(u)$ such that $f(u) \ge g(u)$ for $u \le a$ and $f(u) \le g(u)$ for $u > a$. Prove that $E(X) \le E(Y)$. If in addition $f(u) = g(u) = 0$ for $u < 0$, show that $E(X^n) \le E(Y^n)$ for all positive integers n [49].

15. If the random variable X has values in the interval $[a, b]$, then show that $\text{Var}(X) \le (b-a)^2/4$. (Hints: Reduce to the case $[a, b] = [0, 1]$. If $E(X) = p$, then show that $\text{Var}(X) \le p(1-p)$.)

16. Let X be a random variable with $E(X) = 0$ and $E(X^2) = \sigma^2$. Show that

$$\Pr(X \ge c) \le \frac{a^2 + \sigma^2}{(a+c)^2} \qquad (3.8)$$

for all nonnegative a and c. Prove that the choice $a = \sigma^2/c$ minimizes the right-hand side of (3.8) and that for this choice

$$\Pr(X \ge c) \le \frac{\sigma^2}{\sigma^2 + c^2}.$$

This is Cantelli's inequality [49].

17. Suppose $g(x)$ is a function such that $g(x) \le 1$ for all x and $g(x) \le 0$ for $x \le c$. Demonstrate the inequality

$$\Pr(X \ge c) \ge E[g(X)] \qquad (3.9)$$

for any random variable X [49]. Verify that the polynomial

$$g(x) = \frac{(x-c)(c+2d-x)}{d^2}$$

with $d > 0$ satisfies the stated conditions leading to inequality (3.9). If X is nonnegative with $E(X) = 1$ and $E(X^2) = \beta$ and $c \in (0,1)$, then prove that the choice $d = \beta/(1-c)$ yields

$$\Pr(X \geq c) \; \geq \; \frac{(1-c)^2}{\beta}.$$

Finally, if $E(X^2) = 1$ and $E(X^4) = \beta$, show that

$$\Pr(|X| \geq c) \; \geq \; \frac{(1-c^2)^2}{\beta}.$$

18. Let X be a Poisson random variable with mean λ. Demonstrate that the Chernoff bound

$$\Pr(X \geq c) \; \leq \; \inf_{t>0} e^{-ct} E(e^{tX})$$

amounts to

$$\Pr(X \geq c) \; \leq \; \frac{(\lambda e)^c}{c^c} e^{-\lambda}$$

for any integer $c > \lambda$.

19. Let $B_n f(x) = E[f(S_n/n)]$ denote the Bernstein polynomial of degree n approximating $f(x)$ as discussed in Example 3.5.1. Prove that

(a) $B_n f(x)$ is linear in $f(x)$,
(b) $B_n f(x) \geq 0$ if $f(x) \geq 0$,
(c) $B_n f(x) = f(x)$ if $f(x)$ is linear,
(d) $B_n x(1-x) = \frac{n-1}{n} x(1-x)$.

20. Suppose the function $f(x)$ has continuous derivative $f'(x)$. For $\delta > 0$ show that Bernstein's polynomial satisfies the bound

$$\left| E\left[f\left(\frac{S_n}{n} \right) \right] - f(x) \right| \; \leq \; \delta \|f'\|_\infty + \frac{\|f\|_\infty}{2n\delta^2}.$$

Conclude from this estimate that

$$\left\| E\left[f\left(\frac{S_n}{n} \right) \right] - f \right\|_\infty \; = \; O(n^{-\frac{1}{3}}).$$

21. Let $f(x)$ be a convex function on $[0,1]$. Prove that the Bernstein polynomial of degree n approximating $f(x)$ is also convex. (Hint: Show that

$$\frac{d^2}{dx^2} E\left[f\left(\frac{S_n}{n} \right) \right] \; = \; n(n-1)\left\{ E\left[f\left(\frac{S_{n-2}+2}{n} \right) \right] \right.$$
$$\left. -2 E\left[f\left(\frac{S_{n-2}+1}{n} \right) \right] + E\left[f\left(\frac{S_{n-2}}{n} \right) \right] \right\}.$$

in the notation of Example 3.5.1.)

22. Suppose $1 \le p < \infty$. For a random variable X with $E(|X|^p) < \infty$, define the norm $||X||_p = E(X^p)^{\frac{1}{p}}$. Now prove Minkowski's triangle inequality $||X+Y||_p \le ||X||_p + ||Y||_p$. (Hint: Apply Hölder's inequality to the right-hand side of

$$E(|X + Y|^p) \le E(|X| \cdot |X + Y|^{p-1}) + E(|Y| \cdot |X + Y|^{p-1})$$

and rearrange the result.

23. Suppose X is a random variable satisfying $0 < a \le X \le b < \infty$. Use l'Hôpital's rule to prove that the weighted mean $M(p) = E(X^p)^{\frac{1}{p}}$ is continuous at $p = 0$ if we define $M(0) = e^{E(\ln X)}$.

4
Combinatorics

4.1 Introduction

Combinatorics is the bane of many a student of probability theory. Even elementary combinatorial problems can be frustratingly subtle. The cure for this ill is more exposure, not less. Because combinatorics has so many important applications, serious students of the mathematical sciences neglect it at their peril. Here we explore a few topics in combinatorics that have maximum intersection with probability. Our policy is to assume that readers have a nodding familiarity with combinations and permutations. Based on this background, we discuss inclusion-exclusion (sieve) methods, Stirling numbers of the first and second kind, and the pigeonhole principle. Along the way we meet some applications that we hope will whet readers' appetites for further study. The books [17, 21, 48, 62, 110, 151] are especially recommended.

4.2 Inclusion-Exclusion

The simplest inclusion-exclusion formula is

$$\Pr(A \cup B) \;=\; \Pr(A) + \Pr(B) - \Pr(A \cap B). \qquad (4.1)$$

The probability $\Pr(A \cap B)$ is subtracted from the sum $\Pr(A) + \Pr(B)$ to compensate for doubling counting of the intersection $A \cap B$. As pointed out in Chapter 1, we can derive formula (4.1) by taking expectations of indicator random variables. To derive more general inclusion-exclusion formulas, suppose X is a nonnegative, integer-valued random variable with probability generating function $G(u) = E(u^X)$. In Example 2.4.3 we saw that one can recover the factorial moments of X via

$$E[X(X-1)\cdots(X-j+1)] = \frac{d^j}{du^j}G(1).$$

If X is sufficiently well behaved, say bounded, and if we introduce the binomial moments

$$E\left[\binom{X}{j}\right] = \frac{1}{j!}E[X(X-1)\cdots(X-j+1)] = \frac{1}{j!}\frac{d^j}{du^j}G(1),$$

then we can expand $G(u) = \sum_{k=0}^{\infty}\Pr(X=k)u^k$ about $u=1$ in the Taylor's series

$$
\begin{aligned}
G(u) &= \sum_{j=0}^{\infty}E\left[\binom{X}{j}\right](u-1)^j \\
&= \sum_{j=0}^{\infty}E\left[\binom{X}{j}\right]\sum_{k=0}^{j}(-1)^{j-k}\binom{j}{k}u^k \qquad (4.2)\\
&= \sum_{k=0}^{\infty}u^k\sum_{j=k}^{\infty}(-1)^{j-k}\binom{j}{k}E\left[\binom{X}{j}\right].
\end{aligned}
$$

Equating coefficients of u^k in $G(u)$ therefore gives

$$\Pr(X=k) = \sum_{j=k}^{\infty}(-1)^{j-k}\binom{j}{k}E\left[\binom{X}{j}\right]. \qquad (4.3)$$

Now consider the special case where $X = 1_{A_1} + \cdots + 1_{A_n}$ is a sum of indicator random variables. In view of the trivial identity

$$u^{1_{A_i}} = 1 + 1_{A_i}(u-1),$$

we find that X has probability generating function

$$
\begin{aligned}
G(u) &= E\left\{\prod_{i=1}^{n}[1 + 1_{A_i}(u-1)]\right\} \\
&= \sum_{j=0}^{n}\sum_{|S|=j}E\left(\prod_{i\in S}1_{A_i}\right)(u-1)^j \qquad (4.4)\\
&= \sum_{j=0}^{n}\sum_{|S|=j}\Pr\left(\bigcap_{i\in S}A_i\right)(u-1)^j,
\end{aligned}
$$

where $|S|$ is the number of elements of the subset S of $\{1, \ldots, n\}$. Comparing this expansion with equation (4.2) gives

$$\mathrm{E}\left[\binom{X}{j}\right] = \sum_{|S|=j} \Pr\left(\bigcap_{i \in S} A_i\right),$$

and further comparison with equation (4.3) yields the inclusion-exclusion formula

$$\Pr(X = k) = \sum_{j=k}^{n} (-1)^{j-k} \binom{j}{k} \sum_{|S|=j} \Pr\left(\bigcap_{i \in S} A_i\right) \qquad (4.5)$$

for the probability that exactly k of the events $A_1, \ldots, A_n$ occur. In the sequel, we use the standard notation $p_{[k]}$ to denote $\Pr(X = k)$.

When $k = 0$ in equation (4.5), all of the binomial coefficients $\binom{j}{k} = 1$. Furthermore, the $j = 0$ term in the sum on the right-hand side of equation (4.5) reduces to 1 because only the empty set $\emptyset$ has size 0. Thus,

$$p_{[0]} = 1 + \sum_{j=1}^{n} (-1)^j \sum_{|S|=j} \Pr\left(\bigcap_{i \in S} A_i\right). \qquad (4.6)$$

In many practical examples, the events $A_1, \ldots, A_n$ are exchangeable in the sense that

$$\Pr\left(\bigcap_{i \in S} A_i\right) = \Pr\left(A_1 \cap \cdots \cap A_j\right)$$

for all subsets S of size j. (A similar definition holds for exchangeable random variables.) Because there are $\binom{n}{j}$ such subsets, formula (4.5) reduces to

$$p_{[k]} = \sum_{j=k}^{n} (-1)^{j-k} \binom{j}{k} \binom{n}{j} \Pr\left(A_1 \cap \cdots \cap A_j\right) \qquad (4.7)$$

in the presence of exchangeable events.

We next construct a similar inclusion-exclusion formula for the tail probability $p_{(k)} = \Pr(X \geq k)$ that at least k of the events $A_1, \ldots, A_n$ occur. On one hand,

$$\frac{G(u) - 1}{u - 1} = \sum_{j=1}^{n} \Pr(X = j) \frac{u^j - 1}{u - 1}$$

$$= \sum_{j=1}^{n} \Pr(X = j) \sum_{k=0}^{j-1} u^k$$

$$= \sum_{k=0}^{n-1} u^k \sum_{j=k+1}^{n} \Pr(X = j)$$

$$= \sum_{k=0}^{n-1} u^k \Pr(X > k).$$

On the other hand, equation (4.4) implies that

$$\frac{G(u) - 1}{u - 1} = \sum_{j=1}^{n} \sum_{|S|=j} \Pr\left(\bigcap_{i \in S} A_i\right)(u-1)^{j-1}$$

$$= \sum_{j=1}^{n} \sum_{|S|=j} \Pr\left(\bigcap_{i \in S} A_i\right) \sum_{k=0}^{j-1} (-1)^{j-1-k} \binom{j-1}{k} u^k$$

$$= \sum_{k=0}^{n-1} u^k \sum_{j=k+1}^{n} (-1)^{j-1-k} \binom{j-1}{k} \sum_{|S|=j} \Pr\left(\bigcap_{i \in S} A_i\right).$$

Equating coefficients of u^{k-1} therefore produces

$$p_{(k)} = \sum_{j=k}^{n} (-1)^{j-k} \binom{j-1}{k-1} \sum_{|S|=j} \Pr\left(\bigcap_{i \in S} A_i\right). \qquad (4.8)$$

This formula simplifies to

$$p_{(k)} = \sum_{j=k}^{n} (-1)^{j-k} \binom{j-1}{k-1} \binom{n}{j} \Pr\left(A_1 \cap \cdots \cap A_j\right)$$

in the presence of exchangeable events.

Example 4.2.1 *Fixed Points of a Random Permutation*

Resuming our investigation of Example 2.2.1, let us find the exact distribution of the number of fixed points X of a random permutation π. The pertinent events $A_i = \{\pi : \pi(i) = i\}$ for $1 \le i \le n$ are clearly exchangeable. Furthermore, $\Pr(A_1 \cap \cdots \cap A_j) = (n-j)!/n!$ because the movable integers $\{j+1, \ldots, n\}$ can be permuted in $(n-j)!$ ways. Hence, formula (4.7) gives

$$p_{[k]} = \sum_{j=k}^{n} (-1)^{j-k} \binom{j}{k} \binom{n}{j} \frac{(n-j)!}{n!}$$

$$= \frac{1}{k!} \sum_{j=k}^{n} (-1)^{j-k} \frac{1}{(j-k)!}$$

$$= \frac{1}{k!} \sum_{i=0}^{n-k} (-1)^i \frac{1}{i!}.$$

For $n - k$ reasonably large, the approximation $p_{[k]} \approx e^{-1}/k!$ holds, and this validates our earlier claim that X follows an approximate Poisson distribution with mean 1. ∎

Example 4.2.2 *Euler's Totient Function*

Let n be a positive integer with prime factorization $n = p_1^{m_1} \cdots p_q^{m_q}$. For instance, if $n = 20$, then $n = 2^2 \cdot 5$. If we impose the uniform distribution on the set $\{1, \ldots, n\}$, then it makes sense to ask for the probability $p_{[0]}$ that a random integer N shares no common prime factors with n. Euler considered this problem and gave a lovely formula for his totient function $\varphi(n) = np_{[0]}$. To calculate $\varphi(n)$ via inclusion-exclusion, let A_i be the set of integers divisible by the ith prime p_i in the prime decomposition of n. A little reflection shows that $\Pr(A_i) = \frac{1}{p_i}$ and that in general

$$\Pr\left(\bigcap_{i \in S} A_i\right) = \prod_{i \in S} \frac{1}{p_i}.$$

Hence, equation (4.6) implies

$$\frac{\varphi(n)}{n} = 1 - \sum_i \frac{1}{p_i} + \sum_{i<j} \frac{1}{p_i p_j} - \sum_{i<j<k} \frac{1}{p_i p_j p_k} + \cdots$$

$$= \left(1 - \frac{1}{p_1}\right)\left(1 - \frac{1}{p_2}\right) \cdots \left(1 - \frac{1}{p_q}\right).$$

∎

Example 4.2.3 *0-1 Matrices*

Consider an $m \times n$ random matrix M with entries restricted to the values 0 and 1 [110]. Each entry is determined by flipping an unbiased coin. If the coin lands heads up, then the entry is set to 1; otherwise, it is set to 0. We now ask for the probability $p_{[0]}$ that M possesses no row or column filled entirely with 0's. This problem yields to inclusion-exclusion if we let R_i be the event that row i consists entirely of 0's and C_j be the event that column j consists entirely of 0's. When we intersect s different R_i with t different C_j, the resulting event has probability $1/2^{sn+tm-st}$. Note in this regard that s rows and t columns overlap in st entries; therefore, we must subtract st from $sn + tm$ to avoid double counting of entries. The inclusion-exclusion formula (4.6) now boils down to

$$p_{[0]} = \sum_{s=0}^{m} \sum_{t=0}^{n} (-1)^{s+t} \binom{m}{s} \binom{n}{t} \frac{1}{2^{sn+tm-st}}$$

$$= \frac{1}{2^{mn}} \sum_{s=0}^{m} (-1)^s \binom{m}{s} \sum_{t=0}^{n} \binom{n}{t} (-1)^t 2^{(m-s)(n-t)}$$

$$= \frac{1}{2^{mn}} \sum_{s=0}^{m} (-1)^s \binom{m}{s} \left(2^{m-s} - 1\right)^n$$

because s events R_i can be chosen in $\binom{m}{s}$ ways and t events C_j can be chosen in $\binom{n}{t}$ ways. ∎

In many practical applications, it is cumbersome to calculate all of the terms in the alternating series (4.5). Fortunately, the partial sums

$$\sum_{j=k}^{k+m} (-1)^{j-k} \binom{j}{k} \sum_{|S|=j} \Pr\left(\bigcap_{i \in S} A_i\right)$$

overestimate $p_{[k]}$ for m even and underestimate $p_{[k]}$ for m odd. When $k = 0$, the first two of these Bonferroni inequalities are

$$\Pr\left(A_1 \bigcup \cdots \bigcup A_n\right) = 1 - p_{[0]}$$

$$\leq \sum_{i=1}^{n} \Pr(A_i)$$

and

$$\Pr\left(A_1 \bigcup \cdots \bigcup A_n\right) \geq \sum_{i=1}^{n} \Pr(A_i) - \sum_{i<j} \Pr\left(A_i \bigcap A_j\right) \quad (4.9)$$

as noted in Problem 7. In exactly the same manner, the partial sums

$$\sum_{j=k}^{k+m} (-1)^{j-k} \binom{j-1}{k-1} \sum_{|S|=j} \Pr\left(\bigcap_{i \in S} A_i\right)$$

overestimate $p_{(k)}$ for m even and underestimate $p_{(k)}$ for m odd. Verification of these inequalities can be found in the references [48, 53].

4.3   Applications to Order Statistics

In many practical problems, it is convenient to rearrange n random variables $X_1, X_2, \ldots, X_n$ so that they appear in increasing order. Understanding the marginal distributions and moments of the resulting order statistics $X_{(1)} \leq X_{(2)} \leq \cdots \leq X_{(n)}$ is difficult even when the original random variables are independent and identically distributed. Surprisingly, the principle of inclusion-exclusion sheds considerable light on the subject [8, 15, 36]. To make this claim precise, we need some notation. Denote the distribution function of $X_{(i)}$ by $F_{(i)}(t)$. For an arbitrary subset $S \subset \{1, 2, \ldots, n\}$, let $X_S = \min\{X_j : j \in S\}$ and $X^S = \max\{X_j : j \in S\}$, and designate the corresponding distribution functions $F_S(t)$ and $F^S(t)$, respectively. The following proposition is then true.

Proposition 4.3.1 *The distribution functions of the order statistics $X_{(i)}$ can be expressed as*

$$F_{(i)}(t) = \sum_{j=i}^{n}(-1)^{j-i}\binom{j-1}{i-1}\sum_{|S|=j}F^S(t) \qquad (4.10)$$

$$F_{(n-i+1)}(t) = \sum_{j=i}^{n}(-1)^{j-i}\binom{j-1}{i-1}\sum_{|S|=j}F_S(t), \qquad (4.11)$$

where the sum on S extends over all subsets of $\{1, 2, \ldots, n\}$ with j elements. Consequently, if each X_j possesses a kth absolute moment, then

$$\mathrm{E}[X_{(i)}^k] = \sum_{j=i}^{n}(-1)^{j-i}\binom{j-1}{i-1}\sum_{|S|=j}\mathrm{E}[(X^S)^k] \qquad (4.12)$$

$$\mathrm{E}[X_{(n-i+1)}^k] = \sum_{j=i}^{n}(-1)^{j-i}\binom{j-1}{i-1}\sum_{|S|=j}\mathrm{E}[(X_S)^k]. \qquad (4.13)$$

All of these formulas simplify in the obvious manner if the X_i are exchangeable random variables.

Proof: If we define the events $A_j = \{X_j \le t\}$, then $F_{(i)}(t)$ is the probability that at least i of the n events A_j occur. Equality (4.10) is just a restatement of identity (4.8). To prove equality (4.11), let $Y_j = -X_j$ and apply identity (4.8) to the events $A_j = \{Y_j < -t\} = \{X_j > t\}$. This gives

$$\Pr(X_{(n-i+1)} > t) = \Pr(Y_{(i)} < -t) \qquad (4.14)$$

$$= \sum_{j=i}^{n}(-1)^{j-i}\binom{j-1}{i-1}\sum_{|S|=j}\Pr(\max_{k\in S} -Y_k < -t)$$

$$= \sum_{j=i}^{n}(-1)^{j-i}\binom{j-1}{i-1}\sum_{|S|=j}\Pr(\min_{k\in S} X_k > t).$$

Subtracting the extreme sides of this equation from the constant

$$1 = \sum_{j=i}^{n}(-1)^{j-i}\binom{j-1}{i-1}\sum_{|S|=j}1 \qquad (4.15)$$

gives the final result (4.11). Note that equation (4.15) follows from equation (4.10) by sending t to ∞.

The two moment identities (4.12) and (4.13) follow because two finite measures that share an identical distribution function also share identical moments. (Problem 8 asks the reader to check that all moments in sight exist.) Alternatively, if the X_j are nonnegative, then we can prove identity (4.13) by multiplying both sides of equality (4.14) by kt^{k-1} and integrating as discussed in Example 2.5.1. Identity (4.12) is proved in similar fashion. ∎

4.4   Stirling Numbers

There are two kinds of Stirling numbers [17, 21, 62, 110, 127, 151]. Stirling numbers $\left\{{n \atop k}\right\}$ of the second kind count the number of possible partitions of a set of n objects into k disjoint blocks. For instance, $\left\{{3 \atop 2}\right\} = 3$ because the set $\{1,2,3\}$ can be partitioned into two disjoint blocks in the three ways $\{1,2\} \cup \{3\}$, $\{1,3\} \cup \{2\}$, and $\{2,3\} \cup \{1\}$. We can generate the numbers $\left\{{n \atop k}\right\}$ recursively from the boundary conditions $\left\{{n \atop 1}\right\} = 1$ and $\left\{{n \atop k}\right\} = 0$ for $k > n$ and the recurrence relation

$$\left\{{n \atop k}\right\} = \left\{{n-1 \atop k-1}\right\} + k\left\{{n-1 \atop k}\right\}. \tag{4.16}$$

To prove the recurrence (4.16), imagine adding n to an existing partition of $\{1,\dots,n-1\}$. If the existing partition has $k-1$ blocks, then we must create a new block for n in order to achieve k blocks. If the existing partition has k blocks, then we must add n to one of the k existing blocks. This can be done in k ways. Since none of the other partitions of $\{1,\dots,n-1\}$ can be successfully modified to form k blocks, formula (4.16) is true.

As an application of Stirling numbers of the second kind, consider the problem of throwing n symmetric dice with r faces each. To calculate the probability that k different faces appear when the n dice are thrown, we must first take into account the number of ways $\left\{{n \atop k}\right\}$ that the n dice can be partitioned into k blocks. Once these blocks are chosen, then top-side faces can be assigned to the k blocks in $r(r-1)\cdots(r-k+1)$ ways. Thus, the probability in question amounts to $r^{-n}\left\{{n \atop k}\right\}r(r-1)\cdots(r-k+1)$, which vanishes when $k > \min\{n,r\}$.

In the dice problem, the possible probabilities sum to 1. This establishes the polynomial identity

$$\sum_{k=1}^{n} \left\{{n \atop k}\right\} x(x-1)\cdots(x-k+1) = x^n \tag{4.17}$$

for all positive integers $x = r$ and therefore for all real numbers x. Substituting $-x$ for x in (4.17) gives the similar identity

$$\sum_{k=1}^{n}(-1)^{n-k} \left\{{n \atop k}\right\} x(x+1)\cdots(x+k-1) = x^n. \tag{4.18}$$

Finally, substituting a random variable X for x in equations (4.17) and (4.18) and taking expectations leads to the relations

$$\sum_{k=1}^{n} \left\{{n \atop k}\right\} E[X(X-1)\cdots(X-k+1)] = E(X^n)$$

$$\sum_{k=1}^{n}(-1)^{n-k} \left\{{n \atop k}\right\} E[X(X+1)\cdots(X+k-1)] = E(X^n) \tag{4.19}$$

connecting falling and rising factorial moments to ordinary moments. The former relation is obviously pertinent when we calculate moments by differentiating a probability generating function.

We now turn to Stirling numbers $\begin{bmatrix} n \\ k \end{bmatrix}$ of the first kind. These have a combinatorial interpretation in terms of the cycles of a permutation. Consider the permutation π of $\{1, \ldots, 6\}$ that carries the top row of the matrix

$$\begin{pmatrix} 1 & 2 & 3 & 4 & 5 & 6 \\ 3 & 6 & 5 & 4 & 1 & 2 \end{pmatrix}$$

to its bottom row. We can also represent π by the cycle decomposition $(1, 3, 5), (2, 6), (4)$. The first cycle $(1, 3, 5)$ indicates that π satisfies $\pi(1) = 3$, $\pi(3) = 5$, and $\pi(5) = 1$; the second cycle that $\pi(2) = 6$ and $\pi(6) = 2$; and the third cycle that $\pi(4) = 4$. Note that the order of the cycles is irrelevant in representing π. Also within a cycle, only rotational order is relevant. Thus, the three cycles $(1, 3, 5)$, $(5, 1, 3)$, and $(3, 5, 1)$ are all equivalent. In the preferred or canonical cycle representation of a permutation, the first entry of each cycle is the largest entry of the cycle. The cycles are then ordered so that the successive first entries appear in increasing order. For example, the canonical representation of our given permutation is $(4), (5, 1, 3), (6, 2)$.

The Stirling number $\begin{bmatrix} n \\ k \end{bmatrix}$ counts the number of permutations of $\{1, \ldots, n\}$ with k cycles. These numbers satisfy the boundary conditions $\begin{bmatrix} n \\ k \end{bmatrix} = 0$ for $k > n$ and $\begin{bmatrix} n \\ 1 \end{bmatrix} = (n-1)!$. The former condition is obvious, and the later condition follows once we recall our convention of putting n at the left of the cycle in the canonical representation. All remaining numbers $\begin{bmatrix} n \\ k \end{bmatrix}$ can be generated via the recurrence relation

$$\begin{bmatrix} n \\ k \end{bmatrix} = \begin{bmatrix} n-1 \\ k-1 \end{bmatrix} + (n-1) \begin{bmatrix} n-1 \\ k \end{bmatrix}. \tag{4.20}$$

The proof of (4.20) parallels that of (4.16). The first term on the right counts the number of ways of adding n as a separate cycle to a permutation π of $\{1, \ldots, n-1\}$ with $k-1$ cycles. The second term of the right counts the number of ways of adding n to an existing cycle of a permutation π of $\{1, \ldots, n-1\}$ with k cycles. If π is such a permutation, then we extend π to n by taking $\pi(n) = i$ for $1 \le i \le n-1$. This action conflicts with a current assignment $\pi(j) = i$, so we have to patch things up by defining $\pi(j) = n$. The cycle containing i and j is left intact except for these changes.

We now investigate the number of cycles Y_n in a random permutation π of $\{1, \ldots, n\}$. Clearly, Y_1 is identically 1. If we divide the recurrence (4.20) by $n!$, then we get the recurrence

$$\Pr(Y_n = k) = \frac{1}{n} \Pr(Y_{n-1} = k-1) + \left(1 - \frac{1}{n}\right) \Pr(Y_{n-1} = k).$$

This convolution equation says that $Y_n = Y_{n-1} + Z_n$, where Z_n is independent of Y_{n-1} and follows a Bernoulli distribution with success probability

$1/n$. Proceeding inductively, we conclude that in a distributional sense Y_n can be represented as the sum $Z_1 + \cdots + Z_n$ of n independent Bernoulli random variables with decreasing success probabilities. Problem 12 provides a concrete interpretation of Z_k.

We are now in a position to extract useful information about Y_n. For example, the mean number of cycles is

$$\mathrm{E}(Y_n) \;=\; \sum_{k=1}^{n} \frac{1}{k} \;\approx\; \ln n + \gamma,$$

where $\gamma \approx .5772$ is Euler's constant. Because the probability generating function of Z_n is $\mathrm{E}(x^{Z_n}) = 1 - \frac{1}{n} + \frac{x}{n} = \frac{x+n-1}{n}$, the probability generating function of Y_n is

$$\mathrm{E}(x^{Y_n}) \;=\; \frac{1}{n!} x(x+1) \cdots (x+n-1).$$

In view of the definition of Y_n, we get the interesting identity

$$\begin{aligned} n! \, \mathrm{E}(x^{Y_n}) &= \sum_{k=1}^{n} \begin{bmatrix} n \\ k \end{bmatrix} x^k \\ &= x(x+1) \cdots (x+n-1). \end{aligned} \tag{4.21}$$

This polynomial identity in $x \in [0,1]$ persists for all real x. Substituting $-x$ for x in equation (4.21) yields the dual identity

$$\sum_{k=1}^{n} (-1)^{n-k} \begin{bmatrix} n \\ k \end{bmatrix} x^k \;=\; x(x-1) \cdots (x-n+1). \tag{4.22}$$

Finally, substituting a random variable X for x in equations (4.21) and (4.22) and taking expectations lead to the relations

$$\sum_{k=1}^{n} \begin{bmatrix} n \\ k \end{bmatrix} \mathrm{E}(X^k) \;=\; \mathrm{E}[X(X+1) \cdots (X+n-1)]$$

$$\sum_{k=1}^{n} (-1)^{n-k} \begin{bmatrix} n \\ k \end{bmatrix} \mathrm{E}(X^k) \;=\; \mathrm{E}[X(X-1) \cdots (X-n+1)]$$

connecting ordinary moments to rising and falling factorial moments.

We close this section by giving another combinatorial interpretation of Stirling numbers of the first kind. This interpretation is important in the theory of record values for i.i.d. sequences of random variables [5]. A permutation π is said to possess a left-to-right maximum at i provided $\pi(j) < \pi(i)$ for all $j < i$. The Stirling number $\begin{bmatrix} n \\ k \end{bmatrix}$ also counts the number of permutations of $\{1, \ldots, n\}$ with k left-to-right maxima. To prove this assertion, it suffices to construct a one-to-one correspondence between permutations with k cycles and permutations with k left-to-right maxima. The canonical representation of a permutation achieves precisely this end since the leading number in each cycle is a left-to-right maximum.

4.5   Application to an Urn Model

The family planning model discussed in Example 2.3.3 is a kind of an urn model in which sampling is done with replacement. In other models, sampling without replacement is more appropriate. Such sampling can be realized by starting with n urns and b_j balls in urn j. At each trial a ball is randomly selected from one of the balls currently available and extracted. Under sampling without replacement, this process gradually depletes the supply of balls within the urns. By analogy with the family planning model, it is interesting to set a quota for each urn. When q_j balls from urn j have been drawn, then urn j is said to have reached its quota. Sampling continues until exactly i urns reach their quotas. The trial N_i at which this occurs is a waiting time random variable. Calculation of the moments N_i is challenging. Fortunately, we can apply Proposition 4.3.1 and the magical method of probabilistic embedding [20] .

As a concrete example, consider laundering n pairs of dirty socks. Suppose one removes clean socks from the washing machine one by one. If each pair of socks is distinguishable, let N_1 be the number of clean socks extracted until a pair is found. Blom et al. [21] compute the mean of N_1, a problem originally posed and solved by Friedlen [52]. In the current context, pairs of socks correspond to urns and socks to balls. The quota for the jth pair of socks is $q_j = 2$.

The embedding argument works by imagining the balls in urn j as drawn in turn at times determined by a random sample of size b_j from the uniform distribution on $[0, 1]$. We will refer to such a sample as a uniform process. If the uniform processes for the n different urns are independent, then superimposing them creates a uniform process of size $b = \sum_{j=1}^{n} b_j$ on $[0, 1]$. The original discrete-time urn sampling process is said to be embedded in the superposition process. Observe that the origin of each point in the superposition process is retained.

Thus, each point still corresponds to sampling from a known urn. If we let X_j be the time at which the quota q_j for the jth urn is reached, these attainment times are independent. This feature of embedding can be exploited to good effect.

In the superposition process , the order statistic $X_{(i)}$ is the waiting time until i urns reach their quotas. If N_i is the number of trials until the occurrence of this event, then we need to relate the moments of N_i and $X_{(i)}$. The order statistic $X_{(i)}$ can be represented as

$$X_{(i)} \;=\; \sum_{j=1}^{N_i} Y_j,$$

where $Y_1, Y_2, \ldots, Y_{b+1}$ are the spacings between adjacent points in the superposition process.

It is straightforward to show that the random distance $Z_m = \sum_{j=1}^{m} Y_j$ to the mth point in the superposition process follows a beta distribution with parameters m and $b - m + 1$. Indeed, Z_m is found in the interval $(z, z + dz)$ when one of the b points falls within $(z, z + dz)$, $m - 1$ random points fall to the left of z, and $b - m$ random points fall to the right of $z + dz$. This composite event occurs with approximate probability

$$b \binom{b-1}{m-1} z^{m-1}(1-z)^{b-m+1-1} dz.$$

Dividing this probability by dz and letting dz tend to 0 gives the requisite beta density. In view of this fact, conditioning and Example 2.3.1 yield

$$
\begin{aligned}
\mathrm{E}[X_{(i)}^k] &= \mathrm{E}(Z_{N_i}^k) \\
&= \mathrm{E}\left[\mathrm{E}\left(Z_{N_i}^k \mid N_i\right)\right] \\
&= \frac{1}{(b+1)\cdots(b+k)} \mathrm{E}[N_i \cdots (N_i + k - 1)]. \qquad (4.23)
\end{aligned}
$$

Equation (4.23) gives, for instance, $\mathrm{E}(N_i) = (b+1)\,\mathrm{E}[X_{(i)}]$. Furthermore, we can recover all of the ordinary moments $\mathrm{E}(N_i^k)$ from the ascending factorial moments $\mathrm{E}[N_i \cdots (N_i + k - 1)]$ via equation (4.19).

Formula (4.13) provides a means of computing $\mathrm{E}[X_{(i)}^k]$ exactly. The independence of the urn-specific uniform processes implies

$$\Pr(X_S > t) = \prod_{l \in S}\left[\sum_{m_l=0}^{q_l-1} \binom{b_l}{m_l} t^{m_l}(1-t)^{b_l - m_l}\right].$$

To calculate the expectations $\mathrm{E}[(X_S)^k]$ required by formula (4.13), we define $\mathbf{m} = (m_l)$ to be a multi-index ranging over the Cartesian product set $R = \{\mathbf{m} : 0 \le m_l \le q_l - 1, l \in S\}$. Then with $|\mathbf{m}| = \sum_{l \in S} m_l$, it follows that

$$
\begin{aligned}
\mathrm{E}[(X_S)^k] &= k \int_0^\infty t^{k-1} \Pr(X_S > t)\, dt \\
&= k \sum_{\mathbf{m} \in R} \prod_{l \in S} \binom{b_l}{m_l} \int_0^1 t^{k+|\mathbf{m}|-1}(1-t)^{b-|\mathbf{m}|}\, dt \qquad (4.24) \\
&= k \sum_{\mathbf{m} \in R} \prod_{l \in S} \binom{b_l}{m_l} \frac{\Gamma(k+|\mathbf{m}|)\Gamma(b-|\mathbf{m}|+1)}{\Gamma(k+b+1)},
\end{aligned}
$$

where $\Gamma(u)$ is the gamma function. In the socks in the laundry problem, equation (4.24) reduces to

$$\mathrm{E}[(X_S)^k] = k \sum_{m_1=0}^{1} \cdots \sum_{m_n=0}^{1} \prod_{l=1}^{n} \binom{2}{m_l} \frac{(k+|\mathbf{m}|-1)!(2n-|\mathbf{m}|)!}{(k+2n)!}.$$

$$= k \sum_{i=0}^{n} 2^i \binom{n}{i} \frac{(k+i-1)!(2n-i)!}{(k+2n)!}. \qquad (4.25)$$

For small k and n, we can therefore compute $E(X_{(1)}^k)$ and $E(N_1^k)$ exactly.

For large n it is more instructive to derive asymptotic expressions for $E[X_{(1)}^k]$ by Laplace's method [40]. In the socks in the laundry problem, the independence of the uniform processes entails

$$\Pr[X_{(1)} > t] = \Pr\left(\min_{1 \leq i \leq n} X_i > t \right) = (1 - t^2)^n$$

because any given pair of socks arrives after time t with probability $1 - t^2$. Integrating this tail probability with respect to t produces

$$\begin{aligned}
E[X_{(1)}] &= \int_0^1 (1 - t^2)^n \, dt \\
&= \int_0^1 e^{n \ln(1 - t^2)} \, dt \qquad (4.26) \\
&\asymp \int_0^\infty e^{-nt^2} \, dt \\
&= \frac{\sqrt{\pi}}{2\sqrt{n}}
\end{aligned}$$

and therefore $E(N_1) = (2n + 1) E[X_{(i)}] \asymp \sqrt{\pi n}$. The essence of Laplace's approximation is simply the recognition that the vast majority of the mass of the integral $\int_0^1 (1 - t^2)^n \, dt$ occurs near $t = 0$, where $\ln(1 - t^2) \approx -t^2$. Similar reasoning leads to

$$\begin{aligned}
\mathrm{Var}(N_1) &= (2n + 1)(2n + 2) E[X_{(1)}^2] - E(N_1) - E(N_1)^2 \\
&\asymp (4 - \pi)n.
\end{aligned}$$

4.6 Pigeonhole Principle

The pigeonhole principle is an elementary technique of great beauty and utility [25]. In its simplest form, it deals with p pigeons and b boxes (holes), where $p > b$. If we assign the pigeons to boxes, then some box gets more than one pigeon. A stronger form of the pigeonhole principle deals with n numbers $r_1, \ldots, r_n$. If the sum of $r_1 + \cdots + r_n > mn$, then at least one of the r_i satisfies $r_i > m$.

Example 4.6.1 *Longest Increasing Subsequence*

As an application of the principle, consider a sequence $a_1, \ldots, a_{mn+1}$ of $mn + 1$ distinct real numbers. We claim that a_i contains an increasing subsequence of length $n + 1$ or a decreasing subsequence of length $m + 1$. To apply the pigeonhole principle, we suppose the contrary and label each a_i by the length l_i of the longest increasing subsequence commencing with a_i. For example, if $8, 1, 6, 2, 5, 4, 3$ is the sequence, then a longest increasing subsequence beginning with 1 is $1, 2, 5$. Thus, we label 1 with the number $l_2 = 3$. By assumption, no label can exceed n. Let r_l be the number of l_i satisfying $l_i = l$. Because $r_1 + \cdots + r_n = mn + 1$, at least one of the n summands r_l must exceed m. If $r_l = s > m$, then there are s numbers $a_{i_1}, \ldots, a_{i_s}$ such that each a_{i_j} is the start of a maximal increasing subsequence of length l. Now suppose that $a_{i_j} < a_{i_{j+1}}$ for some j. If this is the case, then by appending a_{i_j} to a maximal increasing subsequence beginning with $a_{i_{j+1}}$, we get an increasing subsequence of length $l + 1$. This contradicts the label of a_{i_j}. Thus, the $s > m$ numbers $a_{i_1}, \ldots, a_{i_s}$ are in decreasing order. In other words, if an increasing subsequence of length $n + 1$ does not exist, then a decreasing subsequence of length $m + 1$ does.

We can put this result to good use in a probabilistic version of the longest increasing subsequence problem. Let $X_1, \ldots, X_n$ be independent random variables uniformly distributed on the interval $[0,1]$. Let I_n and D_n be the random lengths of the longest increasing and decreasing subsequences of $X_1, \ldots, X_n$. These random variables satisfy

$$\sqrt{n} \ \leq \ \max\{I_n, D_n\} \ \leq \ I_n + D_n.$$

Because $\mathrm{E}(I_n) = \mathrm{E}(D_n)$, we therefore conclude that $\mathrm{E}(I_n) \geq \sqrt{n}/2$.

To show that this lower bound is of the correct order of magnitude, we supplement it by a comparable upper bound. We first observe that

$$\Pr(I_n \geq k) \ \leq \ \frac{\binom{n}{k}}{k!}.$$

Indeed, there are $\binom{n}{k}$ subsequences of length k of $X_1, \ldots, X_n$, and each has probability $1/k!$ of being increasing. This inequality comes in handy in the estimate

$$
\begin{aligned}
\mathrm{E}(I_n) \ &\leq \ k + n \Pr(I_n \geq k) \\
&\leq \ k + \frac{n\binom{n}{k}}{k!} \\
&\leq \ k + \frac{n^{k+1}}{(k!)^2}.
\end{aligned}
\tag{4.27}
$$

The latter estimate can be improved by employing Stirling's approximation

$$k! \ \asymp \ \sqrt{2\pi} k^{k+1/2} e^{-k}$$

and choosing k appropriately. The good choice $k = \alpha\sqrt{n}$ yields

$$k + \frac{n^{k+1}}{(k!)^2} \ \approx \ k + \frac{n^{k+1}}{(\sqrt{2\pi} k^{k+1/2} e^{-k})^2}$$

$$= \alpha\sqrt{n} + \frac{e^{2\alpha\sqrt{n}}\sqrt{n}}{2\pi\alpha^2(\alpha\sqrt{n}+1/2)} \qquad (4.28)$$

$$= \alpha\sqrt{n} + \frac{e^{-2\alpha\sqrt{n}(\ln\alpha-1)}\sqrt{n}}{2\pi\alpha}.$$

If we take $\alpha > e$, then $\ln\alpha > 1$, and inequality (4.27) and approximate equality (4.28) together produce $E(I_n) \leq c\sqrt{n}$ for some $c > \alpha$. Thus, $E(I_n)$ is on the order of $\sqrt{n}$ in magnitude. Refinements of these arguments show that $E(I_n)/\sqrt{n}$ tends to a limit as n tends to ∞ [140]. ∎

4.7 Problems

1. If X has discrete geometric density $\Pr(X = k) = pq^k$ for $k \geq 0$, then calculate the binomial moment $E[\binom{X}{j}] = \left(\frac{q}{p}\right)^j$.

2. You have 10 pairs of shoes jumbled together in your closet [48]. Show that if you reach in and randomly pull out four shoes, then the probability of extracting at least one pair is $99/323$.

3. Prove that there are $8! \sum_{k=0}^{8} \frac{(-1)^k}{k!}$ ways of placing eight rooks on a chessboard so that none can take another and none stands on a white diagonal square [48]. (Hint: Think of the rook positions as a random permutation π, and let A_i be the event $\{\pi(i) = i\}$.)

4. A permutation that satisfies the equation $\pi(\pi(i)) = i$ for all i is called an involution [110]. Prove that a random permutation of $\{1, \ldots, n\}$ is an involution with probability

$$\sum_{k=0}^{\lfloor \frac{n}{2} \rfloor} \frac{1}{2^k k!(n-2k)!}.$$

(Hint: An involution has only fixed points and two-cycles. Count the number of involutions and divide by $n!$.)

5. Define q_r to be the probability that in r tosses of two dice each pair $(1,1), \ldots, (6,6)$ appears at least once [48]. Show that

$$q_r = \sum_{k=0}^{6} \binom{6}{k}(-1)^k \left(\frac{36-k}{36}\right)^r.$$

6. Calculate the probability $p_{[k]}$ that exactly k suits are missing in a poker hand [48]. To a good approximation $p_{[0]} = .264$, $p_{[1]} = .588$, $p_{[2]} = .146$, and $p_{[3]} = .002$. (Hint: Each hand has probability $1/\binom{52}{5}$.)

7. Prove the Bonferroni inequality (4.9) by establishing the inequality

$$1_{\cup_{i=1}^n A_i} \geq \sum_{i=1}^n 1_{A_i} - \sum_{i<j} 1_{A_i} 1_{A_j}$$

and then taking expectations.

8. Suppose that each of the random variables $X_1, \ldots, X_n$ of Proposition 4.3.1 satisfies $E(|X_i|^k) < \infty$. Show that all of the expectations $E(|X_{(i)}|^k)$, $E(|X_S|^k)$, and $E(|X^S|^k)$ are finite. (Hints: Bound each random variable in question by $(\sum_{j=1}^n |X_j|)^k$ and apply Minkowski's inequality given in Problem 22 of Chapter 3.)

9. List in canonical form the 11 permutations of $\{1, 2, 3, 4\}$ with 2 cycles.

10. Show that $\begin{bmatrix} n \\ n \end{bmatrix} = \begin{Bmatrix} n \\ n \end{Bmatrix} = 1$, $\begin{bmatrix} n \\ n-1 \end{bmatrix} = \begin{Bmatrix} n \\ n-1 \end{Bmatrix} = \binom{n}{2}$, and $\begin{Bmatrix} n \\ 2 \end{Bmatrix} = 2^{n-1} - 1$.

11. Prove the inequality $\begin{bmatrix} n \\ k \end{bmatrix} \geq \begin{Bmatrix} n \\ k \end{Bmatrix}$ for all n and k by invoking the definitions of the two kinds of Stirling numbers.

12. In our discussion of Stirling numbers of the first kind, we showed that the number of left-to-right maxima Y_n of a random permutation has the decomposition $Y_n = Z_1 + \cdots + Z_n$, where the Z_k are independent Bernoulli variables with decreasing success probabilities. Prove that Z_k can be interpreted as the indicator of the event that position $n-k+1$ is a left-to-right maximum. In other words, Z_k is the indicator of the event $\{\pi(n - k + 1) > \pi(j) \text{ for } 1 \leq j < n - k + 1\}$. Why are the Z_k independent [21]?

13. Let X be the number of fixed points of a random permutation of $\{1, \ldots, n\}$. Demonstrate that

$$E(X^j) = \sum_{k=1}^{\min\{j,n\}} \begin{Bmatrix} j \\ k \end{Bmatrix}.$$

(Hint: Find the binomial moments of X, convert these to factorial moments, and then convert these to ordinary moments [110].)

14. Suppose π is a random permutation of $\{1, \ldots, n\}$. Show that

$$E\left\{ \sum_{j=1}^{n-1} [\pi(j) - \pi(j+1)]^2 \right\} = \binom{n+1}{3}.$$

(Hint: Each term has the same expectation [110].)

15. In the socks in the laundry problem, demonstrate that

$$E(N_1) = \frac{(2^n n!)^2}{(2n)!}.$$

Conclude from this and Stirling's formula that $E(N_1) \asymp \sqrt{\pi n}$. (Hint: Change variables in the first integral of equation (4.26).)

16. Consider a random graph with n nodes. Between every pair of nodes, we independently introduce an edge with probability p. A trio of nodes forms a triangle if each of its three pairs is connected by an edge. If N counts the number of triangles, then demonstrate that $E(N) = \binom{n}{3} p^3$ and $\mathrm{Var}(N) = \binom{n}{3} p^3 (1 - p^3) + \binom{n}{3} 3(n-3)(p^5 - p^6)$.

17. Consider the n-dimensional unit cube $[0,1]^n$. Suppose that each of its $n2^{n-1}$ edges is independently assigned one of two equally likely orientations. Let S be the number of vertices at which all neighboring edges point toward the vertex. Show that S has mean $E(S) = 1$ and variance $\mathrm{Var}(S) = 1 - (n+1)2^{-n}$. When n is large, S follows an approximate Poisson distribution. (Hint: Let X_α be the indicator that vertex α has all of its edges directed toward α. Note that X_α is independent of X_β unless α and β share an edge. If α and β share an edge, then $X_\alpha X_\beta = 0$.)

18. Five points are chosen from an equilateral triangle with sides of length 1. Demonstrate that there exist two points separated by a distance of at most $1/2$.

19. Suppose $n+1$ numbers are chosen from the set $\{1, 2, \ldots, 2n\}$. Show that there is some pair having no common factor other than 1. Show that there is another pair such that one member of the pair is divisible by the other.

20. Consider a graph with more than one node. Let d_i be the degree of node i. Prove at least two d_i coincide.

21. Given n integers $a_1, \ldots, a_n$, demonstrate that there is some sum $\sum_{i=j+1}^{k} a_i$ that is a multiple of n. (Hint: Map each of the $n+1$ partial sums $s_j = \sum_{i=1}^{j} a_i$ into its remainder after division by n.)

5

Combinatorial Optimization

5.1 Introduction

Combinatorial averaging is a supple tool for understanding the solutions of discrete optimization problems. Computer scientists have designed many algorithms to solve such problems. Traditionally, these algorithms have been classified by their worst-case performance. Such an analysis can lead to undue pessimism. The average behavior of an algorithm is usually more relevant. Of course, to evaluate the average complexity of an algorithm, we must have some probability model for generating typical problems on which the algorithm operates. The examples in this chapter on sorting, data compression, and graph coloring illustrate some of the underlying models and the powerful techniques probabilists have created for analyzing algorithms.

Not only is combinatorial averaging helpful in understanding the complexity of algorithms, but it can also yield nonconstructive existence proofs and verify that a proposed solution of a discrete optimization problem is optimal [1, 4, 62, 150]. The former role is just the probabilistic method of combinatorics initiated by Erdös and Rényi. In the probabilistic method, we take a given set of objects, embed it in a probability space, and show that the subset of objects lacking a certain property has probability less than 1. The subset of objects possessing the property must therefore be nonempty. Alternatively, if the property is determined by some number X assigned to each object, then we can view X as a random variable and calculate its expectation. If the property holds for $X \leq c$ and $\mathrm{E}(X) \leq c$, then some object with the property exists. Our treatment of Sperner's theorem illustrates the role of probability in discrete optimization. Finally, we discuss in the current chapter subadditive and superadditive sequences and their application to the longest common subsequence problem. The linear growth in complexity seen in this problem does not always occur, as our concluding example on the Euclidean traveling salesman problem shows.

5.2   Quick Sort

Sorting lists of items such as numbers or words is one of the most thoroughly studied tasks in computer science. It is a pleasant fact that the fastest sorting algorithm can be explained by a probabilistic argument [150]. At the heart of this argument is a recurrence relation specifying the average number of operations encountered in sorting n numbers. In this problem, we can explicitly solve the recurrence relation and estimate the rate of growth of its solution as a function of n.

The quick sort algorithm is based on the idea of finding a splitting entry x_i of a sequence $x_1, \ldots, x_n$ of n distinct numbers in the sense that $x_j < x_i$ for $j < i$ and $x_j > x_i$ for $j > i$. In other words, a splitter x_i is already correctly ordered relative to the rest of the entries of the sequence. Finding a splitter reduces the computational complexity of sorting because it is easier to sort both of the subsequences $x_1, \ldots, x_{i-1}$ and $x_{i+1}, \ldots, x_n$ than it is to sort the original sequence. At this juncture, one can reasonably object that no splitter need exist, and even if one does, it may be difficult to locate. The quick sort algorithm avoids these difficulties by randomly selecting a splitting value and then slightly rearranging the sequence so that this splitting value occupies the correct splitting location.

In the background of quick sort is the probabilistic assumption that all $n!$ permutations of the n values are equally likely. The algorithm begins by randomly selecting one of the n values and moving it to the leftmost or first position of the sequence. Through a sequence of exchanges, this value is then promoted to its correct location. In the probabilistic setting adopted, the correct location of the splitter is uniformly distributed over the n positions of the sequence.

The promotion process works by exchanging or swapping entries to the right of the randomly chosen splitter x_1, which is kept in position 1 until a final swap. Let j be the current position of the sequence as we examine it from left to right. In the sequence up to position j, a candidate position i for the insertion of x_1 must satisfy the conditions $x_k < x_1$ for $1 < k \le i$ and $x_k > x_1$ for $i < k \le j$. Clearly, the choice $i = j$ works when $j = 1$ because then the set $\{k : 1 < k \le i \text{ or } i < k \le j\}$ is empty. Now suppose we examine position $j + 1$. If $x_{j+1} > x_1$, then we keep the current candidate position i. If $x_{j+1} < x_1$, then we swap x_{i+1} and x_{j+1} and replace i by $i+1$.

In either case, the two required conditions imposed on i continue to hold. Thus, we can inductively march from the left end to the right end of the sequence, carrying out a few swaps in the process, so that when $j = n$, the value i marks the correct position to insert x_1. Once this insertion is made, the subsequences $x_1, \ldots, x_{i-1}$ and $x_{i+1}, \ldots, x_n$ can be sorted separately by the same splitting procedure.

Now let e_n be the expected number of operations involved in quick sorting a sequence of n numbers. By convention $e_0 = 0$. If we base our analysis only on how many positions j must be examined at each stage and not on how many swaps are involved, then we can write the recurrence relation

$$
\begin{aligned}
e_n &= n - 1 + \frac{1}{n} \sum_{i=1}^{n} (e_{i-1} + e_{n-i}) \\
&= n - 1 + \frac{2}{n} \sum_{i=1}^{n} e_{i-1}
\end{aligned}
\tag{5.1}
$$

by conditioning on the correct position i of the first splitter.

The recurrence relation (5.1) looks formidable, but a few algebraic maneuvers render it solvable. Multiplying equation (5.1) by n produces

$$
n e_n = n(n-1) + 2 \sum_{i=1}^{n} e_{i-1}.
$$

If we subtract from this the corresponding expression for $(n-1)e_{n-1}$, then we get

$$
n e_n - (n-1)e_{n-1} = 2n - 2 + 2e_{n-1},
$$

which can be rearranged to give

$$
\frac{e_n}{n+1} = \frac{2(n-1)}{n(n+1)} + \frac{e_{n-1}}{n}.
\tag{5.2}
$$

Equation (5.2) can be iterated to yield

$$
\begin{aligned}
\frac{e_n}{n+1} &= 2 \sum_{k=1}^{n} \frac{(k-1)}{k(k+1)} \\
&= 2 \sum_{k=1}^{n} \left(\frac{2}{k+1} - \frac{1}{k} \right) \\
&= 2 \sum_{k=1}^{n} \frac{1}{k} - \frac{4n}{n+1}.
\end{aligned}
$$

Because $\sum_{k=1}^{n} \frac{1}{k}$ approximates $\int_1^n \frac{1}{x} \, dx = \ln n$, it follows that

$$
\lim_{n \to \infty} \frac{e_n}{2n \ln n} = 1.
$$

Quick sort is, indeed, a very efficient algorithm on average. Press et al. [123] provide good computer code implementing it.

5.3 Data Compression and Huffman Coding

Huffman coding is an algorithm for data compression without loss of information [123, 128, 136]. In this section we present the algorithm and prove its optimality in an average sense. To motivate Huffman coding, it is useful to think of an alphabet $\mathcal{A}$ with typical letter $l \in \mathcal{A}$. From previous experience with the alphabet, we can assign a usage probability p_l to l. Inside a computer, we represent l using a bit string s_l, each bit having the value 0 or 1. One possibility is to use bit strings of fixed length to represent all letters. This is inefficient if there is wide variation in the probabilities p_l. Huffman coding uses bit strings of varying length, with frequent letters assigned short strings and infrequent letters assigned long strings.

In addition to the obvious requirement that no two assigned bit strings coincide, we require instantaneous decoding. This is motivated by the necessity of recording words and a sequence of words. Words are separated by spaces, so we enlarge our alphabet to contain a space symbol if necessary. Thus, if we want to encode a message, we do so letter by letter and concatenate the corresponding bit strings. This tactic leads to confusion if we fail to design the bit strings properly. For example, if the alphabet is the ordinary alphabet, we could conceivably assign the letter e the bit string 111 and the letter a the bit string 1110. When we encounter the three bits 111 in the encoded message, we then face the ambiguity of whether we have an e or the start of an a. Consequently, we impose the further constraint that no prefix of a bit string representing a letter coincides with a bit string representing a different letter. We interpret "prefix" to mean either a beginning portion of a bit string or the whole bit string.

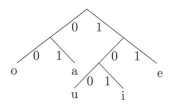

FIGURE 5.1. The Huffman Vowel Tree

Huffman coding solves the instantaneous decoding problem by putting all letters at the bottom of a binary tree. For example, Figure 5.1 shows the Huffman tree corresponding to the alphabet $\mathcal{A} = \{a,e,i,o,u\}$ consisting of the vowels. To construct the bit string for a given letter, we just traverse the tree from the root at its top to the corresponding letter node at its bottom. Each new edge encountered adds a 0 or 1 to the bit string for the letter. Every left edge taken adds a 0, and every right edge taken adds a 1. Thus, we represent the letter o by the bit string 00 and the letter u by the bit string 100 in Figure 5.1.

In an arbitrary encoding of the alphabet $\mathcal{A}$, we also assign each letter l a bit string. The number of bits (or length) in a bit string s is denoted by $\text{len}(s)$. We can view a code S as a random map taking the random letter l to its bit string s_l. Huffman coding minimizes the average length

$$\text{E}[\text{len}(S)] = \sum_{l \in \mathcal{A}} \text{len}(s_l) p_l.$$

A Huffman code is constructed recursively. Consider an alphabet $\mathcal{A}_n$ of n letters $l_1, \ldots, l_n$ arranged so that $p_{l_1} \geq \cdots \geq p_{l_n}$. In the event of one or more ties $p_m = p_{m+1}$, there will be multiple Huffman codings with the same average length. Because we want the most infrequent letters to reside at the bottom of the tree, we build a minitree by joining l_n on the left and l_{n-1} on the right to a parent node above designated $m_{n-1} = \{l_n, l_{n-1}\}$. To node m_{n-1} we attribute probability $p_{l_n} + p_{l_{n-1}}$. We then proceed to construct the Huffman tree for the alphabet $\mathcal{A}_{n-1} = \{l_1, \ldots, l_{n-2}, m_{n-1}\}$. At the final stage of the Huffman algorithm, we have a single letter, which becomes the root of the tree.

For example, the vowels have approximate usage probabilities $p_a = .207$, $p_e = .332$, $p_i = .185$, $p_o = .203$, and $p_u = .073$ in English [128]. Huffman coding first combines u on the left with i on the right into the node $\{u, i\}$ with probability .258. Second, it combines o on the left with a on the right into the node $\{o, a\}$ with probability .410. Third, it combines $\{u, i\}$ on the left with e on the right into the node $\{u, i, e\}$ with probability .590. Finally, it combines $\{o, a\}$ on the left with $\{u, i, e\}$ on the right into the root.

In proving that Huffman coding is optimal, let us simplify notation by identifying the n letters of the alphabet $\mathcal{A}_n$ with the integers $1, \ldots, n$. Under the innocuous assumption $p_1 \geq \cdots \geq p_n$, there are two general methods for improving any instantaneous coding S. First, we can assume that $\text{len}(s_j) \leq \text{len}(s_k)$ whenever $j < k$. If this is not the case, then we can improve S by interchanging s_j and s_k. Second, we can represent any bit string s_j in S by $s_j = (x, y)$, where x is the longest prefix of s_j coinciding with a prefix of any other bit string s_k. If y contains more than just its initial bit y_1, then we can improve S by truncating s_j to (x, y_1).

String truncation has implications for the length of the longest bit string s_n. Suppose that $s_n = (x, y)$ and $\text{len}(s_n) > \text{len}(s_{n-1})$. The longest matching prefix x satisfies $\text{len}(x) < \text{len}(s_{n-1})$; otherwise, x coincides with s_{n-1} or some other bit string having the same length as s_{n-1}. Once we replace s_n by (x, y_1), then we can assume that $\text{len}(s_n) \leq \text{len}(s_{n-1})$. If strict inequality prevails, then we interchange the new s_n with s_{n-1}. If we continue truncating the longest bit string, eventually we reach the point where $s_n = (x, y_1)$ and $\text{len}(s_n) = \text{len}(s_{n-1})$. By definition of x, the bit string $(x, y_1 + 1 \bmod 2)$ also is in S. Performing a final interchange if necessary, we can consequently assume that s_n and s_{n-1} have the same length and differ only in their last bit.

Now consider the Huffman coding H_n of $\mathcal{A}_n$ with h_m denoting the bit string corresponding to m. Huffman's construction replaces the letters n and $n-1$ by a single letter with probability $p_n + p_{n-1}$. If we let h denote the bit string assigned to this new letter, then $\text{len}(h) + 1 = \text{len}(h_n) = \text{len}(h_{n-1})$. The old and new Huffman trees therefore satisfy

$$
\begin{aligned}
\text{E}[\text{len}(H_n)] &= \text{E}[\text{len}(H_{n-1})] - \text{len}(h)(p_n + p_{n-1}) \\
&\quad + [\text{len}(h) + 1]p_n + [\text{len}(h) + 1]p_{n-1} \qquad (5.3) \\
&= \text{E}[\text{len}(H_{n-1})] + (p_n + p_{n-1}).
\end{aligned}
$$

Now consider an alternative coding S_n of $\mathcal{A}_n$. As just demonstrated, we can assume that s_n and s_{n-1} have the same length and differ only in their last bit. The changes necessary to achieve this goal can only decrease the average length of S_n. Without loss of generality, suppose that $s_n = (x, 0)$ and $s_{n-1} = (x, 1)$. Assigning the amalgamation of n and $n-1$ the bit string x and the probability $p_n + p_{n-1}$ leads to a code S_{n-1} on $\mathcal{A}_n$ satisfying

$$
\begin{aligned}
\text{E}[\text{len}(S_n)] &= \text{E}[\text{len}(S_{n-1})] - \text{len}(x)(p_n + p_{n-1}) \\
&\quad + [\text{len}(x) + 1]p_n + [\text{len}(x) + 1]p_{n-1} \qquad (5.4) \\
&= \text{E}[\text{len}(S_{n-1})] + (p_n + p_{n-1}).
\end{aligned}
$$

If we assume by induction on n that $\text{E}[\text{len}(H_{n-1})] \leq \text{E}[\text{len}(S_{n-1})]$, then equations (5.3) and (5.4) prove that $\text{E}[\text{len}(H_n)] \leq \text{E}[\text{len}(S_n)]$. Given the obvious optimality of Huffman coding when $n = 2$, this finishes our inductive argument that Huffman coding minimizes average code length.

5.4   Graph Coloring

In the graph coloring problem, we are given a graph with n nodes and asked to color each node with one color from a palette of k colors [151]. Two adjacent nodes must be colored with different colors. For the sake of convenience, let us label the nodes $1, \ldots, n$ and the colors $1, \ldots, k$. The solution to the well-known four-color problem states that any planar graph can be colored with at most four colors. Roughly speaking, a graph is planar if it can be drawn in two dimensions in such a way that no edges cross. For example, if we wish to color a map of contiguous countries, then countries are nodes, and edges connect adjacent countries. In the general graph coloring problem, the graphs need not be planar.

It turns out that a standard computer science technique called back-tracking will solve every graph coloring problem. The catch is that back-tracking is extremely inefficient for large n on certain worst-case graphs. Nonetheless, the average behavior of backtracking is surprisingly good as n increases. The reason for this good performance is that we can reject the possibility of k-coloring most graphs with n nodes.

To illustrate the backtracking algorithm, consider the simple graph of Figure 5.2 with $n = 4$ nodes. This graph can be colored in several ways with three colors. For instance, one solution is the coloring 1213 that assigns color 1 to node 1, color 2 to node 2, color 1 to node 3, and color 3 to node 4. If carried to completion, the backtracking algorithm will construct all possible colorings. To commence the backtracking algorithm, we assign color 1 to node 1. We then are forced to assign colors 2 or 3 to node 2 to avoid a conflict. In the former case, we represent the partial coloring of the first two nodes by 12. In backtracking we keep growing a partial solution until we reach a full solution or a forbidden color match between two neighboring nodes. When either of these events occur, we backtrack to the first available full or partial solution that we have not previously encountered.

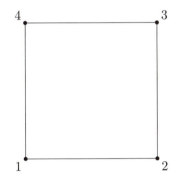

FIGURE 5.2. A Graph with Four Nodes

In our simple example, we extend the partial solution 12 to the larger partial solution 121 by choosing the first available color (color 1) for node 3. From there we extend to the full solution 1212 by choosing the first available color (color 2) for node 4. We then substitute the next available color (color 3) for node 4 to reach the next full solution 1213. At this point, we have exhausted full solutions and are forced to backtrack to node 3 and assign the next available color (color 3) to it. This gives the partial solution 123, which can be grown to the full solution 1232 before backtracking. Figure 5.3 depicts the family of partial and full solutions generated by backtracking with color 1 assigned to node 1. The full backtracking tree with node 1 assigned any of the three available colors is similar to Figure 5.3 but too large to draw.

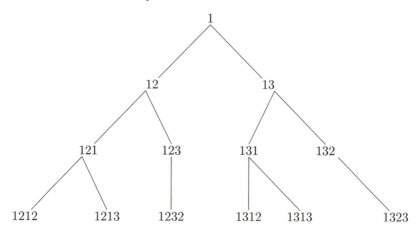

FIGURE 5.3. A Backtracking Tree

In summary, we can depict the functioning of the backtracking algorithm by drawing a backtracking tree with $n + 1$ levels. Level 0 (not shown in Figure 5.3) is a root connected to the k partial solutions $1, \ldots, k$ involving node 1 at level 1. Level l of the backtracking tree contains partial solutions involving nodes 1 through l. Each partial solution is a sequence of length l with entries chosen from the integers 1 to k. The backtracking algorithm finds at least one full solution if and only if the backtracking tree descends all the way to level n.

We now assess the average computational complexity of the backtracking algorithm. To do so, we set up a simple probability model for random graphs with n nodes. There are $\binom{n}{2}$ possible edges in a graph with n nodes and a total of $2^{\binom{n}{2}}$ possible graphs. The uniform distribution on this sample space can be achieved by considering each pair of nodes in turn and independently introducing an edge between the nodes with probability $\frac{1}{2}$. For each given graph G with n nodes, we imagine constructing the corresponding backtrack tree and recording the number $X_l(G)$ of partial solutions at level l. (If $l = n$, the partial solutions are full solutions.) The amount of work done in backtracking is proportional to $\sum_{l=1}^{n} X_l$. With this notation, our goal is to demonstrate the remarkable fact that $\sum_{l=1}^{n} E(X_l)$ is bounded above by a constant that does not depend on n.

To prove this claim, we need to bound $E(X_l)$. Now each partial solution in the backtrack tree at level l represents a proper coloring of the first l nodes of G. There are, of course, k^l possible colorings of l nodes. Instead of trying to estimate the number of colorings compatible with each subgraph on l nodes, let us try to estimate the probability that a random subgraph on l nodes is compatible with a given coloring of the l nodes. Under the uniform distribution, each of the $2^{\binom{l}{2}}$ possible subgraphs on l nodes is equally likely.

Many subgraphs can be eliminated as inconsistent with the given coloring because they involve forbidden edges.

Suppose m_i nodes have color i. Because we can draw an edge only between nodes of different colors, the total number of permitted edges is

$$\sum_{i<j} m_i m_j = \frac{1}{2} \sum_{i=1}^{k} \sum_{j \neq i} m_i m_j$$

$$= \frac{1}{2} \left(\sum_{i=1}^{k} m_i \right)^2 - \frac{1}{2} \sum_{i=1}^{k} m_i^2. \quad (5.5)$$

The variance inequality $\frac{1}{k} \sum_{i=1}^{k} m_i^2 - \left(\frac{1}{k} \sum_{i=1}^{k} m_i \right)^2 \geq 0$ and the counting identity $\sum_{i=1}^{k} m_i = l$ together imply $-\sum_{i=1}^{k} m_i^2 \leq -\frac{l^2}{k}$. Substituting this inequality in equation (5.5) produces the upper bound

$$\sum_{i<j} m_i m_j \leq \frac{l^2}{2} - \frac{l^2}{2k}$$

on the total number of possible edges. Thus, the maximum number of graphs compatible with a given coloring is $2^{l^2/2 - l^2/(2k)}$, and the probability that a random graph is compatible is at most

$$\frac{2^{\frac{l^2}{2} - \frac{l^2}{2k}}}{2^{\binom{l}{2}}} = 2^{\frac{l}{2} - \frac{l^2}{2k}}. \quad (5.6)$$

We now write

$$X_l = \sum_c 1_{A_c},$$

where c is a coloring of the first l nodes and A_c is the event that the underlying random graph is consistent at the first l nodes with the coloring. Taking expectations in this identity and invoking inequality (5.6) yields

$$E(X_l) \leq k^l 2^{\frac{l}{2} - \frac{l^2}{2k}}$$

and therefore the upper bound

$$\sum_{l=1}^{n} E(X_l) \leq \sum_{l=1}^{n} k^l 2^{\frac{l}{2} - \frac{l^2}{2k}} \quad (5.7)$$

on the average number of partial solutions in the backtrack tree. The limit of the series on the right-hand side of inequality (5.7) exists as n tends to ∞ by the ratio test. Indeed, the ratio of term $l+1$ to term l is

$$\frac{k^{l+1} 2^{\frac{l+1}{2} - \frac{(l+1)^2}{2k}}}{k^l 2^{\frac{l}{2} - \frac{l^2}{2k}}} = k 2^{\frac{1}{2} - \frac{2l+1}{2k}}, \quad (5.8)$$

which tends to 0 as l tends to ∞. When $k = 3$, the limit of the series (5.7) is about 197. In other words, the average backtrack tree contains fewer than 197 partial solutions regardless of the size n of the graph. Once again, let us stress that this result is simply a manifestation of the fact that most graphs with n nodes are quickly eliminated as colorable with k colors. ∎

5.5   Point Sets with Only Acute Angles

Consider a finite set of points S in some Euclidean space $\mathbb{R}^d$. Any three points $x, y, z \in S$ determine three angles, depending on which point is taken as apex. For example, taking x as apex produces two vectors $y - x$ and $z - x$ with an angle between them that is obtuse, right, or acute when the inner product $(y - x)^t(z - x)$ is negative, zero, or positive, respectively. It seems likely that at least some of these angles will be obtuse if the number of points is large. For example, Problem 5 asks the reader to check that any five non-collinear points in the plane determine at least one obtuse angle. Asking for only acute angles would seem to diminish the possibilities. Nonetheless, for high-dimensional spaces, it is possible to construct large sets of points with only acute angles [44].

In approaching this problem, we will limit ourselves to sets S contained in the unit cube $\{0, 1\}^d$. This has the advantage of eliminating the possibility of obtuse angles. Indeed, if we express

$$(y - x)^t(z - x) \;=\; \sum_{i=1}^{d}(y_i - x_i)(z_i - x_i),$$

then all of the products in the indicated sum are 0 or 1 because the coordinates x_i, y_i, z_i are chosen from $\{0, 1\}$. When we take $S = \{0, 1\}^d$, we attain a set of maximal size with no obtuse angles [1]. However, many of the angles are right. Let us consider smaller sets $S \subset \{0, 1\}^d$ of size m, where m is to be decided later. Instead of picking S directly, we first construct a set T containing $2m$ random points chosen independently and uniformly from $\{0, 1\}^d$. Some of these points may coincide.

Now consider three points $x, y, z \in T$. Let us call the triple (x, y, z) with apex x a bad triple whenever $(y - x)^t(z - x) = 0$. What is the probability of this happening? For each coordinate i we must have $(y_i - x_i)(z_i - x_i) = 0$. This occurs if either $y_i = x_i$ or $z_i = x_i$ and therefore has probability $\frac{3}{4}$. Because the coordinates are chosen independently, the inner product $(y - x)^t(z - x)$ vanishes with probability $\left(\frac{3}{4}\right)^d$. We now calculate the expected number of bad triples. The total number of triples is $\binom{2m}{3}$. Each triple has three possible choices for its apex. Thus the expected number of bad triples is

$$3\binom{2m}{3}\left(\frac{3}{4}\right)^d \;<\; m(2m)^2\left(\frac{3}{4}\right)^d.$$

We now choose m so that the right-hand side of this inequality is less than m. For example, we can take

$$m = \left\lfloor \frac{1}{2}\left(\frac{2}{\sqrt{3}}\right)^d \right\rfloor.$$

With this choice, there must be at least one configuration T with m or fewer bad triples. For such a T, we throw out the apex of any bad triple. This creates a set S with m or more points and no bad triples. The points of S define only acute angles, no right angles. For example, if $d = 35$, then there is some set S with at least $m = 76$ points defining only acute angles.

5.6 Sperner's Theorem

Consider a family of subsets $\mathcal{F}$ of the set $\{1, \ldots, n\}$ for some positive integer n. In Sperner's theorem, we impose the condition that no two subsets A and B in $\mathcal{F}$ satisfy either $A \subset B$ or $B \subset A$. With this restriction, how many subsets can $\mathcal{F}$ contain?

One extreme case is to take $\mathcal{F}$ to consist of all subsets of $\{1, \ldots, n\}$ having exactly $\lfloor \frac{n}{2} \rfloor$ elements. Because two subsets of the same size either coincide or satisfy the Sperner restriction, it is clear that $\mathcal{F}$ qualifies as a Sperner family. This family contains

$$|\mathcal{F}| = \binom{n}{\lfloor \frac{n}{2} \rfloor}$$

subsets. Following Lubell [110], we now show that this special family contains the maximum possible number of subsets.

Our line of attack proceeds through random permutations. To a given permutation π of $\{1, \ldots, n\}$, there correspond n subsets of the form

$$S(\pi, k) = \{\pi(1), \pi(2), \ldots, \pi(k)\}.$$

These satisfy $S(\pi, k) \subset S(\pi, k + 1)$ for $1 \leq k \leq n - 1$. Now consider the random variable

$$X(\pi) = \sum_{k=1}^{n} 1_{\{S(\pi,k) \in \mathcal{F}\}}$$

defined relative to a Sperner family $\mathcal{F}$. Because at most one of the events $\{S(\pi, k) \in \mathcal{F}\}$ can occur, X must equal either 0 or 1, and $0 \leq \mathrm{E}(X) \leq 1$.

In view of the fact that $X(\pi)$ counts the number of $S(\pi, k)$ in $\mathcal{F}$, we can also write

$$X(\pi) = \sum_{A \in \mathcal{F}} 1_{\{S(\pi,|A|)=A\}},$$

where $|A|$ denotes the number of elements of A. Taking into account the facts that each $S(\pi, |A|)$ is a randomly chosen subset of size $|A|$ and that $\binom{n}{k}$ is maximized by $k = \lfloor \frac{n}{2} \rfloor$, we calculate

$$
\begin{aligned}
\mathrm{E}[X] &= \sum_{A \in \mathcal{F}} \Pr[S(\pi, |A|) = A] \\
&= \sum_{A \in \mathcal{F}} \frac{1}{\binom{n}{|A|}} \\
&\geq \frac{|\mathcal{F}|}{\binom{n}{\lfloor \frac{n}{2} \rfloor}},
\end{aligned}
$$

where $|\mathcal{F}|$ is the size of $\mathcal{F}$. Combining this inequality with our earlier inequality $\mathrm{E}(X) \leq 1$ leads to the desired conclusion $|\mathcal{F}| \leq \binom{n}{\lfloor \frac{n}{2} \rfloor}$.

5.7   Subadditivity and Expectations

Many solutions to hard discrete optimization problems involve complicated random variables whose distributions and moments are nearly impossible to calculate exactly. In such situations, probabilists attempt to pin down the asymptotic behavior of the random variables as the problem size increases. The theory of subadditive sequences constitute one of the most powerful tools for understanding mean behavior.

A sequence $\{a_n\}_{n \geq 1}$ is said to be subadditive if

$$
a_{m+n} \leq a_m + a_n \tag{5.9}
$$

for all positive integers m and n [42, 103, 140]. If the opposite inequality

$$
a_{m+n} \geq a_m + a_n
$$

holds, then the sequence is superadditive . Subadditive and superadditive sequences arise in many combinatorial optimization problems. For example, let X_n denote the minimum effort it takes to solve a random problem of size n. Now suppose that we can decompose a problem of size $m + n$ into two subproblems of size m and n and patch together optimal solutions of these to derive a suboptimal solution of the problem of size $m + n$. If the effort for the concatenated solution is the sum of the efforts for the subsolutions, then the minimal efforts satisfy

$$
X_{m+n} \leq X_m + X_n. \tag{5.10}
$$

In other words, the random sequence X_n is subadditive. Taking expectations in inequality (5.10) demonstrates that the sequence $\mathrm{E}(X_n)$ is also subadditive, provided the expectations exist.

We now prove the remarkable fact that inequality (5.9) implies that

$$\lim_{n\to\infty} \frac{a_n}{n} = \inf_n \frac{a_n}{n} = \gamma. \tag{5.11}$$

The possibility $\gamma = -\infty$ is not ruled out. Consider first the case $\gamma > -\infty$, and set $a_0 = 0$. For any $\epsilon > 0$, we can find a k such that $a_k \leq (\gamma + \epsilon)k$. Because any $m > 0$ can be written as $m = nk + j$ with $0 \leq j < k$, it follows that

$$a_m = a_{nk+j} \leq na_k + a_j \leq (\gamma + \epsilon)nk + \max_{0 \leq l < k} a_l$$

and consequently that

$$\limsup_m \frac{a_m}{m} \leq \gamma + \epsilon \leq \liminf_m \frac{a_m}{m} + \epsilon.$$

By virtue of the arbitrariness of ϵ, this shows that the limit (5.11) exists. The easier case of $\gamma = -\infty$ is left to the reader. A similar result holds for a superadditive sequence.

Example 5.7.1 *Longest Common Subsequence*

A string is a finite sequence of letters taken from some alphabet. For instance in DNA sequence analysis, the relevant alphabet consists of the four letters A, C, T, and G. Two DNA strings sharing an evolutionary history will have long subsequences in common. If we represent two strings of length n by $u_1, \ldots, u_n$ and $v_1, \ldots, v_n$, then the subsequences $u_{i_1}, u_{i_2}, \ldots, u_{i_m}$ and $v_{j_1}, v_{j_2}, \ldots, v_{j_m}$ are shared provided $u_{i_k} = v_{j_k}$ for $1 \leq k \leq m$. Now consider two random strings whose letters are independently and identically distributed. It is important to characterize the random length $M_n = m$ of the longest common subsequence.

Considerable effort has gone into finding $E(M_n)$. We now show that $E(M_n) \asymp \gamma n$ for large n and a constant $\gamma \in [0, 1]$ depending on the letter distribution imposed on the alphabet [140]. This follows readily from the superadditivity property $M_{r+s} \geq M_r + M_s$ derived by concatenating a longest common subsequence drawn from the first block of r pairs of letters with a longest common subsequence drawn from the last block of s pairs of letters. Unfortunately, this argument fails to identify the constant γ. This difficulty plagues all applications of subadditivity and superadditivity. Further problem-specific information must be brought to bear to find γ [140]. For example, Problem 12 provides bounds on γ for the problem of calculating self-avoidance probabilities in a symmetric random walk. ∎

Example 5.7.2 *Euclidean Traveling Salesman Problem*

The average complexity of many combinatorial optimization problems grows at a slower than linear rate, thus falling outside the domain of application of subadditivity. A probabilistic version of the traveling salesman problem furnishes a case in point. In the classical version of the traveling salesman problem, the salesman must visit n towns, starting and ending in his hometown. To minimize his travel time and expense, the salesman takes an optimal route. We defer to Example 7.7.1 the question of how to find such a route.

In the Euclidean, probabilistic version of the problem, n points (cites) $Y_1, \ldots, Y_n$ are drawn uniformly and independently from the unit square [140]. The shortest circuit that the salesman can make through the points is given by the random variable

$$D_n = \min_\sigma \sum_{i=1}^{n} \|Y_{\sigma(i)} - Y_{\sigma(i+1)}\|,$$

where σ denotes a generic permutation of $\{1, \ldots, n\}$, $\sigma(n+1) = \sigma(1)$, and $\|\cdot\|$ is the Euclidean norm. We now demonstrate that the average distance $E(D_n)$ the salesman travels is roughly proportional to $\sqrt{n}$.

One obvious upper bound on $E(D_n)$ is furnished by $M_n = \sup D_n$. It is difficult to calculate M_n, so we will be content with bounding it. We can attack this easier problem by choosing $m = \max\{k \geq 1 : k^2 < n\}$ and dividing the unit square into m^2 nonoverlapping subsquares having sides of length $1/m$. Any two points within one of these subsquares are separated by a distance of at most $\sqrt{2}/m$. Furthermore, $\sqrt{2}/m \leq 2/\sqrt{n}$, as Problem 13 asks the reader to check. Because $m^2 < n$, the pigeonhole principle requires that two of the points, say Y_j and Y_k, fall in the same subsquare. Now consider a minimum length tour of the $n - 1$ points excluding Y_k. If $Y_i \to Y_j$ in this tour, then we can extend the tour to a tour of all n points by replacing the path $Y_i \to Y_j$ by the two paths $Y_i \to Y_k$ and $Y_k \to Y_j$. In view of the triangle inequality

$$\|Y_i - Y_k\| + \|Y_k - Y_j\| \leq \|Y_i - Y_j\| + 2\|Y_k - Y_j\|$$
$$\leq \|Y_i - Y_j\| + \frac{4}{\sqrt{n}},$$

the bounds

$$D_n \leq D_{n-1} + \frac{4}{\sqrt{n}}$$

and

$$M_n \leq M_{n-1} + \frac{4}{\sqrt{n}}$$

hold. If we iterate the last bound and employ $M_0 = 0$, then it is clear that

$$M_n \leq \sum_{i=1}^{n} \frac{4}{\sqrt{i}} \leq 4 \int_0^n \frac{1}{\sqrt{x}} \, dx = 2\sqrt{n}$$

and therefore that $E(D_n) \leq 2\sqrt{n}$.

One can supplement this upper bound with a lower bound of the same order of magnitude. In this case we begin with

$$
\begin{aligned}
E(D_n) &\geq \sum_{i=1}^{n} E(\min_{j \neq i} ||Y_j - Y_i||) \\
&= n E(\min_{j \neq n} ||Y_j - Y_n||) \\
&= n E[E(\min_{j \neq n} ||Y_j - y|| \mid Y_n = y)].
\end{aligned}
$$

To calculate the conditional expectation

$$
E(\min_{j \neq n} ||Y_j - y|| \mid Y_n = y) = E(\min_{j \neq n} ||Y_j - y||),
$$

we use the right-tail probability bound

$$
\Pr(\min_{j \neq n} ||Y_j - y|| \geq r) \geq (1 - \pi r^2)^{n-1}
$$

valid for any y in the unit square. Thus, Example 2.5.1 implies that

$$
\begin{aligned}
E(\min_{j \neq n} ||Y_j - y||) &\geq \int_0^{\frac{1}{\sqrt{\pi}}} (1 - \pi r^2)^{n-1} dr \\
&\approx \int_0^{\frac{1}{\sqrt{\pi}}} e^{-\pi(n-1)r^2} dr \\
&\approx \frac{1}{2} \int_{-\infty}^{\infty} e^{-\pi n r^2} dr \\
&= \frac{1}{2\sqrt{n}}.
\end{aligned}
$$

In summary, we conclude that $E(D_n) \geq \sqrt{n}/2$ to a good approximation for large n. A combination of further theoretical work and numerical experimentation [140] suggests that $\lim_{n \to \infty} D_n/\sqrt{n} \to \gamma$ for some constant $\gamma \in [0.70, 0.73]$. ■

5.8   Problems

1. What is the probability that a random permutation of n distinct numbers contains at least one preexisting splitter? What are the mean and variance of the number of preexisting splitters?

2. Show that the worst case of quick sort takes on the order of n^2 operations.

3. Prove rigorously that $\sum_{k=1}^{n} 1/k \asymp \ln n$ as n tends to ∞.

4. Consider the uniform distribution $p_l = n^{-1}$ on an alphabet $\mathcal{A}$ with n letters. Let $\text{len}(s_l)$ be the number of bits in the bit string s_l representing l under Huffman coding. If $m = \max\{\text{len}(s_l) : l \in \mathcal{A}\}$, then show that $\text{len}(s_l) = m$ or $m - 1$ for all l. If $n = \alpha 2^k$ for $1 < \alpha \le 2$, then determine the number of letters l with $\text{len}(s_l) = j$ for $j = m - 1$ and $j = m$. Use these numbers to calculate $\text{E}[\text{len}(H)]$, where H is a random Huffman bit string.

5. Check that any five non-collinear points in the plane R^2 determine at least one obtuse angle.

6. Let $\|x\|$ be the standard Euclidean norm of a vector $x \in \mathsf{R}^n$. A vector x with $\|x\| = 1$ is said to be a unit vector. For any sequence $x_1, \ldots, x_n$ of n unit vectors from R^n, it is possible to find n numbers $\epsilon_1, \ldots, \epsilon_n$ drawn from $\{-1, 1\}$ such that $\|\epsilon_1 x_1 + \cdots + \epsilon_n x_n\| \le \sqrt{n}$. A different choice of $\epsilon_1, \ldots, \epsilon_n$ yields the reverse inequality [4]. Prove this striking result by setting up a simple probability model. (Hints: Choose the ϵ_i independently from $\{-1, 1\}$ in such a way that $\text{E}(\epsilon_i) = 0$. Now show that the random variable $X = \|\epsilon_1 x_1 + \cdots + \epsilon_n x_n\|^2$ has expectation $\text{E}(X) = n$.)

7. Exactly 10% of the surface of a sphere in R^3 is colored black, and 90% is colored white. Show that it is possible to inscribe a cube in the sphere with all of its vertices colored white [64].

8. Consider a graph with m nodes and n edges. For any set of nodes S, let $X(S)$ be the number of edges with exactly one endpoint in S. Show that $\max_S X(S) \ge n/2$. (Hints: Generate S randomly by independently sampling each node with probability $1/2$. Decompose X as a sum of indicators indexed by the edges.)

9. Consider a family of subsets $\mathcal{F}$ of a set S with the property that each $A \in \mathcal{F}$ has exactly d elements. The family $\mathcal{F}$ is said to be 2-colorable if we can assign one of two colors, say black and white, to each element of S in such a manner that each $A \in \mathcal{F}$ possesses at least one element of each color. Prove that $\mathcal{F}$ is 2-colorable if it contains fewer than 2^{d-1} subsets [1]. (Hints: Randomly color each element of S with one of the two equally likely colors black and white. Let C_A be the event that all elements of $A \in \mathcal{F}$ receive the same color. Show that $\cup_A C_A$ has probability strictly less than 1.)

10. Let $f(t)$ be a nonnegative function on $(0, \infty)$ with $\lim_{t \to 0} f(t) = 0$. If $f(t)$ is subadditive in the sense that $f(s + t) \le f(s) + f(t)$ for all positive s and t, then show that

$$\lim_{t \downarrow 0} \frac{f(t)}{t} = \sup_{t > 0} \frac{f(t)}{t}.$$

(Hint: Demonstrate that

$$p \leq \liminf_{t \downarrow 0} \frac{f(t)}{t} \leq \limsup_{t \downarrow 0} \frac{f(t)}{t}$$

for all $p \in [0, q)$.)

11. A random walk X_n on the integer lattice in R^m is determined by the transition probabilities $\Pr(X_{n+1} = j \mid X_n = i) = q_{j-i}$ and the initial value $X_0 = \mathbf{0}$. Let p_n be the probability that $X_i \neq X_j$ for all pairs $0 \leq i < j \leq n$. In other words, p_n is the probability that the walk avoids itself during its first n steps. Prove that $\lim_{n \to \infty} \frac{1}{n} \ln p_n = \gamma$ for some $\gamma \leq 0$. (Hint: Argue that $p_{m+n} \leq p_m p_n$.)

12. Continuing Problem 11, suppose the random walk is symmetric in the sense that $q_i = 1/(2m)$ if and only if $\|i\| = 1$. Prove that

$$\frac{m^n}{(2m)^n} \leq p_n \leq \frac{2m(2m-1)^{n-1}}{(2m)^n}.$$

Use these inequalities to prove that the constant γ of Problem 11 satisfies $\ln(1/2) \leq \gamma \leq \ln[1 - 1/(2m)]$. (Hints: A random walk is self-avoiding if all its steps are in the positive direction. After its first step, a self-avoiding walk cannot move from its current position back to its previous position.)

13. Suppose $m = \max\{k \geq 1 : k^2 < n\}$. Prove that $\sqrt{2}/m \leq 2/\sqrt{n}$ for all sufficiently large n.

6
Poisson Processes

6.1 Introduction

The Poisson distribution rivals the normal distribution in importance. It occupies this position of eminence because of its connection to Poisson processes [48, 49, 64, 76, 82, 89, 130]. A Poisson process models the formation of random points in space or time. Most textbook treatments of Poisson processes stress one-dimensional processes. This is unfortunate because many of the important applications occur in higher dimensions, and the underlying theory is about as simple there. In this chapter, we emphasize multidimensional Poisson processes, their transformation properties, and computational tools for extracting information about them.

The number of applications of Poisson processes is truly amazing. To give just a few examples, physicists use them to describe the emission of radioactive particles, astronomers to account for the distribution of stars, communication engineers to model the arrival times of telephone calls at an exchange, radiologists to reconstruct medical images in emission and transmission tomography, and ecologists to test for the random location of plants. Almost equally important, Poisson processes can provide a theoretical perspective helpful in complex probability calculations that have no obvious connection to random points. We will visit a few applications of both types as we proceed.

6.2   The Poisson Distribution

It is helpful to begin our exposition of Poisson processes with a brief review of the Poisson distribution. Readers will recall that a Poisson random variable $Z \geq 0$ has discrete density $\Pr(Z = k) = e^{-\mu}\frac{\mu^k}{k!}$ with mean and variance μ and probability generating function $E(t^Z) = e^{\mu(t-1)}$. Furthermore, if $Z_1, \ldots, Z_m$ are independent Poisson random variables, then the sum $Z = \sum_{k=1}^{m} Z_k$ is also a Poisson random variable. Less well known is the next proposition.

Proposition 6.2.1 *Suppose a Poisson random variable Z with mean μ represents the number of outcomes from some experiment. Let each outcome be independently classified in one of m categories, the kth of which occurs with probability p_k. Then the number of outcomes Z_k falling in category k is Poisson distributed with mean $\mu_k = p_k\mu$. Furthermore, the random variables $Z_1, \ldots, Z_m$ are independent. Conversely, if $Z = \sum_{k=1}^{m} Z_k$ is a sum of independent Poisson random variables Z_k with means $\mu_k = p_k\mu$, then conditional on $Z = n$, the vector $(Z_1, \ldots, Z_m)$ follows a multinomial distribution with n trials and cell probabilities $p_1, \ldots, p_m$.*

Proof: If $n = n_1 + \cdots + n_m$, then

$$
\begin{aligned}
\Pr(Z_1 = n_1, \ldots, Z_m = n_m) &= e^{-\mu}\frac{\mu^n}{n!}\binom{n}{n_1, \ldots, n_m}\prod_{k=1}^{m} p_k^{n_k} \\
&= \prod_{k=1}^{m} e^{-\mu_k}\frac{\mu_k^{n_k}}{n_k!} \\
&= \prod_{k=1}^{m} \Pr(Z_k = n_k).
\end{aligned}
$$

To prove the converse, divide the last string of equalities by the probability $\Pr(Z = n) = e^{-\mu}\frac{\mu^n}{n!}$. ∎

In practice, it is useful to extend the definition of a Poisson random variable to include the limiting cases $X \equiv 0$ and $X \equiv \infty$ with corresponding means 0 and ∞.

6.3   Characterization and Construction

A Poisson process involves points randomly scattered in some measurable region S of m-dimensional space $\mathbb{R}^m$. To formalize the notion that the points are completely random but concentrated on average more in some regions rather than in others, we introduce several postulates involving an intensity function $\lambda(x) \geq 0$ on S. Postulate (c) below uses the notation $o(\mu)$, which is to be interpreted as a generic error term satisfying $\lim_{\mu \to 0} \frac{o(\mu)}{\mu} = 0$.

(a) There exists a sequence of disjoint subregions S_n satisfying $S = \bigcup_n S_n$ and $\int_{S_n} \lambda(x)\, dx < \infty$ for all n.

(b) The probability $p_k(\mu)$ that a region $T \subset S$ contains k random points depends only on the mass $\mu = \int_T \lambda(x)\, dx$ of T. If $\mu = \infty$, then all $p_k(\mu) = 0$, and the given region possesses an infinite number of points.

(c) The numbers of random points in disjoint regions are independent.

(d) The first two probabilities $p_0(\mu)$ and $p_1(\mu)$ have the asymptotic values

$$
\begin{aligned}
p_0(\mu) &= 1 - \mu + o(\mu) \\
p_1(\mu) &= \mu + o(\mu)
\end{aligned}
\tag{6.1}
$$

as μ tends to 0. Thus, $p_k(\mu) = o(\mu)$ for all $k > 1$.

Proposition 6.3.1 *Based on postulates (a) through (d), the number of random points in a region with mass μ has the Poisson distribution*

$$
p_k(\mu) = e^{-\mu} \frac{\mu^k}{k!}.
$$

Proof: We first remark that for any region T with $\int_T \lambda(x)\, dx < \infty$ and any μ in the interval $[0, \int_T \lambda(x)\, dx]$, there exists a region $R \subset T$ with mass $\mu = \int_R \lambda(x)\, dx$. To verify this assertion, we need only consider regions of the form $R = T \cap B_t$, where B_t is the ball $\{x \in \mathbb{R}^m : \|x\| \le t\}$. The function $t \mapsto \int_{T \cap B_t} \lambda(x)\, dx$ is continuous from the right by the dominated convergence theorem. It is continuous from the left because the surface of B_t has volume 0 and therefore $\int_{\{x : \|x\| = t\}} \lambda(x)\, dx = 0$. The intermediate value theorem now gives the desired conclusion.

Let μ and $d\mu$ represent the masses of two nonoverlapping regions. By the preceding comments, $d\mu$ can be made as small as we please. The equality

$$
\begin{aligned}
p_0(\mu + d\mu) &= p_0(\mu) p_0(d\mu) \\
&= p_0(\mu)[1 - d\mu + o(d\mu)]
\end{aligned}
$$

follows immediately from the postulates and can be rearranged to give the difference quotient

$$
\frac{p_0(\mu + d\mu) - p_0(\mu)}{d\mu} = -p_0(\mu) + o(1).
$$

Taking limits as $d\mu$ tends to 0 produces the ordinary differential equation $p_0'(\mu) = -p_0(\mu)$ with solution $p_0(\mu) = e^{-\mu}$ satisfying the initial condition $p_0(0) = 1$.

For $k \geq 1$, we again invoke the postulates and execute the expansion

$$
\begin{aligned}
p_k(\mu + d\mu) &= p_k(\mu)p_0(d\mu) + p_{k-1}(\mu)p_1(d\mu) + \sum_{j=2}^{k} p_{k-j}(\mu)p_j(d\mu) \\
&= p_k(\mu)[1 - d\mu + o(d\mu)] + p_{k-1}(\mu)[d\mu + o(d\mu)] \\
&\quad + \sum_{j=2}^{k} p_{k-j}(\mu)o(d\mu).
\end{aligned}
$$

Rearrangement of this approximation yields the difference quotient

$$
\frac{p_k(\mu + d\mu) - p_k(\mu)}{d\mu} = -p_k(\mu) + p_{k-1}(\mu) + o(1)
$$

and ultimately the ordinary differential equation $p_k'(\mu) = -p_k(\mu) + p_{k-1}(\mu)$ with initial condition $p_k(0) = 0$. The transformed function $q_k(\mu) = p_k(\mu)e^{\mu}$ satisfies the simpler ordinary differential equation $q_k'(\mu) = q_{k-1}(\mu)$ with initial condition $q_k(0) = 0$. If the $p_k(\mu)$ are Poisson, then $q_k(\mu) = \frac{\mu^k}{k!}$ should hold. This formula is certainly true for $q_0(\mu) = 1$. Assuming that it is true for $q_{k-1}(\mu)$, we see that

$$
\begin{aligned}
q_k(\mu) &= \int_0^{\mu} q_{k-1}(u)\, du \\
&= \int_0^{\mu} \frac{u^{k-1}}{(k-1)!}\, du \\
&= \frac{\mu^k}{k!}
\end{aligned}
$$

has the necessary value to advance the inductive argument and complete the proof.   ∎

At this junction, several remarks are in order. First, the proposition is less than perfectly rigorous because we have only considered derivatives from the right. A better proof under less restrictive conditions is given in reference [89]. Second, $\mu = \int_T \lambda(x)\, dx$ is the expected number of random points in the region T. Third, only a finite number of random points can occur in any S_n. Fourth, if $\int_T \lambda(x)\, dx = \infty$, then an infinite number of random points occur in T. Fifth, because every $y \in S$ has mass $\int_{\{y\}} \lambda(x)\, dx = 0$, the probability that a random point coincides with y is 0. Sixth, no two random points ever coincide. Seventh, the approximations (6.1) are consistent with the final result $p_k(\mu) = e^{-\mu}\frac{\mu^k}{k!}$. Eighth and finally, if the intensity function $\lambda(x)$ is constant, then the Poisson process is said to be homogeneous.

We now turn the question of Proposition 6.3.1 around and ask how can one construct a Poisson process with a given intensity function $\lambda(x)$? This question has more than theoretical interest because we often need to simulate a Poisson process on a computer. Briefly, we attack the problem by independently generating random points in each of the disjoint regions S_n. The number of points N_{S_n} to be scattered in S_n follows a Poisson distribution with mean $\mu_n = \int_{S_n} \lambda(x)\,dx$. Once we sample N_{S_n}, then we independently distribute the corresponding N_{S_n} points X_{ni} one by one over the region S_n according to the probability measure

$$\Pr(X_{ni} \in R) \quad = \quad \frac{1}{\mu_n} \int_R \lambda(x)\,dx.$$

This procedure incorporates the content of Proposition 6.2.1. The resulting union of random points $\Pi = \bigcup_n \bigcup_{i=1}^{N_{S_n}} X_{ni}$ constitutes one realization from the required Poisson process.

6.4   One-Dimensional Processes

When a Poisson process occurs on a subset of the real line, it is often convenient to refer to time instead of space and events instead of points. Consider a homogeneous Poisson process on $[0, \infty)$ with intensity λ. Let T_k be the waiting time until the kth event after time 0. The interarrival time between events $k-1$ and k equals $W_k = T_k - T_{k-1}$ for $k > 1$. By convention $W_1 = T_1$.

Proposition 6.4.1 *The random waiting time T_k has a gamma distribution with density $\lambda \frac{(\lambda t)^{k-1}}{(k-1)!} e^{-\lambda t}$. Furthermore, the interarrival times W_k are independent and exponentially distributed with intensity λ.*

Proof: The event $T_k > t$ is equivalent to the event that $k-1$ or fewer random points fall in $[0, t]$. Hence,

$$\Pr(T_k \le t) \quad = \quad 1 - \Pr(T_k > t)$$
$$= \quad 1 - \sum_{j=0}^{k-1} e^{-\lambda t} \frac{(\lambda t)^j}{j!}.$$

Differentiating this distribution function with respect to t gives the density function

$$-\sum_{j=0}^{k-1} j \lambda e^{-\lambda t} \frac{(\lambda t)^{j-1}}{j!} + \sum_{j=0}^{k-1} \lambda e^{-\lambda t} \frac{(\lambda t)^j}{j!} \quad = \quad \lambda \frac{(\lambda t)^{k-1}}{(k-1)!} e^{-\lambda t}.$$

This proves the first claim.

Assume for the moment that the second claim is true. Because the matrix of the linear transformation

$$T_1 \quad = \quad W_1$$

$$T_2 = W_1 + W_2$$

$$\vdots$$

$$T_n = W_1 + W_2 + \cdots + W_n$$

taking $(W_1, \ldots, W_n)$ to $(T_1, \ldots, T_n)$ is lower triangular with 1's down its diagonal, it has Jacobian 1. The change of variables formula (1.11) therefore implies that the random vector $(T_1, \ldots, T_n)$ has density

$$f_n(t_1, \ldots, t_n) = \prod_{i=1}^{n} \lambda e^{-\lambda w_i} = \lambda^n e^{-\lambda t_n}$$

on the region $\Gamma_n = \{0 \leq t_1 \leq \cdots \leq t_n\}$. Conversely, if $(T_1, \ldots, T_n)$ possesses the density $f_n(t_1, \ldots, t_n)$, then applying the inverse transformation shows that the W_k are independent and exponentially distributed with intensity λ.

Now let $F_n(t_1, \ldots, t_n)$ be the distribution function corresponding to $f_n(t_1, \ldots, t_n)$. Integrating the obvious identity

$$\begin{aligned} f_n(s_1, \ldots, s_n) &= \lambda^n e^{-\lambda s_n} \\ &= \lambda e^{-\lambda s_1} f_{n-1}(s_2 - s_1, \ldots, s_n - s_1) \end{aligned}$$

over the intersection $\Gamma_n \cap \{s_1 \leq t_1, \ldots, s_n \leq t_n\}$ yields the identity

$$F_n(t_1, \ldots, t_n) = \int_0^{t_1} \lambda e^{-\lambda s_1} F_{n-1}(t_2 - s_1, \ldots, t_n - s_1) \, ds_1 \quad (6.2)$$

recursively determining the distribution functions $F_n(t_1, \ldots, t_n)$ starting with $F_1(t_1) = 1 - e^{-\lambda t_1}$. If $G_n(t_1, \ldots, t_n)$ denotes the actual distribution function of $(T_1, \ldots, T_n)$, then our strategy is to show that $G_n(t_1, \ldots, t_n)$ satisfies identity (6.2). Given the fact that $G_1(t_1) = 1 - e^{-\lambda t_1}$, induction on n then shows that $G_n(t_1, \ldots, t_n)$ and $F_n(t_1, \ldots, t_n)$ coincide.

To verify that $G_n(t_1, \ldots, t_n)$ satisfies identity (6.2), we first note that

$$\Pr(T_1 \leq t_1, \ldots, T_n \leq t_n) = \Pr(\cap_{i=1}^{n}\{N_{t_i} \geq i\}). \quad (6.3)$$

We also note the identity

$$\begin{aligned} \frac{(\lambda t)^j}{j!} e^{-\lambda t} &= \frac{\lambda^j}{(j-1)!} e^{-\lambda t} \int_0^t s^{j-1} ds \\ &= \frac{\lambda^j}{(j-1)!} e^{-\lambda t} \int_0^t (t-s)^{j-1} ds \\ &= \int_0^t \lambda e^{-\lambda s} \frac{[\lambda(t-s)]^{j-1}}{(j-1)!} e^{-\lambda(t-s)} ds. \end{aligned}$$

Consequently, if $1 \leq j_1 \leq \cdots \leq j_n$, then

$$
\begin{aligned}
\Pr(\cap_{i=1}^{n}\{N_{t_i} = j_i\}) &= \Pr(N_{t_1} = j_1)\Pr(\cap_{i=2}^{n}\{N_{t_i} - N_{t_{i-1}} = j_i - j_{i-1}\}) \\
&= \int_0^{t_1} \lambda e^{-\lambda s}\frac{[\lambda(t_1 - s)]^{j_1 - 1}}{(j_1 - 1)!}e^{-\lambda(t_1 - s)} \\
&\quad \times \Pr(\cap_{i=2}^{n}\{N_{t_i - s} - N_{t_{i-1} - s} = j_i - j_{i-1}\})\, ds \\
&= \int_0^{t_1} \lambda e^{-\lambda s}\Pr(\cap_{i=1}^{n}\{N_{t_i - s} = j_i - 1\})\, ds.
\end{aligned}
$$

Summing this equality over the intersection of the sets $\{j_1 \leq \cdots \leq j_n\}$ and $\{j_1 \geq 1, \dots, j_n \geq n\}$ produces

$$
\Pr(\cap_{i=1}^{n}\{N_{t_i} \geq i\}) = \int_0^{t_1} \lambda e^{-\lambda s}\Pr(\cap_{i=2}^{n}\{N_{t_i - s} \geq i - 1\})\, ds,
$$

which in light of identity (6.3) is just a disguised form of identity (6.2). ∎

This proposition implies that generating a sequence of exponentially distributed interarrival times $W_1, W_2, \dots$ and extracting the corresponding waiting times $T_k = \sum_{j=1}^{k} W_j$ from it provides another method of constructing a homogeneous Poisson process on $[0, \infty)$.

The exponential distribution has an important "lack of memory" property. If X is exponentially distributed with intensity λ, then

$$
\begin{aligned}
\Pr(X > t + h \mid X > t) &= \frac{\Pr(X > t + h)}{\Pr(X > t)} \\
&= \frac{e^{-\lambda(t+h)}}{e^{-\lambda t}} \\
&= e^{-\lambda h} \\
&= \Pr(X > h).
\end{aligned}
$$

Lack of memory characterizes the exponential.

Proposition 6.4.2 *Suppose X is a random variable with values in $(0, \infty)$ and satisfying $\Pr(X > t + h) = \Pr(X > t)\Pr(X > h)$ for all positive h and t. Then X is exponentially distributed.*

Proof: If we let $g(t) = \Pr(X > t)$, then $g(0) = 1$ and $g(t)$ satisfies the familiar functional equation (2.5). Given differentiability of $g(t)$, the reasoning in Section 2.4.7 leads to the solution $g(t) = e^{-\lambda t}$. Here $\lambda > 0$ because $g(t)$ tends to 0 as t tends to ∞. ∎

Example 6.4.1 *Waiting Time Paradox*

Buses arrive at a bus stop at random times according to a Poisson process with intensity λ. I arrive at time t and ask how much time $E(W)$ on average I will have to wait for the next bus. To quote Feller [49], there are two mutually contradictory responses:

(a) "The lack of memory of the Poisson process implies that the distribution of my waiting time should not depend on the epoch of my arrival. In this case, $E(W) = 1/\lambda$."

(b) "The epoch of my arrival is chosen at random in the interval between two consecutive buses, and for reasons of symmetry my expected waiting time should be half the expected time between two consecutive buses, that is $E(W) = 1/(2\lambda)$."

Answer (a) is correct; answer (b) neglects the fact that I am more likely to arrive during a long interval than a short interval. This is the paradox of length-biased sampling. In fact, the random length of the interval capturing my arrival is distributed as the sum of two independent exponentially distributed random variables with intensity λ. This assertion is clear if I arrive at time 0, and it continues to hold for any other time t because a homogeneous Poisson process is stationary and possesses no preferred time origin. ∎

Example 6.4.2 *Order Statistics from an Exponential Sample*

The lack of memory property of the exponential distribution makes possible an easy heuristic derivation of a convenient representation of the order statistics $X_{(1)} \leq \cdots \leq X_{(n)}$ from an independent sample $X_1, \ldots, X_n$ of exponentially distributed random variables with common intensity λ [49]. From the calculation $\Pr(X_{(1)} \geq x) = \prod_{j=1}^{n} \Pr(X_j \geq x) = e^{-n\lambda x}$, we find that $X_{(1)}$ is exponentially distributed with intensity $n\lambda$. Because of the lack of memory property of the exponential, the $n-1$ random points to the right of $X_{(1)}$ provide an exponentially distributed sample of size $n-1$ starting at $X_{(1)}$. Duplicating our argument for $X_{(1)}$, we find that the difference $X_{(2)} - X_{(1)}$ is independent of $X_{(1)}$ and exponentially distributed with intensity $(n-1)\lambda$. Arguing inductively, we now see that $Z_1 = X_{(1)}$ and the differences $Z_{k+1} = X_{(k+1)} - X_{(k)}$ are independent and that Z_k is exponentially distributed with intensity $(n-k+1)\lambda$. Problem 2 provides the moments of $X_{(j)}$ based on the representation $X_{(j)} = \sum_{k=1}^{j} Z_k$. ∎

6.5 Transmission Tomography

The purpose of transmission tomography is to reconstruct the local attenuation properties of the object being imaged. Attenuation is to be roughly equated with density. In medical applications, material such as bone is dense and stops or deflects X-rays better than soft tissue. With enough radiation, even small gradations in soft tissue can be detected. The traditional method of image reconstruction in transmission tomography relies on Fourier analysis and the Radon transform [71]. An alternative to this deterministic approach is to pose an explicitly Poisson process model that

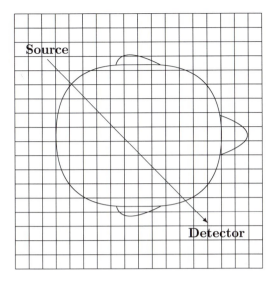

FIGURE 6.1. Cartoon of Transmission Tomography

permits parameter estimation by maximum likelihood [98]. The MM algorithm presented in Chapter 3 immediately suggests itself in this context.

The stochastic model depends on dividing the object of interest into small nonoverlapping regions of constant attenuation called pixels. Typically the pixels are squares. The attenuation attributed to pixel j constitutes parameter θ_j of the model. Since there may be thousands of pixels, implementation of maximum likelihood algorithms such as scoring or Newton's method is out of the question. Each observation Y_i is generated by beaming a stream of X-rays or high-energy photons from an X-ray source toward some detector on the opposite side of the object. The observation (or projection) Y_i counts the number of photons detected along the ith line of flight. Figure 6.1 shows one such projection line beamed through a cartoon of the human head. Naturally, only a fraction of the photons are successfully transmitted from source to detector. If l_{ij} is the length of the segment of projection line i intersecting pixel j, then we claim that the probability of a photon escaping attenuation along projection line i is the exponentiated line integral $\exp(-\sum_j l_{ij}\theta_j)$.

This result can be demonstrated by considering a Poisson process along projection i, starting with the source as origin. Each random point corresponds to a possible attenuation event. The first attenuation event encountered stops or deflects the photon and thus prevents it from being detected. The intensity of the attenuation process is determined locally by the attenuation coefficient of the surrounding pixel. It follows that a photon escapes attenuation with Poisson probability $\exp(-\sum_j l_{ij}\theta_j)$. Example 8.7.2

continues this discussion from the perspective of continuous-time Markov chains.

Of course, a second Poisson process is lurking in the background. In the absence of the intervening object, the number of photons generated and ultimately detected follows a Poisson distribution. Let the mean of this distribution be d_i for projection line i. Since Proposition 6.2.1 implies that random thinning of a Poisson random variable gives a Poisson random variable, the number Y_i is Poisson distributed with mean $d_i \exp(-\sum_j l_{ij}\theta_j)$. Owing to the Poisson nature of X-ray generation, the different projections will be independent even if collected simultaneously. This fact enables us to write the loglikelihood of the observed data $Y_i = y_i$ as the finite sum

$$L(\theta) \;=\; \sum_i \left[-d_i e^{-\sum_j l_{ij}\theta_j} - y_i \sum_j l_{ij}\theta_j + y_i \ln d_i - \ln y_i! \right]. \quad (6.4)$$

Omitting irrelevant constants, we can rewrite the loglikelihood (6.4) more succinctly as

$$L(\theta) \;=\; -\sum_i f_i(l_i^t \theta),$$

where $f_i(s) = d_i e^{-s} + y_i s$ and $l_i^t \theta = \sum_j l_{ij}\theta_j$ is the inner product of the attenuation parameter vector θ and the vector of intersection lengths l_i for projection i.

Following the lead of De Pierro [37] in emission tomography, one can devise an MM algorithm based on a convexity argument [99]. First define admixture constants

$$\alpha_{ij} \;=\; \frac{l_{ij}\theta_j^n}{l_i^t \theta^n}. \quad (6.5)$$

Since $\sum_j \alpha_{ij} = 1$ and each $f_i(s)$ is strictly convex, the inequality

$$
\begin{aligned}
L(\theta) &\;=\; -\sum_i f_i\left(\sum_j \alpha_{ij} \frac{\theta_j}{\theta_j^n} l_i^t \theta^n \right) \\
&\;\geq\; -\sum_i \sum_j \alpha_{ij} f_i\left(\frac{\theta_j}{\theta_j^n} l_i^t \theta^n \right) \quad (6.6) \\
&\;=\; Q(\theta \mid \theta^n)
\end{aligned}
$$

holds. Furthermore, equality occurs when $\theta_j = \theta_j^n$ for all j. Thus, the surrogate function $Q(\theta \mid \theta^n)$ minorizes $L(\theta)$. By construction, maximizing $Q(\theta \mid \theta^n)$ separates into a sequence of one-dimensional maximization problems, each of which can be solved approximately by one step of Newton's method as noted in Problem 7.

The images produced by maximum likelihood estimation in transmission tomography look grainy. Geman and McClure [58] recommend incorporating a Gibbs prior that enforces image smoothness. A Gibbs prior $\pi(\theta)$ can be written as

$$\ln \pi(\theta) \quad = \quad -\gamma \sum_{\{j,k\}\epsilon N} w_{jk}\psi(\theta_j - \theta_k),$$

where γ and the weights w_{jk} are positive constants, N is a set of unordered pairs $\{j, k\}$ defining a neighborhood system, and $\psi(r)$ is called a potential function. For instance, if the pixels are squares, we might define the weights by $w_{jk} = 1$ for orthogonal nearest neighbors and $w_{jk} = 1/\sqrt{2}$ for diagonal nearest neighbors. The constant γ scales the overall strength assigned to the prior. To achieve a smooth image with good resolution, we maximize the log posterior function $L(\theta) + \ln \pi(\theta)$ rather than $L(\theta)$.

Choice of the potential function $\psi(r)$ is the most crucial feature of the Gibbs prior. It is convenient to assume that $\psi(r)$ is even and strictly convex. Strict convexity leads to strict concavity of the $L(\theta) + \ln \pi(\theta)$ and permits simple modification of the MM algorithm based on the $Q(\theta \mid \theta^n)$ function defined by inequality (6.6). Many potential functions exist satisfying these conditions. One simple example is $\psi(r) = r^2$. Because this choice tends to deter the formation of boundaries, Green [63] has suggested the gentler alternative $\psi(r) = \ln[\cosh(r)]$, which grows for large $|r|$ linearly rather than quadratically.

One adverse consequence of introducing a prior is that it couples the parameters in the maximization step of the MM algorithm for finding the posterior mode. One can decouple the parameters by exploiting the convexity and evenness of the potential function $\psi(r)$ through the inequality

$$\psi(\theta_j - \theta_k) \quad = \quad \psi\left(\frac{1}{2}\left[2\theta_j - \theta_j^n - \theta_k^n\right] + \frac{1}{2}\left[-2\theta_k + \theta_j^n + \theta_k^n\right]\right)$$
$$\leq \quad \frac{1}{2}\psi(2\theta_j - \theta_j^n - \theta_k^n) + \frac{1}{2}\psi(2\theta_k - \theta_j^n - \theta_k^n),$$

which is strict unless $\theta_j + \theta_k = \theta_j^n + \theta_k^n$ [37]. This inequality allows us to redefine the minorizing function as

$$Q(\theta \mid \theta^n) \quad = \quad -\sum_i \sum_j \alpha_{ij} f_i\left(\frac{\theta_j}{\theta_j^n} l_i^t \theta^n\right)$$
$$-\frac{\gamma}{2} \sum_{\{j,k\}\epsilon N} w_{jk}[\psi(2\theta_j - \theta_j^n - \theta_k^n) + \psi(2\theta_k - \theta_j^n - \theta_k^n)],$$

where $f_i(s) = d_i e^{-s} + y_i s$ and the admixture constants a_{ij} are given by (6.5). The parameters are once again separated in the M step, and maximizing $Q(\theta \mid \theta^n)$ drives the logposterior uphill and eventually leads to the posterior mode (maximum).

6.6 Mathematical Applications

Poisson processes not only offer realistic models for scientific phenomena, but they also provide devices for solving certain problems in combinatorics and probability theory [2, 13, 16, 21]. The Poisson strategies at the heart of the next two examples succeed by replacing dependent random variables by closely related independent random variables.

Example 6.6.1 *Schrödinger's Method*

Schrödinger's method is a technique for solving occupancy problems in multinomial sampling. Consider a multinomial sample with m equally likely categories and n trials. If we desire the probability of the event A_n that all m categories are occupied, then we can use an inclusion-exclusion argument. Alternatively in Schrödinger's method, we assume that the number of trials N is a Poisson random variable with mean λ. According to Proposition 6.2.1, this assumption decouples the categories in the sense that the numbers of outcomes falling in the different categories are independent Poisson random variables with common mean λ/m. In the Poisson setting, the probability of the event A that all m categories are occupied satisfies

$$
\begin{aligned}
e^{\lambda} \Pr(A) &= e^{\lambda}\left(1 - e^{-\frac{\lambda}{m}}\right)^{m} \\
&= \left(e^{\frac{\lambda}{m}} - 1\right)^{m} \\
&= \sum_{j=0}^{m}\binom{m}{j}(-1)^{j} e^{(m-j)\lambda/m} \\
&= \sum_{j=0}^{m}\binom{m}{j}(-1)^{j}\sum_{n=0}^{\infty}\left(1 - \frac{j}{m}\right)^{n}\frac{\lambda^{n}}{n!} \qquad (6.7)\\
&= \sum_{n=0}^{\infty}\frac{\lambda^{n}}{n!}\sum_{j=0}^{m}\binom{m}{j}(-1)^{j}\left(1 - \frac{j}{m}\right)^{n}.
\end{aligned}
$$

On the other hand, conditioning on N produces

$$
\Pr(A) = \sum_{n=0}^{\infty}\Pr(A_n)e^{-\lambda}\frac{\lambda^{n}}{n!}. \qquad (6.8)
$$

We now multiply equation (6.8) by e^{λ} and equate the result to equation (6.7). Because the coefficients of λ^{n} must match, the conclusion

$$
\Pr(A_n) = \sum_{j=0}^{m}\binom{m}{j}(-1)^{j}\left(1 - \frac{j}{m}\right)^{n}
$$

follows immediately. This Poisson randomization technique extends to more complicated occupancy problems [13]. ∎

Example 6.6.2 *Poissonization in the Family Planning Model 2.3.3*

Poisson processes come into play in this model when we embed the births to the couple at the random times determined by a Poisson process on $[0, \infty)$ of unit intensity. Hence, on average n births occur during $[0, n]$ for any positive integer n. When births are classified by sex, then as suggested by Proposition 6.2.1 and discussed in more detail in the next section, male births and female births form two independent Poisson processes. Let T_s and T_d be the continuously distributed waiting times until the birth of s sons and d daughters, respectively. The waiting time until the quota of at least s sons and d daughters is reached is $T_{sd} = \max\{T_s, T_d\}$. Now independence of T_s and T_d and Example 2.5.1 entail

$$
\begin{aligned}
\mathrm{E}(T_{sd}) &= \int_0^\infty [1 - \Pr(T_{sd} \le t)]\, dt \\
&= \int_0^\infty [1 - \Pr(T_s \le t)\Pr(T_d \le t)]\, dt \\
&= \int_0^\infty \{1 - [1 - \Pr(T_s > t)][1 - \Pr(T_d > t)]\}\, dt \quad (6.9) \\
&= \int_0^\infty \Pr(T_s > t)\, dt + \int_0^\infty \Pr(T_d > t)\, dt \\
&\quad - \int_0^\infty \Pr(T_s > t)\Pr(T_d > t)\, dt.
\end{aligned}
$$

Proposition 6.4.1 implies that $\Pr(T_s > t) = \sum_{k=0}^{s-1} \frac{(pt)^k}{k!} e^{-pt}$ and similarly for $\Pr(T_d > t)$. Combining these facts with the identity

$$
\int_0^\infty t^n e^{-rt}\, dt = \frac{n!}{r^{n+1}}
$$

and equation (6.9) leads to the conclusion that

$$
\begin{aligned}
\mathrm{E}(T_{sd}) &= \int_0^\infty \sum_{k=0}^{s-1} \frac{(pt)^k}{k!} e^{-pt}\, dt + \int_0^\infty \sum_{l=0}^{d-1} \frac{(qt)^l}{l!} e^{-qt}\, dt \\
&\quad - \int_0^\infty \sum_{k=0}^{s-1} \sum_{l=0}^{d-1} \frac{p^k}{k!} \frac{q^l}{l!} t^{k+l} e^{-t}\, dt \quad (6.10) \\
&= \sum_{k=0}^{s-1} \frac{p^k}{p^{k+1}} + \sum_{l=0}^{d-1} \frac{q^l}{q^{l+1}} - \sum_{k=0}^{s-1} \sum_{l=0}^{d-1} \binom{k+l}{k} p^k q^l \\
&= \frac{s}{p} + \frac{d}{q} - \sum_{k=0}^{s-1} \sum_{l=0}^{d-1} \binom{k+l}{k} p^k q^l.
\end{aligned}
$$

Having calculated $E(T_{sd})$, we now show that $E(N_{sd}) = E(T_{sd})$ by considering the random sum

$$T_{sd} = \sum_{k=1}^{N_{sd}} W_k,$$

where the W_k are the independent exponential waiting times between successive births. Because $E(W_1) = 1$ and N_{sd} is independent of the W_k, Example 2.4.4 implies $E(T_{sd}) = E(N_{sd}) E(W_1) = E(N_{sd})$. Alternatively, readers can check the equality $E(N_{sd}) = E(T_{sd})$ by verifying that formula (6.10) for $E(T_{sd})$ satisfies the same boundary conditions and the same recurrence relation as $E(N_{sd})$.   ■

6.7   Transformations

In this section, we informally discuss various ways of constructing new Poisson processes from old ones. (Detailed proofs of all assertions made here can be found in reference [89].) For instance, suppose Π is a Poisson process on the region $S \subset \mathsf{R}^m$. If T is a measurable subset of S, then the random points Π_T falling in T clearly satisfy the postulates (a) through (d) of a Poisson process. The Poisson process Π_T is called the restriction of Π to T. Similarly, if Π_1 and Π_2 are two independent Poisson processes on S, then the union $\Pi = \Pi_1 \bigcup \Pi_2$ is a Poisson process called the superposition of Π_1 and Π_2. The intensity function $\lambda(x)$ of Π is the sum $\lambda_1(x) + \lambda_2(x)$ of the intensity functions of Π_1 and Π_2.

In some circumstances, one can create a new Poisson process $T(\Pi)$ from an existing Poisson process by transforming the underlying space $U \subset \mathsf{R}^m$ to a new space $V \subset \mathsf{R}^n$ via a measurable map $T : U \mapsto V$. In this paradigm, a random point $X \in \Pi$ is sent into the new random point $T(X)$. To prevent random points from piling up on one another, we must impose some restriction on the map $T(x)$. We can achieve this goal and avoid certain measure-theoretic subtleties by requiring the existence of an intensity function $\lambda_T(y)$ such that

$$\int_A \lambda_T(y)\, dy = \int_{T^{-1}(A)} \lambda(x)\, dx \qquad (6.11)$$

for all measurable $A \subset V$. Equality (6.11) is just another way of stating that the expected number $E(N_A)$ of transformed random points on A matches the expected number of random points $E(N_{T^{-1}(A)})$ on the inverse image $T^{-1}(A) = \{x \in S : T(x) \in A\}$ of A. Because the inverse image operation sends disjoint regions B and C into disjoint inverse regions $T^{-1}(B)$ and $T^{-1}(C)$, the number of transformed random points N_B and N_C in B and C enjoy the crucial independence property of a Poisson process. A possible difficulty in the construction of $\lambda_T(y)$ lies in finding a sequence of subregions V_n such that $V = \bigcup_n V_n$ and $\int_{V_n} \lambda_T(y)\,dy < \infty$. The broader definition of a Poisson process adopted in reference [89] solves this apparent problem.

Two special cases of formula (6.11) cover most applications. In the first case, the transformation $T(x)$ is continuously differentiable and invertible. When this is true, the change of variables formula (1.11) implies that

$$\lambda_T(y) \quad = \quad \lambda \circ T^{-1}(y) |\det dT^{-1}(y)|.$$

Of course, invertibility presupposes that the dimensions m and n match. In the second case, suppose that $U = \mathsf{R}^m$ and $V = \mathsf{R}^n$ with $n < m$. Consider the projection $T(x_1, \ldots, x_m) = (x_1, \ldots, x_n)$ of a point x onto its first n coordinates. If a Poisson process on R^m has intensity function $\lambda(x_1, \ldots, x_m)$, then the projected Poisson process on R^n has the intensity function

$$\lambda_T(x_1, \ldots, x_n) \quad = \quad \int \cdots \int \lambda(x_1, \ldots, x_m)\,dx_{n+1} \cdots dx_m$$

created by integrating over the last $m - n$ variables. Note that the multidimensional integral defining $\lambda_T(x_1, \ldots, x_n)$ can be infinite on a finite region $A \subset \mathsf{R}^n$. When this occurs, the projected Poisson process attributes an infinite number of random points to A. This phenomenon crops up when $\lambda(x) \equiv 1$.

Example 6.7.1 *Polar Coordinates*

Let $T(x_1, x_2)$ be the map taking each point $(x_1, x_2) \in U = \mathsf{R}^2$ to its polar coordinates (r, θ). The change of variables formula

$$\int \int_A \lambda(r \cos \theta, r \sin \theta) r \, dr \, d\theta \quad = \quad \int \int_{T^{-1}(A)} \lambda(x_1, x_2)\,dx_1\,dx_2$$

shows that the intensity function $\lambda(x_1, x_2)$ is transformed into the intensity function $\lambda(r \cos \theta, r \sin \theta) r$. If $\lambda(x_1, x_2)$ is a function of $r = \sqrt{x_1^2 + x_2^2}$ alone and we further project onto the r coordinate of (r, θ), then the doubly transformed Poisson process has intensity $2\pi r \lambda(r)$ on the interval $[0, \infty)$. Readers can exploit this fact in solving Problem 1.   ■

6.8   Marking and Coloring

Our final construction involves coloring and marking. In coloring, we randomly assign a color to each random point in a Poisson process, with probability p_k attributed to color k. Expanding on Proposition 6.2.1, we can assert that the random points of different colors form independent Poisson processes. If $\lambda(x)$ is the intensity function of the overall process, then $p_k \lambda(x)$ is the intensity function of the Poisson process for color k.

In marking, we generalize this paradigm in two ways. First, we replace colors by points y in some arbitrary marking space M. Second, we allow the selection procedure assigning a mark $y \in M$ to a point $x \in S$ to depend on x. Thus, we select y according to the probability density $p(y \mid x)$, which we now assume to be a continuous density for the sake of consistency with our slightly restricted definition of a Poisson process. Marking is still carried out independently from point to point. The marking theorem [89] says that the pairs (X, Y) of random points X and their associated marks Y generate a Poisson process on the product space $S \times M$ with intensity function $\lambda(x)p(y \mid x)$. Thus, the number of pairs falling in the region $R \subset S \times M$ is $\int \int_R \lambda(x)p(y \mid x) \, dy \, dx$. If we combine marking with projection onto the marking space, then we can assert that the random marks Y constitute a Poisson process with intensity $\int \lambda(x)p(y \mid x) \, dx$.

Example 6.8.1 *New Cases of AIDS*

In a certain country, new HIV viral infections occur according to a Poisson process with intensity $\lambda(t)$ that varies with time t. Given someone is infected at t, he or she lives a random length of time $Y_t \geq 0$ until the onset of AIDS. Suppose that the latency period (mark) Y_t has a density function $p(y \mid t)$. If the Y_t are assigned independently from person to person, then the pairs (T, Y_T), of random infection times T and associated latency periods Y_T constitute a Poisson process concentrated in the upper-half plane of $\mathbb{R}^2$. The intensity function of this two-dimensional Poisson process is given by the product $\lambda(t)p(y \mid t)$.

If we apply the mapping procedure to the function $f(t, y) = t + y = u$, then we infer that the random onset times $U = T + Y_T$ of AIDS determine a Poisson process. It follows immediately that the numbers of new AIDS cases arising during disjoint time intervals are independent. The equation

$$\int \int_{\{t+y \in A\}} \lambda(t)p(y \mid t) \, dt \, dy \; = \; \int_A \int_{-\infty}^{u} \lambda(t)p(u - t \mid t) \, dt \, du$$

identifies $\int_{-\infty}^{u} \lambda(t)p(u - t \mid t) \, dt$ as the intensity function of the Poisson process. It is of some interest to calculate the expected number $E(N_{[c,d]})$ of AIDS cases during $[c, d]$ given explicit models for $\lambda(t)$ and $p(y \mid t)$. Under exponential growth of the HIV epidemic, $\lambda(t) = \alpha e^{\beta t}$ for positive constants α and β. The model $p(y \mid t) = \gamma e^{-\gamma(y-\delta)} 1_{\{y \geq \delta\}}$ incorporates an absolute delay δ during which the immune system weakens before the onset of AIDS. After this waiting period, there is a constant hazard rate γ for the appearance of AIDS in an infected person. With these assumptions

$$\int_{-\infty}^{u} \lambda(t)p(u - t \mid t) \, dt \; = \; \int_{-\infty}^{u-\delta} \alpha e^{\beta t} \gamma e^{-\gamma(u-t-\delta)} \, dt$$

$$= \; \frac{\alpha \gamma}{\beta + \gamma} e^{\beta(u-\delta)},$$

and

$$
\begin{aligned}
\mathrm{E}(N_{[c,d]}) &= \frac{\alpha\gamma}{\beta+\gamma} \int_c^d e^{\beta(u-\delta)}\,du \\
&= \frac{\alpha\gamma}{\beta(\beta+\gamma)}[e^{\beta(d-\delta)} - e^{\beta(c-\delta)}].
\end{aligned}
$$

■

6.9 Campbell's Moment Formulas

In many Poisson process models, we are confronted with the task of evaluating moments of random sums of the type

$$
S = \sum_{X \in \Pi} f(X),
$$

where X ranges over the random points of a process Π and $f(x)$ is a deterministic measurable function. Campbell devised elegant formulas for precisely this purpose [89]. It is easiest to derive Campbell's formulas when $f(x) = \sum_{j=1}^m c_j 1_{A_j}$ is a simple function defined by a partition $A_1, \ldots, A_m$ of the underlying space. If N_{A_j} counts the number of random points in A_j, then by virtue of the disjointness of the sets A_j, we can write

$$
S = \sum_{j=1}^m c_j N_{A_j}.
$$

This representation makes it clear that

$$
\begin{aligned}
\mathrm{E}(S) &= \sum_{j=1}^m c_j\,\mathrm{E}(N_{A_j}) \\
&= \sum_{j=1}^m c_j \int_{A_j} \lambda(x)\,dx \\
&= \int f(x)\lambda(x)\,dx,
\end{aligned}
\tag{6.12}
$$

where $\lambda(x)$ is the intensity function of Π. Similar reasoning leads to the formulas

$$
\begin{aligned}
\mathrm{E}(e^{itS}) &= \exp\left\{ \int [e^{itf(x)} - 1]\lambda(x)\,dx \right\} \\
\mathrm{E}(u^S) &= \exp\left\{ \int [u^{f(x)} - 1]\lambda(x)\,dx \right\}
\end{aligned}
\tag{6.13}
$$

for the characteristic function of S and for the probability generating function of S when $f(x)$ is nonnegative and integer valued.

If we have a second random sum $T = \sum_{X \in \Pi} g(X)$ defined by a simple function $g(x) = \sum_{k=1}^{n} d_k 1_{B_k}$, then it is often useful to calculate $\mathrm{Cov}(S, T)$. Toward this end, note that

$$
\begin{aligned}
N_{A_j} &= N_{A_j \setminus B_k} + N_{A_j \cap B_k} \\
N_{B_k} &= N_{B_k \setminus A_j} + N_{A_j \cap B_k}.
\end{aligned}
$$

Because the numbers of random points occurring on disjoint sets are independent and Poisson distributed, these decompositions produce

$$
\mathrm{Cov}(N_{A_j}, N_{B_k}) = \mathrm{Var}(N_{A_j \cap B_k}) = \mathrm{E}(N_{A_j \cap B_k}) = \int_{A_j \cap B_k} \lambda(x)\, dx.
$$

It follows that

$$
\begin{aligned}
\mathrm{Cov}(S, T) &= \sum_{j=1}^{m} \sum_{k=1}^{n} c_j d_k \, \mathrm{Cov}(N_{A_j}, N_{B_k}) \\
&= \sum_{j=1}^{m} \sum_{k=1}^{n} c_j d_k \int_{A_j \cap B_k} \lambda(x)\, dx \qquad (6.14) \\
&= \int f(x) g(x) \lambda(x)\, dx.
\end{aligned}
$$

The special choice $g(x) = f(x)$ yields

$$
\mathrm{Var}(S) = \int f(x)^2 \lambda(x)\, dx. \qquad (6.15)
$$

Campbell's formulas (6.12), (6.13), (6.14), and (6.15) extend to more general functions $f(x)$ and $g(x)$ by passing to appropriate limits [89].

Example 6.9.1 *An Astronomical Application*

Suppose stars occur in the universe $U \subset \mathbf{R}^3$ according to a Poisson process with intensity function $\lambda(x)$. Furthermore, assume each star radiates light at a level y chosen independently from a probability density $p(y)$. The marked Poisson process Π of pairs (X, Y) of random locations and radiation levels has intensity function $\lambda(x)p(y)$. At the center of the universe, the incoming radiation has level

$$
S = \sum_{(X,Y) \in \Pi} \frac{Y}{\|X\|^2},
$$

where $\|x\|$ is the Euclidean distance of x from the origin, and where we assume that the light radiated by different stars acts additively. Campbell's formula (6.12) implies

$$
\begin{aligned}
\mathrm{E}(S) &= \int_U \int \frac{y}{\|x\|^2} \lambda(x) p(y) \, dy \, dx \\
&= \int y p(y) \, dy \int_U \frac{1}{\|x\|^2} \lambda(x) \, dx.
\end{aligned}
$$

Given this naive physical model and a constant $\lambda(x)$, passage to spherical coordinates shows that it is possible for the three-dimensional integral $\int_U \|x\|^{-2} \lambda(x) \, dx$ to diverge on an unbounded set such as $U = \mathsf{R}^3$. The fact that we are not blinded by light on a starlit night suggests that U is bounded.   ∎

6.10 Problems

1. Consider a Poisson process in the plane with constant intensity λ. Find the distribution and density function of the distance from the origin of the plane to the nearest random point. What is the mean of this distribution? (Hint: Using Example 6.7.1, you should calculate the distribution function $1 - e^{-\lambda \pi r^2}$.)

2. In the context of Example 6.4.2, show that the jth order statistic $X_{(j)}$ had mean and variance

$$
\mathrm{E}(X_{(j)}) = \sum_{k=1}^{j} \frac{1}{\lambda(n - k + 1)}
$$

$$
\mathrm{Var}(X_{(j)}) = \sum_{k=1}^{j} \frac{1}{\lambda^2 (n - k + 1)^2}.
$$

3. Continuing Problem 2, prove that $X_{(j)}$ has distribution and density functions

$$
F_{(j)}(x) = \sum_{k=j}^{n} \binom{n}{k} (1 - e^{-\lambda x})^k e^{-(n-k)\lambda x}
$$

$$
f_{(j)}(x) = n \binom{n-1}{j-1} (1 - e^{-\lambda x})^{j-1} e^{-(n-j)\lambda x} \lambda e^{-\lambda x}.
$$

(Hint: Ignore the representation of Example 6.4.2 and reason directly.)

4. In the context of Example 6.4.2, suppose you observe $X_{(1)}, \ldots, X_{(r)}$ and wish to estimate λ^{-1} by a linear combination $S = \sum_{i=1}^{r} \alpha_i X_{(i)}$. Demonstrate that $\mathrm{Var}(S)$ is minimized subject to $\mathrm{E}(S) = \lambda^{-1}$ by taking $\alpha_j = 1/r$ for $1 \le j < r$ and $\alpha_r = (n - r + 1)/r$ [49].

5. For a fixed positive integer n, we define the segmental functions $_n\alpha_j(x)$ of x as the finite Fourier transform coefficients

$$_n\alpha_j(x) = \frac{1}{n}\sum_{k=0}^{n-1} e^{xu_n^k}u_n^{-jk},$$

where $u_n = e^{2\pi i/n}$ is the nth principal root of unity. These functions generalize the hyperbolic trig functions $\cosh(x)$ and $\sinh(x)$. Prove the following assertions:

(a) $_n\alpha_j(x) = {_n\alpha_{j+n}}(x)$.

(b) $_n\alpha_j(x+y) = \sum_{k=0}^{n-1} {_n\alpha_k}(x){_n\alpha_{j-k}}(y)$.

(c) $_n\alpha_j(x) = \sum_{k=0}^{\infty} \frac{x^{j+kn}}{(j+kn)!}$ for $0 \le j \le n-1$.

(d) $\frac{d}{dx}\left[{_n\alpha_j}(x)\right] = {_n\alpha_{j-1}}(x)$.

(e) $\lim_{x\to\infty} e^{-x} {_n\alpha_j}(x) = \frac{1}{n}$.

(f) In a Poisson process of intensity 1, $e^{-x} {_n\alpha_j}(x)$ is the probability that the number of random points on $[0,x]$ equals j modulo n.

(g) Relative to this Poisson process, let N_x count every nth random point on $[0,x]$. Then N_x has probability generating function

$$P(s) = e^{-x}\sum_{j=0}^{n-1} s^{-\frac{j}{n}} {_n\alpha_j}(s^{\frac{1}{n}}x).$$

(h) Furthermore, N_x has mean

$$E(N_x) = \frac{x}{n} - \frac{e^{-x}}{n}\sum_{j=0}^{n-1} j{_n\alpha_j}(x).$$

(i) $\lim_{x\to\infty}\left[E(N_x) - \frac{x}{n}\right] = -\frac{n-1}{2n}$.

6. Show that the loglikelihood (6.4) for the transmission tomography model is concave. State a necessary condition for strict concavity in terms of the number of pixels and the number of projections. Prove that the sufficient conditions mentioned in the text guarantee that the logposterior function $L(\theta) + \ln\pi(\theta)$ is strictly concave.

7. In the absence of a Gibbs smoothing prior, show that one step of Newton's method leads to the approximate MM update

$$\theta_j^{n+1} = \theta_j^n \frac{\sum_i l_{ij}[d_i e^{-l_i^t\theta^n}(1+l_i^t\theta^n) - y_i]}{\sum_i l_{ij}l_i^t\theta^n d_i e^{-l_i^t\theta^n}}$$

in the transmission tomography model.

8. Under the assumptions of Problem 7, demonstrate that the exact solution of the one-dimensional equation

$$\frac{\partial}{\partial\theta_j}Q(\theta\mid\theta^n) \;=\; 0$$

exists and is positive when $\sum_i l_{ij}d_i > \sum_i l_{ij}y_i$. Why would this condition usually obtain in practical implementations of transmission tomography?

9. Prove that the function $\psi(r) = \ln[\cosh(r)]$ is even, strictly convex, infinitely differentiable, and asymptotic to $|r|$ as $|r| \to \infty$.

10. In the family planning model of Example 2.3.3, we showed how to compute the probability R_{sd} that the couple reach their quota of s sons before their quota of d daughters. Deduce the formula

$$R_{sd} \;=\; \sum_{k=0}^{d-1}\binom{s+k-1}{s-1}p^s q^k$$

by viewing the births to the couple as occurring at the times of a Poisson process with unit intensity. Can you also derive this formula by counting all possible successful sequences of births? (Hint: The final birth in a successful sequence is a son.)

11. In the family planning model of Example 6.6.2, let M_{sd} be the number of children born when the family first attains either its quota of s sons or d daughters. Show that

$$\mathrm{E}(M_{sd}) \;=\; \mathrm{E}(\min\{T_s,T_d\}) \;=\; \sum_{k=0}^{s-1}\sum_{l=0}^{d-1}\binom{k+l}{k}p^k q^l.$$

Note that the formulas for $\mathrm{E}(\max\{T_s,T_d\})$ and $\mathrm{E}(\min\{T_s,T_d\})$ together yield

$$\mathrm{E}(\min\{T_s,T_d\})+\mathrm{E}(\max\{T_s,T_d\}) \;=\; \mathrm{E}(T_s)+\mathrm{E}(T_d). \quad (6.16)$$

Prove the general identity

$$\mathrm{E}(\min\{X,Y\})+\mathrm{E}(\max\{X,Y\}) \;=\; \mathrm{E}(X)+\mathrm{E}(Y)$$

for any pair of random variables X and Y with finite expectations. Finally, argue that

$$\mathrm{E}(M_{sd}) \;=\; d\sum_{k=0}^{s-1}\binom{d+k}{k}p^k q^d + s\sum_{l=0}^{d-1}\binom{s+l}{l}p^s q^l \quad (6.17)$$

by counting all possible successful sequences of births that lead to either the daughter quota or the son quota being fulfilled first. Combining equations (6.16) and (6.17) permits us to write

$$E(N_{sd}) = \frac{s}{p} + \frac{d}{q} - d\sum_{k=0}^{s-1}\binom{d+k}{k}p^k q^d - s\sum_{l=0}^{d-1}\binom{s+l}{l}p^s q^l,$$

replacing a double sum with two single sums.

12. Suppose you randomly drop n balls into m boxes. Assume that a ball is equally likely to land in any box. Use Schrödinger's method to prove that each box receives an even number of balls with probability

$$e_n = \frac{1}{2^m}\sum_{j=0}^{m}\binom{m}{j}\left(1 - \frac{2j}{m}\right)^n$$

and an odd number of balls with probability

$$o_n = \frac{1}{2^m}\sum_{j=0}^{m}\binom{m}{j}(-1)^j\left(1 - \frac{2j}{m}\right)^n.$$

(Hint: The even terms of e^t sum to $\frac{1}{2}(e^t + e^{-t})$ and the odd terms to $\frac{1}{2}(e^t - e^{-t})$.)

13. Continuing Problem 12, show that the probability that exactly j boxes are empty is

$$\binom{m}{j}\sum_{k=1}^{m-j}\binom{m-j}{k}(-1)^{m-j-k}\left(\frac{k}{m}\right)^n.$$

14. A one-way highway extends from 0 to ∞. Cars enter at position 0 at times s determined by a Poisson process on $[0, t]$ with constant intensity λ. Each car is independently assigned a velocity v from a density $g(v)$ on $[0, \infty)$. Demonstrate that the number of cars located in the interval (a, b) at time t has a Poisson distribution with mean $\lambda\int_0^t[G(\frac{b}{t-s}) - G(\frac{a}{t-s})]\,ds$, where $G(v)$ is the distribution function of $g(v)$ [130].

15. Suppose we generate random circles in the plane by taking their centers (x, y) to be the random points of a Poisson process of constant intensity λ. Each center we independently mark with a radius r sampled from a probability density $g(r)$ on $[0, \infty)$. If we map each random triple (X, Y, R) to the point $U = \sqrt{X^2 + Y^2} - R$, then show that the random points so generated constitute a Poisson process with intensity

$$\eta(u) = 2\pi\lambda\int_0^\infty (r + u)_+ g(r)\,dr.$$

Conclude from this analysis that the number of random circles that overlap the origin is Poisson with mean $\lambda\pi \int_0^\infty r^2 g(r)\, dr$ [143].

16. Continuing Problem 15, perform the same analysis in three dimensions for spheres. Conclude that the number of random spheres that overlap the origin is Poisson with mean $\frac{4\lambda\pi}{3} \int_0^\infty r^3 g(r)\, dr$ [143].

17. If $f(x)$ be a simple function and Π is a Poisson process with intensity function $\lambda(x)$, then demonstrate the formulas in equation (6.13) for the characteristic function and generating function of the random sum S.

18. A train departs at time $t > 0$. During the interval $[0, t]$, passengers arrive at the depot at times T determined by a Poisson process with constant intensity λ. The total waiting time passengers spend at the depot is $W = \sum_T (t - T)$. Show that W has mean $E(W) = \frac{\lambda t^2}{2}$ and variance $\text{Var}(W) = \frac{\lambda t^3}{3}$ by invoking Campbell's formulas (6.12) and (6.15) [130].

19. Claims arrive at an insurance company at the times T of a Poisson process with constant intensity λ on $[0, \infty)$. Each time a claim arrives, the company pays S dollars, where S is independently drawn from a probability density $g(s)$ on $[0, \infty)$. Because of inflation and the ability of the company to invest premiums, the longer a claim is delayed, the less it costs the company. If a claim is discounted at rate β, then show that the company's ultimate liability $L = \sum_T S e^{-\beta T}$ has mean and variance

$$
\begin{aligned}
E(L) &= \frac{\lambda}{\beta} \int_0^\infty s g(s)\, ds \\
\text{Var}(L) &= \frac{\lambda}{2\beta} \int_0^\infty s^2 g(s)\, ds.
\end{aligned}
$$

(Hints: The random pairs (T, S) constitute a marked Poisson process. Use Campbell's formulas (6.12) and (6.15).)

7
Discrete-Time Markov Chains

7.1 Introduction

Applied probability thrives on models. Markov chains are one of the richest sources of good models for capturing dynamical behavior with a large stochastic component [18, 19, 48, 64, 82, 83, 93]. In this chapter we give a few examples and a quick theoretical overview of discrete-time Markov chains. The highlight of our theoretical development, Proposition 7.4.1, relies on a coupling argument. Because coupling is one of the most powerful and intuitively appealing tools available to probabilists, we examine a few of its general applications as well. We also stress reversible Markov chains. Reversibility permits explicit construction of the long-run or equilibrium distribution of a chain when such a distribution exists. Chapter 8 will cover continuous-time Markov chains.

7.2 Definitions and Elementary Theory

For the sake of simplicity, we will only consider chains with a finite or countable number of states [18, 48, 64, 82, 83]. The movement of such a chain from epoch to epoch (equivalently generation to generation) is governed by its transition probability matrix $P = (p_{ij})$. This matrix is infinite dimensional when the number of states is infinite. If Z_n denotes the state of the chain at epoch n, then $p_{ij} = \Pr(Z_n = j \mid Z_{n-1} = i)$. As a consequence, every entry of P satisfies $p_{ij} \geq 0$, and every row of P satisfies

$\sum_j p_{ij} = 1$. Implicit in the definition of p_{ij} is the fact that the future of the chain is determined by its present regardless of its past. This Markovian property is expressed formally by the equation

$$\Pr(Z_n = i_n \mid Z_{n-1} = i_{n-1}, \ldots, Z_0 = i_0) \quad = \quad \Pr(Z_n = i_n \mid Z_{n-1} = i_{n-1}).$$

The n-step transition probability $p_{ij}^{(n)} = \Pr(Z_n = j \mid Z_0 = i)$ is given by the entry in row i and column j of the matrix power P^n. This follows because the decomposition

$$p_{ij}^{(n)} \quad = \quad \sum_{i_1} \cdots \sum_{i_{n-1}} p_{ii_1} \cdots p_{i_{n-1}j}$$

over all paths $i \to i_1 \to \cdots \to i_{n-1} \to j$ of n steps corresponds to $n-1$ matrix multiplications. If the chain tends toward stochastic equilibrium, then the limit of $p_{ij}^{(n)}$ as n increases should exist independently of the starting state i. In other words, the matrix powers P^n should converge to a matrix with identical rows. Denoting the common limiting row by π, we deduce that $\pi = \pi P$ from the calculation

$$\begin{pmatrix} \pi \\ \vdots \\ \pi \end{pmatrix} \quad = \quad \lim_{n \to \infty} P^{n+1}$$

$$= \quad \left(\lim_{n \to \infty} P^n \right) P$$

$$= \quad \begin{pmatrix} \pi \\ \vdots \\ \pi \end{pmatrix} P.$$

Any probability distribution π on the states of the chain satisfying the condition $\pi = \pi P$ is termed an equilibrium (or stationary) distribution of the chain. The jth component

$$\pi_j \quad = \quad \sum_i \pi_i p_{ij} \tag{7.1}$$

of the equation $\pi = \pi P$ suggests a balance between the probabilistic flows into and out of state j. Indeed, if the left-hand side of equation (7.1) represents the probability of being in state j at the current epoch, then the right-hand side represents the probability of being in state j at the next epoch. At equilibrium, these two probabilities must match. For finite-state chains, equilibrium distributions always exist [48, 64]. The real issue is uniqueness.

Probabilists have attacked the uniqueness problem by defining appropriate ergodic conditions. For finite-state Markov chains, two ergodic assumptions are invoked. The first is aperiodicity; this means that the greatest common divisor of the set $\{n \geq 1 : p_{ii}^{(n)} > 0\}$ is 1 for every state i. Aperiodicity trivially holds when $p_{ii} > 0$ for all i. The second ergodic assumption is irreducibility; this means that for every pair of states (i, j), there exists a positive integer n_{ij} such that $p_{ij}^{(n_{ij})} > 0$. In other words, every state is reachable from every other state. Said yet another way, all states communicate. For a finite-state irreducible chain, Problem 1 states that the integer n_{ij} can be chosen independently of the particular pair (i, j) if and only if the chain is also aperiodic. Thus, we can merge the two ergodic assumptions into the single assumption that some power P^n has all entries positive. Under this single ergodic condition, we show in Proposition 7.4.1 that a unique equilibrium distribution π exists and that $\lim_{n \to \infty} p_{ij}^{(n)} = \pi_j$. Because all states communicate, the entries of π are necessarily positive.

Equally important is the ergodic theorem [48, 64]. This theorem permits one to run a chain and approximate theoretical means by sample means. More precisely, let $f(z)$ be some real-valued function defined on the states of an ergodic chain. Then given that Z_i is the state of the chain at epoch i and π is the equilibrium distribution, we have

$$\lim_{n \to \infty} \frac{1}{n} \sum_{i=0}^{n-1} f(Z_i) = E_\pi[f(Z)] = \sum_z \pi_z f(z).$$

This result generalizes the law of large numbers for independent sampling and has important applications in Markov chain Monte Carlo methods as discussed later in this chapter.

In many Markov chain models, the equilibrium distribution satisfies the stronger condition

$$\pi_j p_{ji} = \pi_i p_{ij} \tag{7.2}$$

for all pairs (i, j). If this is the case, then the probability distribution π is said to satisfy detailed balance, and the Markov chain, provided it is irreducible, is said to be reversible. Summing equation (7.2) over i yields the equilibrium condition (7.1). Thus, detailed balance implies balance. Irreducibility is imposed as part of reversibility to guarantee that π is unique and has positive entries. Given the latter condition, detailed balance implies that $p_{ij} > 0$ if and only if $p_{ji} > 0$.

If $i_1, \ldots, i_m$ is any sequence of states in a reversible chain, then detailed balance also entails

$$\pi_{i_1} p_{i_1 i_2} = \pi_{i_2} p_{i_2 i_1}$$
$$\pi_{i_2} p_{i_2 i_3} = \pi_{i_3} p_{i_3 i_2}$$
$$\vdots$$
$$\pi_{i_{m-1}} p_{i_{m-1} i_m} = \pi_{i_m} p_{i_m i_{m-1}}$$
$$\pi_{i_m} p_{i_m i_1} = \pi_{i_1} p_{i_1 i_m}.$$

Multiplying these equations together and canceling the common positive factor $\pi_{i_1} \cdots \pi_{i_m}$ from both sides of the resulting equality give Kolmogorov's circulation criterion [86]

$$p_{i_1 i_2} p_{i_2 i_3} \cdots p_{i_{m-1} i_m} p_{i_m i_1} \;=\; p_{i_1 i_m} p_{i_m i_{m-1}} \cdots p_{i_3 i_2} p_{i_2 i_1}. \qquad (7.3)$$

Conversely, suppose an irreducible Markov chain satisfies Kolmogorov's criterion and the condition that $p_{ij} > 0$ if and only if $p_{ji} > 0$. We can prove that the chain is reversible by explicitly constructing the equilibrium distribution and showing that it satisfies detailed balance. The idea behind the construction is to choose some arbitrary reference state i and to pretend that π_i is given. If j is another state, let $i \to i_1 \to \cdots \to i_m \to j$ be any path leading from i to j. Then the formula

$$\pi_j \;=\; \pi_i \frac{p_{i i_1} p_{i_1 i_2} \cdots p_{i_m j}}{p_{j i_m} p_{i_m i_{m-1}} \cdots p_{i_1 i}} \qquad (7.4)$$

defines π_j. A straightforward application of Kolmogorov's criterion (7.3) shows that definition (7.4) does not depend on the particular path chosen from i to j. To validate detailed balance, suppose that k is adjacent to j. Then $i \to i_1 \to \cdots \to i_m \to j \to k$ furnishes a path from i to k through j. It follows from (7.4) that

$$\begin{aligned}
\pi_k &= \pi_i \frac{p_{i i_1} p_{i_1 i_2} \cdots p_{i_m j} p_{jk}}{p_{j i_m} p_{i_m i_{m-1}} \cdots p_{i_1 i} p_{kj}} \\
&= \pi_j \frac{p_{jk}}{p_{kj}},
\end{aligned}$$

which is obviously equivalent to detailed balance. In general, the value of π_i is determined by the requirement that $\sum_j \pi_j = 1$. For a chain with a finite number of states, we can guarantee this condition by replacing π by $\tilde{\pi}$ with components

$$\tilde{\pi}_j \;=\; \frac{\pi_j}{\sum_k \pi_k}.$$

In practice, explicit calculation of the sum $\sum_k \pi_k$ may be nontrivial. For a chain with an infinite number of states, in contrast, it may impossible to renormalize the π_j defined by equation (7.4) so that $\sum_j \pi_j = 1$. This situation occurs in Example 7.3.1 in the next section.

7.3   Examples

Here are a few examples of discrete-time chains classified according to the concepts just introduced. If possible, the unique equilibrium distribution is identified.

Example 7.3.1 *Random Walk on a Graph*

Consider a connected graph with node set N and edge set E. The number of edges $d(v)$ incident on a given node v is called the degree of v. Owing to the connectedness assumption, $d(v) > 0$ for all $v \in N$. Now define the transition probability matrix $P = (p_{uv})$ by

$$p_{uv} = \begin{cases} \frac{1}{d(u)} & \text{for } \{u, v\} \in E \\ 0 & \text{for } \{u, v\}, \notin E. \end{cases}$$

This Markov chain is irreducible because of the connectedness assumption. It is also aperiodic unless the graph is bipartite. (A graph is said to be bipartite if we can partition its node set into two disjoint subsets F and M, say females and males, such that each edge has one node in F and the other node in M.) If V has m edges, then the equilibrium distribution π of the chain has components $\pi_v = \frac{d(v)}{2m}$. It is trivial to show that this choice of π satisfies detailed balance.

One hardly needs this level of symmetry to achieve detailed balance. For instance, consider a random walk on the nonnegative integers with neighboring integers connected by an edge. For $i > 0$, let

$$p_{ij} = \begin{cases} q_i, & j = i - 1 \\ r_i, & j = i \\ p_i, & j = i + 1 \\ 0, & \text{otherwise} \end{cases}$$

and $p_{00} = r_0$ and $p_{01} = p_0$. With state 0 as a reference state, Kolmogorov's formula (7.4) becomes

$$\pi_i = \pi_0 \prod_{j=1}^{i} \frac{p_{j-1}}{q_j}.$$

This definition of π_i clearly satisfies detailed balance. However, because the state space is infinite, we must impose the additional constraint

$$\sum_{i=0}^{\infty} \prod_{j=1}^{i} \frac{p_{j-1}}{q_j} < \infty$$

to achieve a legitimate equilibrium distribution. When this condition holds, we define

$$\pi_i = \frac{\prod_{j=1}^{i} \frac{p_{j-1}}{q_j}}{\sum_{k=0}^{\infty} \prod_{j=1}^{k} \frac{p_{j-1}}{q_j}} \tag{7.5}$$

and eliminate the unknown π_0. For instance, if all $p_j = p$ and all $q_j = q$, then $p < q$ is a necessary and sufficient condition for the existence of an equilibrium distribution. When $p < q$, formula (7.5) implies $\pi_i = \frac{(q-p)p^i}{q^{i+1}}$.

∎

Example 7.3.2 *Wright-Fisher Model of Genetic Drift*

Consider a population of m organisms from some animal or plant species. Each member of this population carries two genes at some genetic locus, and these genes take two forms (or alleles) labeled a_1 and a_2. At each generation, the population reproduces itself by sampling $2m$ genes with replacement from the current pool of $2m$ genes. If Z_n denotes the number of a_1 alleles at generation n, then it is clear that the Z_n constitute a Markov chain with binomial transition probability matrix

$$p_{jk} = \binom{2m}{k}\left(\frac{j}{2m}\right)^k\left(1 - \frac{j}{2m}\right)^{2m-k}.$$

This chain is reducible because once one of the states 0 or $2m$ is reached, the corresponding allele is fixed in the population, and no further variation is possible. An infinity of equilibrium distributions π exist. Each one is characterized by $\pi_0 = \alpha$ and $\pi_{2m} = 1 - \alpha$ for some $\alpha \in [0, 1]$.   ∎

Example 7.3.3 *Ehrenfest's Model of Diffusion*

Consider a box with m gas molecules. Suppose the box is divided in half by a rigid partition with a very small hole. Molecules drift aimlessly around each half until one molecule encounters the hole and passes through. Let Z_n be the number of molecules in the left half of the box at epoch n. If epochs are timed to coincide with molecular passages, then the transition matrix of the chain is

$$p_{jk} = \begin{cases} 1 - \frac{j}{m} & \text{for } k = j + 1 \\ \frac{j}{m} & \text{for } k = j - 1 \\ 0 & \text{otherwise.} \end{cases}$$

This chain is a random walk with finite state space. It is periodic with period 2, irreducible, and reversible with equilibrium distribution

$$\pi_j = \binom{m}{j}\left(\frac{1}{2}\right)^m.$$

The binomial form of π_j follows from either equation (7.2) or equation (7.5).   ∎

Example 7.3.4 *Discrete Renewal Process*

Many treatments of Markov chain theory depend on a prior development of renewal theory. Here we reverse the logical flow and consider a discrete renewal process as a Markov chain. A renewal process models repeated visits to a special state [48, 64]. Shortly after entering the state, the process leaves and eventually returns for the first time after $j > 0$ steps with probability f_j. The return times following different visits are independent.

In modeling this behavior by a Markov chain, we let Z_n denote the number of additional epochs left after epoch n until a return to the special state occurs. The renewal mechanism generates the transition matrix with entries

$$p_{ij} = \begin{cases} f_{j+1}, & i = 0 \\ 1, & i > 0 \text{ and } j = i - 1 \\ 0, & i > 0 \text{ and } j \neq i - 1. \end{cases}$$

In order for $\sum_j p_{0j} = 1$, we must have $f_0 = 0$. If $f_j = 0$ for $j > m$, then the chain has m states; otherwise, it has an infinite number of states. Because the chain always ratchets downward from $i > 0$ to $i-1$, it is both irreducible and irreversible. State 0, and therefore the whole chain, is aperiodic if and only if the set $\{j : f_j > 0\}$ has greatest common divisor 1. This number theoretic fact is covered in Appendix A of reference [137] and the appendix of reference [19].

One of the primary concerns in renewal theory is predicting what fraction of epochs are spent in the special state. This problem is solved by finding the equilibrium probability π_0 of state 0 in the associated Markov chain. Assuming the chain is aperiodic and the mean $\mu = \sum_i i f_i$ is finite, we can easily calculate the equilibrium distribution. The balance conditions defining equilibrium are

$$\pi_j = \pi_{j+1} + \pi_0 f_{j+1}.$$

One can demonstrate by induction that the unique solution to this system of equations is given by

$$\pi_j = \pi_0 \left(1 - \sum_{i=1}^{j} f_i \right) = \pi_0 \sum_{i=j+1}^{\infty} f_i$$

subject to

$$1 = \sum_{i=0}^{\infty} \pi_i$$

$$= \pi_0 \sum_{i=0}^{\infty} \sum_{i=j+1}^{\infty} f_i$$

$$= \pi_0 \mu.$$

Here we assume an infinite number of states and invoke Example 2.5.1. It follows that $\pi_0 = \mu^{-1}$ and $\pi_j = \mu^{-1} \sum_{i=j+1}^{\infty} f_i$. ∎

Example 7.3.5 *Card Shuffling and Random Permutations*

Imagine a deck of cards labeled $1, \ldots, m$. The cards taken from top to bottom provide a permutation σ of these m labels. The usual method of shuffling cards, the so-called riffle shuffle, is difficult to analyze probabilistically. A far simpler shuffle is the top-in shuffle [3, 38]. In this shuffle, one takes the top card on the deck and moves it to a random position in the deck. Of course, if the randomly chosen position is the top position, then the deck suffers no change. Repeated applications of the top-in shuffle constitute a Markov chain. This chain is aperiodic and irreducible. Aperiodicity is obvious because the deck can remain constant for an arbitrary number of shuffles. Irreducibility is slightly more subtle. Suppose we follow the card originally at the bottom of the deck. Cards inserted below it occur in completely random order. Once the original bottom card reaches the top of the deck and is moved, then the whole deck is randomly rearranged. This argument shows that all permutations are ultimately equally likely and can be reached from any starting permutation. We extend this analysis in Example 7.4.3. Finally, the chain is irreversible. For example, if a deck of seven cards is currently in the order $\sigma = (4, 7, 5, 2, 3, 1, 6)$, equating left to top and right to bottom, then inserting the top card 4 in position 3 produces $\eta = (7, 5, 4, 2, 3, 1, 6)$. Clearly, it is impossible to return from η to σ by moving the new top card 7 to another position. Under reversibility, each individual step of a Markov chain can be reversed. ∎

7.4 Coupling

In this section we undertake an investigation of the convergence of a ergodic Markov chain to its equilibrium. Our method of attack exploits a powerful proof technique known as coupling. By definition, two random variables or stochastic processes are coupled if they reside on the same probability space [106, 107]. As a warm-up, we illustrate coupling arguments by two examples having little to do with Markov chains.

Example 7.4.1 *Correlated Random Variables*

Suppose X is a random variable and the functions $f(x)$ and $g(x)$ are both increasing or both decreasing. If the random variables $f(X)$ and $g(X)$ have finite second moments, then it is reasonable to conjecture that $\text{Cov}[f(X), g(X)] \geq 0$. To prove this fact by coupling, consider a second random variable Y independent of X but sharing the same distribution. If $f(x)$ and $g(x)$ are both increasing or both decreasing, then the product $[f(X) - f(Y)][g(X) - g(Y)] \geq 0$. Hence,

$$
\begin{aligned}
0 &\leq \text{E}\{[f(X) - f(Y)][g(X) - g(Y)]\} \\
&= \text{E}[f(X)g(X)] + \text{E}[f(Y)g(Y)] - \text{E}[f(X)]\,\text{E}[g(Y)] - \text{E}[f(Y)]\,\text{E}[g(X)] \\
&= 2\,\text{Cov}[f(X), g(X)].
\end{aligned}
$$

When one of the two functions is increasing and the other is decreasing, the same proof with obvious modifications shows that $\text{Cov}[f(X), g(X)] \leq 0$. ∎

Example 7.4.2 *Monotonicity in Bernstein's Approximation*

In Example 3.5.1, we considered Bernstein's proof of the Weierstrass approximation theorem. When the continuous function $f(x)$ being approximated is increasing, it is plausible that the approximating polynomial

$$\mathrm{E}\left[f\left(\frac{S_n}{n}\right)\right] \;=\; \sum_{k=0}^{n} f\left(\frac{k}{n}\right)\binom{n}{k} x^k (1-x)^{n-k}.$$

is as well [107]. To prove this assertion by coupling, imagine scattering n points randomly on the unit interval. If $x \leq y$ and we interpret S_n as the number of points less than or equal to x and T_n as the number of points less than or equal to y, then these two binomially distributed random variables satisfy $S_n \leq T_n$. The desired inequality

$$\mathrm{E}\left[f\left(\frac{S_n}{n}\right)\right] \;\leq\; \mathrm{E}\left[f\left(\frac{T_n}{n}\right)\right]$$

now follows directly from the assumption that f is increasing. ∎

Our coupling proof of the convergence of an ergodic Markov chain depends on quantifying the distance between the distributions π_X and π_Y of two integer-valued random variables X and Y. One candidate distance is the total variation norm

$$\begin{aligned}
\|\pi_X - \pi_Y\|_{TV} &= \sup_{A \subset \mathcal{Z}} |\Pr(X \in A) - \Pr(Y \in A)| \\
&= \frac{1}{2}\sum_{k} |\Pr(X = k) - \Pr(Y = k)|, \qquad (7.6)
\end{aligned}$$

where A ranges over all subsets of the integers $\mathcal{Z}$ [38]. Problem 13 asks the reader to check that these two definitions of the total variation norm are equivalent. The coupling bound

$$\begin{aligned}
\|\pi_X - \pi_Y\|_{TV} &= \sup_{A \subset \mathcal{Z}} |\Pr(X \in A) - \Pr(Y \in A)| \\
&= \sup_{A \subset \mathcal{Z}} |\Pr(X \in A, X = Y) + \Pr(X \in A, X \neq Y) \\
&\qquad - \Pr(Y \in A, X = Y) - \Pr(Y \in A, X \neq Y)| \quad (7.7) \\
&= \sup_{A \subset \mathcal{Z}} |\Pr(X \in A, X \neq Y) - \Pr(Y \in A, X \neq Y)| \\
&\leq \sup_{A \subset \mathcal{Z}} \mathrm{E}(1_{\{X \neq Y\}} |1_A(X) - 1_A(Y)|) \\
&\leq \Pr(X \neq Y)
\end{aligned}$$

has many important applications.

In our convergence proof, we actually consider two random sequences X_n and Y_n and a random stopping time T such that $X_n = Y_n$ for all $n \geq T$. The bound

$$\Pr(X_n \neq Y_n) \;\leq\; \Pr(T > n)$$

suggests that we study the tail probabilities $\Pr(T > n)$. As we shall see in Proposition 7.4.1, if $\mathcal{F}_n$ is the σ-algebra of events determined by the X_i and Y_i with $i \leq n$, then $\{T \leq n\} \in \mathcal{F}_n$ and

$$\Pr(T \leq n + r \mid \mathcal{F}_n) \;\geq\; \epsilon \tag{7.8}$$

for some $\epsilon > 0$ and $r \geq 1$ and all n. These assumptions imply the inequality

$$
\begin{aligned}
\Pr(T > n + r) &= \mathrm{E}(1_{\{T > n + r\}}) \\
&= \mathrm{E}(1_{\{T > n\}} 1_{\{T > n + r\}}) \\
&= \mathrm{E}[1_{\{T > n\}}\, \mathrm{E}(1_{\{T > n + r\}} \mid \mathcal{F}_n)] \\
&\leq \mathrm{E}[1_{\{T > n\}} (1 - \epsilon)] \\
&= (1 - \epsilon)\Pr(T > n),
\end{aligned}
$$

which can be iterated to produce

$$\Pr(T > kr) \;\leq\; (1 - \epsilon)^k. \tag{7.9}$$

In the last step of the iteration, we must take $\mathcal{F}_0$ to be the trivial σ-algebra consisting of the null event and the whole sample space. From inequality (7.9) it is immediately evident that $\Pr(T < \infty) = 1$.

   With these preliminaries out of the way, we now turn to proving convergence based on a standard coupling argument [106, 129].

Proposition 7.4.1 *Every finite-state ergodic Markov chain has a unique equilibrium distribution π. Furthermore, the n-step transition probabilities $p_{ij}^{(n)}$ satisfy $\lim_{n \to \infty} p_{ij}^{(n)} = \pi_j$.*

Proof: Without loss of generality, we identify the states of the chain with the integers $\{1, \dots, m\}$. From the inequality

$$
\begin{aligned}
p_{ij}^{(n)} &= \sum_k p_{ik} p_{kj}^{(n-1)} \\
&\leq \max_l p_{lj}^{(n-1)} \sum_k p_{ik} \\
&= \max_l p_{lj}^{(n-1)}
\end{aligned}
$$

involving the n-step transition probabilities, we immediately deduce that $\max_i p_{ij}^{(n)}$ is decreasing in n. Likewise, $\min_i p_{ij}^{(n)}$ is increasing in n. If

$$\lim_{n \to \infty} |p_{uj}^{(n)} - p_{vj}^{(n)}| \;=\; 0$$

for all initial states u and v, then the gap between $\lim_{n \to \infty} \min_i p_{ij}^{(n)}$ and $\lim_{n \to \infty} \max_i p_{ij}^{(n)}$ is 0. This forces the existence of $\lim_{n \to \infty} p_{ij}^{(n)} = \pi_j$, which we identify as the equilibrium distribution of the chain.

We now construct two coupled chains X_n and Y_n on $\{1, \ldots, m\}$ that individually move according to the transition matrix $P = (p_{ij})$. The X chain starts at u and the Y chain at $v \neq u$. These two chains move independently until the first epoch $T = n$ at which $X_n = Y_n$. Thereafter, they move together. The pair of coupled chains has joint transition matrix

$$p_{(ij),(kl)} = \begin{cases} p_{ik}p_{jl} & \text{if } i \neq j \\ p_{ik} & \text{if } i = j \text{ and } k = l \\ 0 & \text{if } i = j \text{ and } k \neq l \end{cases}.$$

By definition it is clear that the probability that the coupled chains occupy the same state at epoch r is at least as great as the probability that two completely independent chains occupy the same state at epoch r. Invoking the ergodic assumption and choosing r so that some power P^r has all of its entries bounded below by a positive constant ϵ, it follows that

$$\Pr(T \leq r \mid X_0 = u, Y_0 = v) \geq \sum_k p_{uk}^{(r)} p_{vk}^{(r)} \geq \epsilon \sum_k p_{vk}^{(r)} = \epsilon.$$

Exactly the same reasoning demonstrates that every rth epoch the two chains have a chance of colliding of at least ϵ, regardless of their starting positions r epochs previous. In other words, inequality (7.8) holds.

Because $\Pr(T > n)$ is decreasing in n, we now harvest the bound

$$\Pr(T > n) \leq (1 - \epsilon)^{\lfloor \frac{n}{r} \rfloor}$$

from inequality (7.9). Combining this bound with the coupling inequality (7.7) yields

$$\frac{1}{2} \sum_j |p_{uj}^{(n)} - p_{vj}^{(n)}| = \|\pi_{X_n} - \pi_{Y_n}\|_{TV}$$
$$\leq \Pr(X_n \neq Y_n)$$
$$\leq \Pr(T > n) \tag{7.10}$$
$$\leq (1 - \epsilon)^{\lfloor \frac{n}{r} \rfloor}.$$

In view of the fact that u and v are arbitrary, this concludes the proof that the $\lim_{n \to \infty} p_{ij}^{(n)} = \pi_j$ exists. ∎

In the next example, we concoct a different kind of waiting time $T < \infty$ connected with a Markov chain X_n. If at epoch T the chain achieves its equilibrium distribution π and X_T is independent of T, then T is said to be a strong stationary time. When a strong stationary time exists, it gives considerable insight into the rate of convergence of the underlying chain [3, 38]. In view of the fact that X_n is at equilibrium when $T \leq n$, we readily deduce the total variation bound

$$\|\pi_{X_n} - \pi\|_{TV} \leq \Pr(T > n). \tag{7.11}$$

Example 7.4.3 *A Strong Stationary Time for Top-in Shuffling*

In top-in card shuffling, let $T - 1$ be the epoch at which the original bottom card reaches the top of the deck. At epoch T the bottom card is reinserted, and the deck achieves equilibrium. For our purposes it is fruitful to view T as the sum $T = S_1 + S_2 + \cdots + S_m$ of independent geometrically distributed random variables. If $T_0 = 0$ and T_i is the first epoch at which i total cards occur under the bottom card, then the waiting time $S_i = T_i - T_{i-1}$ is geometrically distributed with success probability $\frac{i}{m}$. Because $E(S_i) = \frac{m}{i}$, the strong stationary time T has mean

$$E(T) = \sum_{i=1}^{m} \frac{m}{i} \approx m \ln m.$$

To exploit the bound (7.11), we consider a random variable T^* having the same distribution as T but generated by a different mechanism. Suppose we randomly drop balls into m equally likely boxes. If we let T^* be the first trial at which no box is empty, then it is clear that we can decompose $T^* = S_m^* + S_{m-1}^* + \cdots + S_1^*$, where S_i^* is the number of trials necessary to go from i empty boxes to $i-1$ empty boxes. Once again the S_i^* are independent and geometrically distributed. This perspective makes it simple to bound $\Pr(T > n)$. Indeed, if A_i is the event that box i is empty after n trials, then

$$
\begin{aligned}
\Pr(T > n) &= \Pr(T^* > n) \\
&\leq \sum_{i=1}^{m} \Pr(A_i) \\
&= m\left(1 - \frac{1}{m}\right)^n \\
&\leq me^{-\frac{n}{m}}.
\end{aligned}
\tag{7.12}
$$

Here we invoke the inequality $\ln(1 - x) \leq -x$ for $x \in [0, 1)$.

Returning to the top-in shuffle problem, we now combine inequality (7.12) with inequality (7.11). If $n = (1 + \epsilon)m \ln m$, we deduce that

$$\|\pi_{X_n} - \pi\|_{TV} \leq me^{-\frac{(1+\epsilon)m \ln m}{m}} = \frac{1}{m^\epsilon},$$

where π is the uniform distribution on permutations of $\{1, \ldots, m\}$. This shows that if we wait much beyond the mean of T, then top-in shuffling will completely randomize the deck. Hence, the mean $E(T) \approx m \ln m$ serves as a fairly sharp cutoff for equilibrium. Some statistical applications of these ideas appear in reference [104].   ■

7.5 Hitting Probabilities and Hitting Times

Consider a Markov chain X_k with state space $\{1,\ldots,n\}$ and transition matrix $P = (p_{ij})$. Suppose that we can divide the states into a transient set $B = \{1,\ldots,m\}$ and an absorbing set $A = \{m+1,\ldots,n\}$ such that $p_{ij} = 0$ for every $i \in A$ and $j \in B$ and such that every $i \in B$ leads to at least one $j \in A$. Then every realization of the chain starting in B is eventually trapped in A. It is often of interest to find the probability h_{ij} that the chain started at $i \in B$ enters A at $j \in A$. The $m \times (n-m)$ matrix of hitting probabilities $H = (h_{ij})$ can be found by solving the system of equations

$$h_{ij} \;=\; p_{ij} + \sum_{k=1}^{m} p_{ik} h_{kj}$$

derived by conditioning on the next state visited by the chain starting from state i. We can summarize this system as the matrix equation $H = R + QH$ by decomposing P into the block matrix

$$P \;=\; \begin{pmatrix} Q & R \\ 0 & S \end{pmatrix},$$

where Q is $m \times m$, R is $m \times (n-m)$, and S is $(n-m) \times (n-m)$. If I is the $m \times m$ identity matrix, then the formal solution of our system of equations is $H = (I - Q)^{-1}R$. To prove that the indicated matrix inverse exists, we turn to a simple proposition.

Proposition 7.5.1 *Suppose that $\lim_{k \to \infty} Q^k = \mathbf{0}$, where $\mathbf{0}$ is the $m \times m$ zero matrix. Then $(I - Q)^{-1}$ exists and equals $\lim_{l \to \infty} \sum_{k=0}^{l} Q^k$.*

Proof: By assumption $\lim_{k \to \infty}(I - Q^k) = I$. Because the determinant function is continuous and $\det I = 1$, it follows that $\det(I - Q^k) \neq 0$ for k sufficiently large. Taking determinants in the identity

$$(I - Q)(I + Q + \cdots + Q^{k-1}) \;=\; I - Q^k$$

yields

$$\det(I - Q)\det(I + Q + \cdots + Q^{k-1}) \;=\; \det(I - Q^k).$$

Thus, $\det(I - Q) \neq 0$, and $I - Q$ is nonsingular. Finally, the power series expansion for $I - Q$ follows from taking limits in

$$I + Q + \cdots + Q^{k-1} \;=\; (I - Q)^{-1}(I - Q^k).$$

To apply Proposition 7.5.1, we need to interpret the entries of the matrix $Q^k = (q_{ij}^{(k)})$. A moment's reflection shows that

$$q_{ij}^{(k)} = \sum_{l=1}^{m} q_{il}^{(k-1)} p_{lj}$$

is just the probability that the chain passes from i to j in k steps. Note that the sum defining $q_{ij}^{(k)}$ stops at $l = m$ because once the chain leaves the transient states, it can not reenter them. This fact also makes it intuitively obvious that $\lim_{k\to\infty} q_{ij}^{(k)} = 0$. To verify this limit, it suffices to prove that the chain leaves the transient states after a finite number of steps. Suppose on the contrary that the chain wanders from transient state to transient state forever. In this case, the chain visits some transient state i an infinite number of times. However, i leads to an absorbing state $j > m$ along a path of positive probability. One of these visits to i must successfully take the path to j. This argument can be tightened by defining a first passage time T to the transient states and invoking inequalities (7.8) and (7.9).

In much the same way that we calculate hitting probabilities, we can calculate the mean number of epochs t_{ij} that the chain spends in transient state j prior to absorption starting from transient state i. These expectations satisfy the system of equations

$$t_{ij} = 1_{\{j=i\}} + \sum_{k=1}^{m} p_{ik} t_{kj},$$

which reads as $T = I + QT$ in matrix form. The solution $T = (I - Q)^{-1}$ can be used to write the mean hitting time vector t with ith entry $t_i = \sum_j t_{ij}$ as $t = (I - Q)^{-1}\mathbf{1}$, where $\mathbf{1}$ is the vector with all entries 1. Finally, if f_{ij} is the probability of ever reaching transient state j starting from transient state i, we can rearrange the identity $t_{ij} = f_{ij} t_{jj}$ to yield the simple formula $f_{ij} = t_{ij}/t_{jj}$ for f_{ij}.

Example 7.5.1 *A Illness-Death Cancer Model*

Figure 7.1 depicts a naive Markov chain model for cancer morbidity and mortality [50]. The two transient states 1 (healthy) and 2 (cancerous) lead to the absorbing states 3 (death from other causes) and 4 (death from cancer). A brief calculation shows that

$$Q = \begin{pmatrix} p_{11} & p_{12} \\ p_{21} & p_{22} \end{pmatrix}, \qquad R = \begin{pmatrix} p_{13} & 0 \\ 0 & p_{24} \end{pmatrix},$$

and

$$(I - Q)^{-1} = \frac{1}{(1 - p_{11})(1 - p_{22}) - p_{12} p_{21}} \begin{pmatrix} 1 - p_{22} & p_{12} \\ p_{21} & 1 - p_{11} \end{pmatrix}.$$

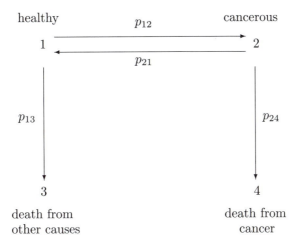

FIGURE 7.1. An Illness-Death Markov Chain

These are precisely the ingredients necessary to calculate the hitting probabilities $H = (I - Q)^{-1}R$ and mean hitting times $t = (I - Q)^{-1}\mathbf{1}$.   ■

7.6   Markov Chain Monte Carlo

The Markov chain Monte Carlo (MCMC) revolution sweeping statistics is drastically changing how statisticians perform integration and summation. In particular, the Metropolis algorithm and Gibbs sampling make it straightforward to construct a Markov chain that samples from a complicated conditional distribution. Once a sample is available, then according to the ergodic theorem, any conditional expectation can be approximated by forming its corresponding sample average. The implications of this insight are profound for both classical and Bayesian statistics. As a bonus, trivial changes to the Metropolis algorithm yield simulated annealing, a general-purpose algorithm for solving difficult combinatorial optimization problems.

Our limited goal in this section is to introduce a few of the major MCMC themes. One issue of paramount importance is how rapidly the underlying chains reach equilibrium. This is the Achilles heel of the whole business and not just a mathematical nicety. Unfortunately, probing this delicate issue is scarcely possible in the confines of a brief overview. We analyze one example to give a feel for the power of coupling arguments. Readers interested in pursuing MCMC methods and related method of simulated annealing further will enjoy the pioneering articles [55, 57, 69, 90, 113], the elementary surveys [27, 29], and the books [56, 59, 142].

7.6.1   The Hastings-Metropolis Algorithm

The Hastings-Metropolis algorithm is a device for constructing a Markov chain with a prescribed equilibrium distribution π on a given state space [69, 113]. Each step of the chain is broken into two stages, a proposal stage and an acceptance stage. If the chain is currently in state i, then in the proposal stage a new destination state j is proposed according to a probability density $q_{ij} = q(j \mid i)$. In the subsequent acceptance stage, a random number is drawn uniformly from $[0, 1]$ to determine whether the proposed step is actually taken. If this number is less than the Hastings-Metropolis acceptance probability

$$a_{ij} = \min\left\{\frac{\pi_j q_{ji}}{\pi_i q_{ij}}, 1\right\}, \tag{7.13}$$

then the proposed step is taken. Otherwise, the proposed step is declined, and the chain remains in place. Problem 19 indicates that equation (7.13) defines the most generous acceptance probability consistent with the given proposal mechanism.

Like most good ideas, the Hastings-Metropolis algorithm has undergone successive stages of abstraction and generalization. For instance, Metropolis et al. [113] considered only symmetric proposal densities with $q_{ij} = q_{ji}$. In this case the acceptance probability reduces to

$$a_{ij} = \min\left\{\frac{\pi_j}{\pi_i}, 1\right\}. \tag{7.14}$$

In this simpler setting it is clear that any proposed destination j with $\pi_j > \pi_i$ is automatically accepted. In applying either formula (7.13) or formula (7.14), it is noteworthy that the π_i need only be known up to a multiplicative constant.

To prove that π is the equilibrium distribution of the chain constructed from the Hastings-Metropolis scheme (7.13), it suffices to check that detailed balance holds. If π puts positive weight on all points of the state space, then we must require the inequalities $q_{ij} > 0$ and $q_{ji} > 0$ to be simultaneously true or simultaneously false if detailed balance is to have any chance of holding. Now suppose without loss of generality that the fraction

$$\frac{\pi_j q_{ji}}{\pi_i q_{ij}} \leq 1$$

for some $j \neq i$. Then detailed balance follows immediately from

$$\pi_i q_{ij} a_{ij} = \pi_i q_{ij} \frac{\pi_j q_{ji}}{\pi_i q_{ij}}$$

$$= \pi_j q_{ji}$$

$$= \pi_j q_{ji} a_{ji}.$$

Besides checking that π is the equilibrium distribution, we should also be concerned about whether the Hastings-Metropolis chain is irreducible and aperiodic. Aperiodicity is the rule because the acceptance-rejection step allows the chain to remain in place. Problem 20 states a precise result and a counterexample. Irreducibility holds provided the entries of π are positive and the proposal matrix $Q = (q_{ij})$ is irreducible.

Example 7.6.1 *Random Walk on a Subset of the Integers*

Random walk sampling occurs when the proposal density $q_{ij} = q_{j-i}$ for some density q_k. This construction requires that the sample space be closed under subtraction. If $q_k = q_{-k}$, then the Metropolis acceptance probability (7.14) applies.   ∎

Example 7.6.2 *Independence Sampler*

If the proposal density $q_{ij} = q_j$, then candidate points are drawn independently of the current point. To achieve quick convergence of the chain, q_i should be close to π_i for most i. Furthermore, the ratio q_i/π_i should not be allowed to become too small. Indeed, if it is exceptionally small for a given state i, then it is exceptionally hard to exit i.   ∎

7.6.2 Gibbs Sampling

The Gibbs sampler is a special case of the Hastings-Metropolis algorithm for Cartesian product state spaces [55, 57, 59]. Suppose that each sample point $i = (i_1, \ldots, i_m)$ has m components. The Gibbs sampler updates one component of i at a time. If the component is chosen randomly and resampled conditional on the remaining components, then the acceptance probability is 1. To prove this assertion, let i_c be the uniformly chosen component, and denote the remaining components by $i_{-c} = (i_1, \ldots, i_{c-1}, i_{c+1}, \ldots, i_m)$. If j is a neighbor of i reachable by changing only component i_c, then $j_{-c} = i_{-c}$. For such a neighbor j, the proposal probability

$$q_{ij} = \frac{1}{m} \cdot \frac{\pi_j}{\sum_{\{k:k_{-c}=i_{-c}\}} \pi_k}$$

satisfies $\pi_i q_{ij} = \pi_j q_{ji}$, and the ratio appearing in the acceptance probability (7.13) is 1.

In contrast to random sampling of components, we can repeatedly cycle through the components in some fixed order, say $1, 2, \ldots, m$. If the transition matrix for changing component c while leaving other components unaltered is $P^{(c)}$, then the transition matrices for random sampling and sequential (or cyclic) sampling are $R = \frac{1}{m} \sum_c P^{(c)}$ and $S = P^{(1)} \cdots P^{(m)}$, respectively. Because each $P^{(c)}$ satisfies $\pi P^{(c)} = \pi$, we have $\pi R = \pi$ and $\pi S = \pi$ as well. Thus, π is the unique equilibrium distribution for R or S if either is irreducible. However as pointed out in Problem 21, R satisfies detailed balance while S ordinarily does not.

Example 7.6.3 *Ising Model*

Consider m elementary particles equally spaced around the boundary of the unit circle. Each particle c can be in one of two magnetic states—spin up with $i_c = 1$ or spin down with $i_c = -1$. The Gibbs distribution

$$\pi_i \;\propto\; e^{\beta \sum_d i_d i_{d+1}} \tag{7.15}$$

takes into account nearest-neighbor interactions in the sense that states like $(1,1,1,\ldots,1,1,1)$ are favored and states like $(1,-1,1,\ldots,1,-1,1)$ are shunned for $\beta > 0$. (Note that in equation (7.15) the index $m+1$ of i_{m+1} is interpreted as 1.) Specification of the normalizing constant (or partition function)

$$Z \;=\; \sum_i e^{\beta \sum_d i_d i_{d+1}}$$

is unnecessary to carry out Gibbs sampling. If we elect to resample component c, then the choices $j_c = -i_c$ and $j_c = i_c$ are made with respective probabilities

$$\frac{e^{\beta(-i_{c-1}i_c - i_c i_{c+1})}}{e^{\beta(i_{c-1}i_c + i_c i_{c+1})} + e^{\beta(-i_{c-1}i_c - i_c i_{c+1})}} \;=\; \frac{1}{e^{2\beta(i_{c-1}i_c + i_c i_{c+1})} + 1}$$

$$\frac{e^{\beta(i_{c-1}i_c + i_c i_{c+1})}}{e^{\beta(i_{c-1}i_c + i_c i_{c+1})} + e^{\beta(-i_{c-1}i_c - i_c i_{c+1})}} \;=\; \frac{1}{1 + e^{-2\beta(i_{c-1}i_c + i_c i_{c+1})}}.$$

When the number of particles m is even, the odd-numbered particles are independent given the even-numbered particles, and vice versa. This fact suggests alternating between resampling all odd-numbered particles and resampling all even-numbered particles. Such multi-particle updates take longer to execute but create more radical rearrangements than single-particle updates. ∎

7.6.3   *Convergence of the Independence Sampler*

For the independence sampler, it is possible to give a coupling bound on the rate of convergence to equilibrium [108]. Suppose that $X_0, X_1, \ldots$ represents the sequence of states visited by the independence sampler starting from $X_0 = x_0$. We couple this Markov chain to a second independence sampler $Y_0, Y_1, \ldots$ starting from the equilibrium distribution π. By definition, each Y_k has distribution π. The two chains are coupled by a common proposal stage and a common uniform deviate U sampled in deciding whether to accept the common proposed point. They differ in having different acceptance probabilities. If $X_n = Y_n$ for some n, then $X_k = Y_k$ for all $k \geq n$. Let T denote the random epoch when X_n first meets Y_n and the X chain attains equilibrium.

The importance ratios $w_j = \pi_j/q_j$ determine what proposed points are accepted. Without loss of generality, assume that the states of the chain are numbered $1,\ldots,m$ and that $w_1 \geq w_j$ for all j. If $X_n = x \neq y = Y_n$, then according to equation (7.13) the next proposed point is accepted by both chains with probability

$$\sum_{j=1}^{m} q_j \min\left\{\frac{w_j}{w_x},\frac{w_j}{w_y},1\right\} \;=\; \sum_{j=1}^{m} \pi_j \min\left\{\frac{1}{w_x},\frac{1}{w_y},\frac{1}{w_j}\right\}$$

$$\geq \;\frac{1}{w_1}.$$

In other words, at each trial the two chains meet with at least probability $1/w_1$. This translates into the tail probability bound

$$\Pr(T > n) \;\leq\; \left(1 - \frac{1}{w_1}\right)^n. \tag{7.16}$$

By the same type of reasoning that led to inequality (7.10), we deduce the further bound

$$\begin{aligned}
\|\pi_{X_n} - \pi\|_{TV} &\leq \Pr(X_n \neq Y_n) \\
&= \Pr(T > n) \\
&\leq \left(1 - \frac{1}{w_1}\right)^n
\end{aligned}$$

on the total variation distance of X_n from equilibrium.

7.7 Simulated Annealing

In simulated annealing we are interested in finding the most probable state of a Markov chain [90, 123]. If this state is k, then $\pi_k > \pi_i$ for all $i \neq k$. To accentuate the weight given to state k, we can replace the equilibrium distribution π by a distribution putting probability

$$\pi_i^{(\tau)} \;=\; \frac{\pi_i^{1/\tau}}{\sum_j \pi_j^{1/\tau}}$$

on state i. Here τ is a positive parameter traditionally called temperature. With a symmetric proposal density, the distribution $\pi_i^{(\tau)}$ can be attained by running a chain with Metropolis acceptance probability

$$a_{ij} \;=\; \min\left\{\left(\frac{\pi_j}{\pi_i}\right)^{1/\tau}, 1\right\}. \tag{7.17}$$

In simulated annealing, the chain is run with τ gradually decreasing to 0 rather than with τ fixed. If τ starts out large, then in the early steps of simulated annealing, almost all proposed steps are accepted, and the chain broadly samples the state space. As τ declines, fewer unfavorable steps are taken, and the chain eventually settles on some nearly optimal state. With luck, this state is k or a state equivalent to k if several states are optimal. Simulated annealing is designed to mimic the gradual freezing of a substance into a crystalline state of perfect symmetry and hence minimum energy.

Example 7.7.1 *The Traveling Salesman Problem*

As discussed in Example 5.7.2, a salesman must visit m towns, starting and ending in his hometown. Given fixed distances d_{ij} between every pair of towns i and j, in what order should he visit the towns to minimize the length of his circuit? This problem belongs to the class of NP-complete problems; these have deterministic solutions that are conjectured to increase in complexity at an exponential rate in m.

In the simulated annealing approach to the traveling salesman problem, we assign to each permutation $\sigma = (\sigma_1, \ldots, \sigma_m)$ a cost $c_\sigma = \sum_{i=1}^{m} d_{\sigma_i, \sigma_{i+1}}$, where $\sigma_{m+1} = \sigma_1$. Defining $\pi_\sigma \propto e^{-c_\sigma}$ turns the problem of minimizing the cost into one of finding the most probable permutation σ. In the proposal stage of simulated annealing, we randomly select two indices $i \neq j$ and reverse the block of integers beginning at σ_i and ending at σ_j in the current permutation $(\sigma_1, \ldots, \sigma_m)$. This proposal is accepted with probability (7.17) depending on the temperature τ. In *Numerical Recipes'* [123] simulated annealing algorithm for the traveling salesman problem, τ is lowered in multiplicative decrements of 10% after every $100m$ epochs or every $10m$ accepted steps, whichever comes first.   ∎

7.8   Problems

1. Demonstrate that a finite-state Markov chain is ergodic (irreducible and aperiodic) if and only if some power P^n of the transition matrix P has all entries positive. (Hints: For sufficiency, show that if some power P^n has all entries positive, then P^{n+1} has all entries positive. For necessity, note that $p_{ij}^{(r+s+t)} \geq p_{ik}^{(r)} p_{kk}^{(s)} p_{kj}^{(t)}$, and use the number theoretic fact that the set $\{s : p_{kk}^{(s)} > 0\}$ contains all sufficiently large positive integers s if k is aperiodic. See Appendix A of reference [137] or the appendix of reference [19] for the requisite number theory.)

2. Show that Kolmogorov's criterion (7.3) implies that definition (7.4) does not depend on the particular path chosen from i to j.

3. In the Bernoulli-Laplace model, we imagine two boxes with m particles each. Among the $2m$ particles there are b black particles and

w white particles, where $b + w = 2m$ and $b \leq w$. At each epoch one particle is randomly selected from each box, and the two particles are exchanged. Let Z_n be the number of black particles in the first box. Is the corresponding chain irreducible, aperiodic, or reversible? Show that its equilibrium distribution is hypergeometric.

4. In Example 7.3.1, show that the chain is aperiodic if and only if the underlying graph is not bipartite.

5. Consider the $n!$ different permutations $\sigma = (\sigma_1, \ldots, \sigma_n)$ of the set $\{1, \ldots, n\}$ equipped with the uniform distribution $\pi_\sigma = 1/n!$ [38]. Declare a permutation w to be a neighbor of σ if there exist two indices $i \neq j$ such that $w_i = \sigma_j$, $w_j = \sigma_i$, and $w_k = \sigma_k$ for $k \notin \{i, j\}$. How many neighbors does a permutation σ possess? Show how the set of permutations can be made into a reversible Markov chain using the construction of Example 7.3.1. Is the underlying graph bipartite? If we execute one step of the chain by randomly choosing two indices i and j and switching σ_i and σ_j, how can we slightly modify the chain so that it is aperiodic?

6. Let P be the transition matrix and π the equilibrium distribution of a reversible Markov chain with n states. Define an inner product $\langle u, v \rangle_\pi$ on complex column vectors u and v with n components by

$$\langle u, v \rangle_\pi = \sum_i u_i \pi_i v_i^*.$$

Verify that P satisfies the self-adjointness condition

$$\langle Pu, v \rangle_\pi = \langle u, Pv \rangle_\pi,$$

and conclude by standard arguments that P has only real eigenvalues.

7. In Example 7.4.1, suppose that $f(x)$ is strictly increasing and $g(x)$ is increasing. Show that $\mathrm{Cov}[f(X), g(X)] = 0$ occurs if and only if $\Pr[g(X) = c] = 1$ for some constant c. (Hint: For necessity, examine the proof of the example and show that $\mathrm{Cov}[f(X), g(X)] = 0$ entails $\Pr[g(X) = g(Y)] = 1$ and therefore $\mathrm{Var}[g(X) - g(Y)] = 0$.)

8. Suppose that X follows the hypergeometric distribution

$$\Pr(X = i) = \frac{\binom{r}{i}\binom{n-r}{m-i}}{\binom{n}{m}}.$$

Let Y follow the same hypergeometric distribution except that $r + 1$ replaces r. Give a coupling proof that $\Pr(X \geq k) \leq \Pr(Y \geq k)$ for all k. (Hint: Consider an urn with r red balls, 1 white ball, and $n - r - 1$ black balls. If we draw m balls from the urn without replacement, then X is the number of red balls drawn, and Y is the number of red or white balls drawn.)

9. Let X be a binomially distributed random variable with n trials and success probability p. Show by a coupling argument that $\Pr(X \geq k)$ is increasing in n for fixed p and k and in p for fixed n and k.

10. Let Y be a Poisson random variable with mean λ. Demonstrate that $\Pr(Y \geq k)$ is increasing in λ for k fixed. (Hint: If $\lambda_1 < \lambda_2$, then construct coupled Poisson random variables Y_1 and Y_2 with means λ_1 and λ_2 such that $Y_1 \leq Y_2$.)

11. Consider a random graph with n nodes. Between every pair of nodes, independently introduce an edge with probability p. If $c(p)$ denotes the probability that the graph is connected, then it is intuitively clear that $c(p)$ is increasing in p. Give a coupling proof of this fact.

12. Consider a random walk on the integers $0, \ldots, m$ with transition probabilities

$$
p_{ij} \;=\; \begin{cases} q_i & j = i - 1 \\ 1 - q_i & j = i + 1 \end{cases}
$$

for $i = 1, \ldots, m - 1$ and $p_{00} = p_{mm} = 1$. All other transition probabilities are 0. Eventually the walk gets trapped at 0 or m. Let f_i be the probability that the walk is absorbed at 0 starting from i. Show that f_i is an increasing function of the entries of $q = (q_1, \ldots, q_{m-1})$. (Hint: Let q and q^* satisfy $q_i \leq q_i^*$ for $i = 1, \ldots, m - 1$. Construct coupled walks X_n and Y_n based on q and q^* such that $X_0 = Y_0 = i$ and such that at the first step $Y_1 \leq X_1$. This requires coordinating the first step of each chain. If $X_1 > Y_1$, then run the X_n chain until it reaches either m or Y_1. In the latter case, take another coordinated step of the two chains.)

13. Show that the two definitions of the total variation norm given in (7.6) coincide.

14. Let X have a Bernoulli distribution with success probability p and Y a Poisson distribution with mean p. Prove the total variation inequality

$$
\| \pi_X - \pi_Y \| \;\leq\; p^2 \tag{7.18}
$$

involving the distributions π_X and π_Y of X and Y.

15. Suppose the integer-valued random variables U_1, U_2, V_1, and V_2 are such that U_1 and U_2 are independent and V_1 and V_2 are independent. Demonstrate that

$$
\| \pi_{U_1 + U_2} - \pi_{V_1 + V_2} \| \;\leq\; \| \pi_{U_1} - \pi_{V_1} \| + \| \pi_{U_2} - \pi_{V_2} \|. \tag{7.19}
$$

16. Let $Z_0, Z_1, Z_2, \ldots$ be a realization of a finite-state ergodic chain. If we sample every kth epoch, then show (a) that the sampled chain $Z_0, Z_k, Z_{2k}, \ldots$ is ergodic, (b) that it possesses the same equilibrium distribution as the original chain, and (c) that it is reversible if the original chain is. Thus, based on the ergodic theorem, we can estimate theoretical means by sample averages using only every kth epoch of the original chain.

17. Take three numbers x_1, x_2, and x_3 and form the successive running averages $x_n = (x_{n-3} + x_{n-2} + x_{n-1})/3$ starting with x_4. Prove that

$$\lim_{n \to \infty} x_n = \frac{x_1 + 2x_2 + 3x_3}{6}.$$

18. Consider a random walk on the integers $\{0, 1, \ldots, n\}$. States 0 and n are absorbing in the sense that $p_{00} = p_{nn} = 1$. If i is a transient state, then the transition probabilities $p_{i,i+1} = p$ and $p_{i,i-1} = q$, where $p + q = 1$. Verify that the hitting probabilities are

$$h_{in} = \begin{cases} \frac{(\frac{q}{p})^i - 1}{(\frac{q}{p})^n - 1}, & p \neq q \\ \frac{i}{n}, & p = q \end{cases}$$

and the mean hitting times are

$$t_i = \begin{cases} \frac{n}{p-q} \frac{(\frac{q}{p})^i - 1}{(\frac{q}{p})^n - 1} - \frac{i}{p-q}, & p \neq q \\ i(n - i), & p = q. \end{cases}$$

(Hint: First argue that

$$t_i = 1 + \sum_{k=1}^{m} p_{ik} t_k$$

in the notation of Section 7.5.)

19. An acceptance function $a : (0, \infty) \mapsto [0, 1]$ satisfies the functional identity $a(x) = xa(1/x)$. Prove that the detailed balance condition

$$\pi_i q_{ij} a_{ij} = \pi_j q_{ji} a_{ji}$$

holds if the acceptance probability a_{ij} is defined by

$$a_{ij} = a\left(\frac{\pi_j q_{ji}}{\pi_i q_{ij}}\right)$$

in terms of an acceptance function $a(x)$. Check that the Barker function $a(x) = x/(1 + x)$ qualifies as an acceptance function and that any acceptance function is dominated by the Metropolis acceptance function in the sense that $a(x) \leq \min\{x, 1\}$ for all x.

20. The Metropolis acceptance mechanism (7.14) ordinarily implies aperiodicity of the underlying Markov chain. Show that if the proposal distribution is symmetric and if some state i has a neighboring state j such that $\pi_i > \pi_j$, then the period of state i is 1, and the chain, if irreducible, is aperiodic. For a counterexample, assign probability $\pi_i = 1/4$ to each vertex i of a square. If the two vertices adjacent to a given vertex i are each proposed with probability $1/2$, then show that all proposed steps are accepted by the Metropolis criterion and that the chain is periodic with period 2.

21. Consider the Cartesian product space $\{0, 1\} \times \{0, 1\}$ equipped with the probability distribution

$$(\pi_{00}, \pi_{01}, \pi_{10}, \pi_{11}) \;=\; \left(\frac{1}{2}, \frac{1}{4}, \frac{1}{8}, \frac{1}{8}\right).$$

Demonstrate that sequential Gibbs sampling does not satisfy detailed balance by showing that $\pi_{00} s_{00,11} \neq \pi_{11} s_{11,00}$, where $s_{00,11}$ and $s_{11,00}$ are entries of the matrix S for first resampling component one and then resampling component two.

22. It is known that every planar graph can be colored by four colors [25]. Design, program, and test a simulated annealing algorithm to find a four coloring of any planar graph. (Suggestions: Represent the graph by a list of nodes and a list of edges. Assign to each node a color represented by a number between 1 and 4. The cost of a coloring is the number of edges with incident nodes of the same color. In the proposal stage of the simulated annealing solution, randomly choose a node, randomly reassign its color, and recalculate the cost. If successful, simulated annealing will find a coloring with the minimum cost of 0.)

8
Continuous-Time Markov Chains

8.1 Introduction

This chapter introduces the subject of continuous-time Markov chains [18, 41, 48, 64, 82, 83, 93, 117]. In practice, continuous-time chains are more useful than discrete-time chains. For one thing, continuous-time chains permit variation in the waiting times for transitions between neighboring states. For another, they avoid the annoyances of periodic behavior. Balanced against these advantages is the disadvantage of a more complex theory involving linear differential equations. The primary distinction between the two types of chain is the substitution of transition intensities for transition probabilities. Once one grasps this difference, it is straightforward to formulate relevant continuous-time models. Implementing such models numerically and understanding them theoretically then requires the matrix exponential function. The chapter concludes with a discussion of Kendall's birth-death-immigration process, which involves an infinite number of states and transition intensities that depend on time.

8.2 Finite-Time Transition Probabilities

Just as with a discrete-time chain, the behavior of a continuous-time chain is described by an indexed family Z_t of random variables giving the state occupied by the chain at each time t. Now, however, the index t ranges over the nonnegative real numbers rather than the nonnegative integers. Of fundamental theoretical importance are the finite-time transition probabilities $p_{ij}(t) = \Pr(Z_{s+t} = j \mid Z_s = i)$ for all $s, t \geq 0$. We shall see momentarily how these probabilities can be found by solving a matrix differential equation.

The perspective of competing risks sharpens our intuitive understanding of how a continuous-time chain operates. Imagine that a particle executes a Markov chain by moving from state to state. If the particle is currently in state i, then each neighboring state independently beckons the particle to switch positions. The intensity of the temptation exerted by state j is the constant λ_{ij}. In the absence of competing temptations, the particle waits an exponential length of time T_{ij} with intensity λ_{ij} before moving to state j. Taking into account competing independent temptations, the particle moves at the moment $T_i = \min_{j \neq i} T_{ij}$, which is exponentially distributed with intensity $\lambda_i = \sum_{j \neq i} \lambda_{ij}$. Of course, exponentially distributed waiting times are inevitable in a Markovian model; otherwise, the intensity of leaving state i would depend on the past history of waiting in i. Once the particle decides to leave i, it moves to j with probability $q_{ij} = \lambda_{ij}/\lambda_i$.

An important consequence of these assumptions is that the destination state D_i is chosen independently of the waiting time T_i. Indeed, conditioning on the value of T_{ik} gives

$$
\begin{aligned}
\Pr(D_i = k, T_i \geq t) &= \Pr(T_{ik} \geq t, T_{ij} > T_{ik} \text{ for } j \neq k) \\
&= \int_t^\infty \lambda_{ik} e^{-\lambda_{ik} s} \Pr(T_{ij} > T_{ik} \text{ for } j \neq k \mid T_{ik} = s) \, ds \\
&= \int_t^\infty \lambda_{ik} e^{-\lambda_{ik} s} \prod_{j \notin \{i,k\}} e^{-\lambda_{ij} s} ds \\
&= \int_t^\infty \lambda_{ik} e^{-\lambda_i s} ds \\
&= q_{ik} e^{-\lambda_i t} \\
&= \Pr(D_i = k) \Pr(T_i \geq t).
\end{aligned}
$$

Not only does this calculation establish the independence of D_i and T_i, but it also validates their claimed marginal distributions. If we ignore the times at which transitions occur, the sequence of transitions in a continuous-time chain determines a discrete-time chain with transition probability matrix $Q = (q_{ij})$.

At this juncture, it is helpful to pause and consider the nature of the finite-time transition matrix $P(t) = [p_{ij}(t)]$ for small times $t > 0$. Suppose the chain starts in state i at time 0. Because it will be in state i at time t if it never leaves in the interim, we have

$$
p_{ii}(t) \geq e^{-\lambda_i t} = 1 - \lambda_i t + o(t).
$$

The chain can reach a destination state $j \neq i$ if it makes a one-step transition to j sometime during $[0, t]$ and remains there for the duration of the interval. Given that $\lambda_i q_{ij} = \lambda_{ij}$, this observation leads to the inequality

$$p_{ij}(t) \geq (1 - e^{-\lambda_i t})q_{ij}e^{-\lambda_j t} = \lambda_{ij}t + o(t).$$

If there are only a finite number of states, the sum of these inequalities satisfies

$$1 = \sum_j p_{ij}(t) \geq 1 + o(t),$$

and therefore equality must hold in each of the participating inequalities to order $o(t)$. In view of the approximations embodied in the formula $p_{ij}(t) = \lambda_{ij}t + o(t)$, the transition intensities λ_{ij} are also termed infinitesimal transition probabilities.

8.3   Derivation of the Backward Equations

We now show that the finite-time transition probabilities $p_{ij}(t)$ satisfy a system of ordinary differential equations called the backward equations. The integral form of this system amounts to

$$p_{ij}(t) = 1_{\{j=i\}}e^{-\lambda_i t} + \int_0^t \lambda_i e^{-\lambda_i s} \sum_{k \neq i} q_{ik}p_{kj}(t-s)\,ds. \qquad (8.1)$$

The first term on the right of equation (8.1) represents the probability that a particle initially in state i remains there throughout the period $[0, t]$. Of course, this is only possible when $j = i$. The integral contribution on the right of equation (8.1) involves conditioning on the time s of the first departure from state i. If state k is chosen as the destination for this departure, then the particle ends up in state j at time t with probability $p_{kj}(t-s)$.

Multiplying equation (8.1) by $e^{\lambda_i t}$ yields

$$e^{\lambda_i t}p_{ij}(t) = 1_{\{j=i\}} + \int_0^t \lambda_i e^{\lambda_i(t-s)} \sum_{k \neq i} q_{ik}p_{kj}(t-s)\,ds$$

$$= 1_{\{j=i\}} + \int_0^t \lambda_i e^{\lambda_i s} \sum_{k \neq i} q_{ik}p_{kj}(s)\,ds \qquad (8.2)$$

after an obvious change of variables. Because all of the finite-time transition probabilities satisfy $|p_{ik}(s)| \leq 1$ and all rows of the matrix Q satisfy $\sum_{k \neq i} q_{ik} = 1$, the integrand $\lambda_i e^{\lambda_i s} \sum_{k \neq i} q_{ik}p_{kj}(s)$ is bounded on every finite interval. It follows that both its integral and $p_{ij}(t)$ are continuous in t. Given continuity of the $p_{ij}(t)$, the integrand $\lambda_i e^{\lambda_i s} \sum_{k \neq i} q_{ik}p_{kj}(s)$ is continuous, being the limit of a uniformly converging series of continuous functions. The fundamental theorem of calculus therefore implies that $p_{ij}(t)$ is differentiable. Taking derivatives in equation (8.2) produces

$$\lambda_i e^{\lambda_i t} p_{ij}(t) + e^{\lambda_i t} p'_{ij}(t) \;=\; \lambda_i e^{\lambda_i t} \sum_{k \neq i} q_{ik} p_{kj}(t).$$

After straightforward rearrangement using $\lambda_{ii} = -\lambda_i$, we arrive at the differential form

$$p'_{ij}(t) \;=\; \sum_k \lambda_{ik} p_{kj}(t) \tag{8.3}$$

of the backward equations.

For a chain with a finite number of states, the backward equation (8.3) can be summarized in matrix notation by introducing the two matrices $P(t) = [p_{ij}(t)]$ and $\Lambda = (\lambda_{ij})$. The backward equations in this notation become

$$\begin{aligned} P'(t) &= \Lambda P(t) \tag{8.4} \\ P(0) &= I, \end{aligned}$$

where I is the identity matrix. The solution of the initial value problem (8.4) is furnished by the matrix exponential [74, 93]

$$\begin{aligned} P(t) &= e^{t\Lambda} \\ &= \sum_{k=0}^{\infty} \frac{1}{k!} (t\Lambda)^k. \tag{8.5} \end{aligned}$$

One can check this fact formally by differentiating the series expansion (8.5) term by term. Probabilists call Λ the infinitesimal generator or infinitesimal transition matrix of the process. Because $\lambda_{ii} = -\sum_{j \neq i} \lambda_{ij}$, all row sums of Λ are 0, and the column vector $\mathbf{1}$ is an eigenvector of Λ with eigenvalue 0.

8.4   Equilibrium Distributions and Reversibility

A probability distribution $\pi = (\pi_i)$ on the states of a continuous-time Markov chain is a row vector whose components satisfy $\pi_i \geq 0$ for all i and $\sum_i \pi_i = 1$. If

$$\pi P(t) = \pi \tag{8.6}$$

holds for all $t \geq 0$, then π is said to be an equilibrium distribution for the chain. Written in components, the eigenvector equation (8.6) reduces to $\sum_i \pi_i p_{ij}(t) = \pi_j$. For small t and a finite number of states, the series expansion (8.5) implies that equation (8.6) can be rewritten as

$$\pi(I + t\Lambda) + o(t) = \pi.$$

This approximate form makes it obvious that $\pi\Lambda = \mathbf{0}$ is a necessary condition for π to be an equilibrium distribution. Here $\mathbf{0}$ is a row vector of zeros. Multiplying equation (8.5) on the left by π shows that $\pi\Lambda = \mathbf{0}$ is also a sufficient condition for π to be an equilibrium distribution. In components this necessary and sufficient condition is equivalent to the balance equation

$$\sum_{j\neq i} \pi_j \lambda_{ji} = \pi_i \sum_{j\neq i} \lambda_{ij} \tag{8.7}$$

for all i. If all of the states of a finite-state Markov chain communicate, then the chain is said to be irreducible, and there is a unique equilibrium distribution π. Moreover, each row of $P(t)$ approaches π as t tends to ∞.

One particularly simple method of proving convergence to the equilibrium distribution is to construct a Liapunov function. A Liapunov function steadily declines along a trajectory of the chain until reaching its minimum at the equilibrium distribution. Let $q(t) = [q_j(t)]$ denote the distribution of the chain at time t. The relative information

$$H(t) = \sum_k q_k(t) \ln \frac{q_k(t)}{\pi_k} = \sum_k \pi_k \frac{q_k(t)}{\pi_k} \ln \frac{q_k(t)}{\pi_k}$$

furnishes one Liapunov function exploiting the strict convexity of the function $h(u) = u \ln u$ [86]. Proof that $H(t)$ is a Liapunov function hinges on the Chapman-Kolmorgorov relation

$$q_k(t+d) = \sum_j q_j(t) p_{jk}(d)$$

for t and d nonnegative. This equation simply says that the process must pass through some intermediate state j at time t enroute to state k at time $t+d$.

We now fix d and define $\alpha_{kj} = \pi_j p_{jk}(d)/\pi_k$. Because each α_{jk} is nonnegative and $\sum_j \alpha_{kj} = 1$, we deduce that

$$H(t+d) = \sum_k \pi_k h\left[\frac{q_k(t+d)}{\pi_k}\right]$$

$$= \sum_k \pi_k h\left[\frac{\sum_j q_j(t) p_{jk}(d)}{\pi_k}\right]$$

$$= \sum_k \pi_k h\left[\sum_j \alpha_{kj} \frac{q_j(t)}{\pi_j}\right]$$

$$\leq \sum_j \sum_k \pi_k \alpha_{kj} h \left[\frac{q_j(t)}{\pi_j} \right] \qquad (8.8)$$

$$= \sum_j \sum_k \pi_j p_{jk}(t) h \left[\frac{q_j(t)}{\pi_j} \right]$$

$$= H(t),$$

with strict inequality unless $q(t) = \pi$. Note here the implicit assumption that all entries of π are positive. This holds because all states communicate.

To prove that $\lim_{t \to \infty} q(t) = \pi$, we demonstrate that all limit points of the trajectory $q(t)$ coincide with π. Certainly at least one limit point exists when the number of states is finite because then $q(t)$ belongs to a compact (closed and bounded) set. Furthermore, $H(t)$ monotonely decreases to a finite limit c. If $\lim_{n \to \infty} q(t_n) = \nu$, then the equality

$$q_k(t_n + d) = \sum_j q_j(t_n) p_{jk}(d)$$

implies that $\lim_{n \to \infty} q(t_n + d) = \omega$ exists as well. We now take limits in inequality (8.8) along the sequence t_n. In view of the continuity of $H(t)$ and its convergence to c, this gives

$$c = \sum_k \pi_k h \left(\frac{\omega_k}{\pi_k} \right) \leq \sum_k \pi_k h \left(\frac{\nu_k}{\pi_k} \right) = c. \qquad (8.9)$$

Rederivation of inequality (8.8) with ω_k substituting for $q_k(t + d)$ and ν_j substituting for $q_j(t)$ shows that strict inequality holds in (8.9) whenever $\nu \neq \pi$, contradicting the evident equality throughout. It follows that $\nu = \pi$ and that π is the limit of $q(t)$. For other proofs of convergence to the equilibrium distribution, see the references [93, 117].

Fortunately as pointed out in Problem 4, the annoying feature of periodicity present in the discrete-time theory disappears in the continuous-time theory. The definition and properties of reversible chains carry over directly from discrete time to continuous time provided we substitute transition intensities for transition probabilities [86]. For instance, the detailed balance condition becomes

$$\pi_i \lambda_{ij} = \pi_j \lambda_{ji} \qquad (8.10)$$

for all pairs $i \neq j$. Kolmogorov's circulation criterion for reversibility continues to hold. When it is true, the equilibrium distribution is constructed from the transition intensities exactly as in discrete time, substituting transition intensities for transition probabilities.

It helps to interpret equations (8.7) and (8.10) as probabilistic flows. Imagine a vast number of independent particles executing the same Markov chain. If we station ourselves at some state i, particles are constantly entering and leaving the state. Viewed from a distance, this particle swarm

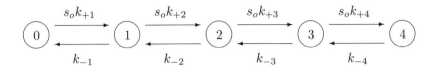

FIGURE 8.1. A Markov Chain Model for Oxygen Attachment to Hemoglobin

looks like a fluid flowing into and out of i. If we let the ensemble evolve, then it eventually reaches an equilibrium where the flows into and out of i match. Equation (8.7) is the quantification of this balance. With a reversible process, the flow from state i to state j must eventually match the flow from j to i; otherwise, the process reversed in time could be distinguished from the original process. Equation (8.10) is the quantification of this detailed balance.

8.5 Examples

Here are a few examples of continuous-time Markov chains.

Example 8.5.1 *Oxygen Attachment to Hemoglobin*

A hemoglobin molecule has four possible sites to which oxygen (O_2) can attach. If the concentration s_o of O_2 is high compared to that of hemoglobin, then we can model the number of sites occupied on a single hemoglobin molecule as a continuous-time Markov chain [132]. Figure 8.1 depicts the model. In the figure, each arc is labeled by a transition intensity and each state by the circled number of O_2 molecules attached to the hemoglobin molecule. The forward rates $s_o k_{+j}$ incorporate the concentration of O_2. The higher the concentration of O_2, the more frequently successful collisions occur between O_2 and the hemoglobin attachment sites. Because this chain is reversible, we can calculate its equilibrium distribution starting from the reference state 0 as $\pi_i = \pi_0 s_o^i \prod_{j=1}^i k_{+j}/k_{-j}$. If each site operated independently, then we could postulate per site association and disassociation intensities of $s_o k_+$ and k_-, respectively. Under the independent-site hypothesis, $\lambda_{i,i+1} = (4-i)s_o k_+$ and $\lambda_{i,i-1} = ik_-$. ∎

Example 8.5.2 *Kimura's DNA Substitution Model*

Kimura has suggested a model for base-pair substitution in molecular evolution [88, 105]. Recall that DNA is a long, double polymer constructed from the four bases (or nucleotides) adenine, guanine, cytosine, and thymine. These bases are abbreviated A, G, C, and T, respectively. Two of the bases are purines (A and G), and two are pyrimidines (C and

T). The two strands of DNA form a double helix containing complementary hereditary information in the sense that A and T and C and G always pair across strands. For example, if one strand contains the block –ACCGT– of bases, then the other strand contains the complementary block –TGGCA– of bases taken in reverse order. Thus, one is justified in following the evolutionary development of one strand and ignoring the other strand.

Kimura defines a continuous-time Markov chain with the four states A, G, C, and T that captures the evolutionary history of a species at a single position (or site) along a DNA strand. Mutations occur from time to time that change the base at the site. Let λ_{ij} be the intensity at which base i mutates to base j. Kimura radically simplifies these intensities. Let us write $i \simeq j$ if i and j are both purines or both pyrimidines and $i \not\simeq j$ if one is a purine and the other is a pyrimidine. Then Kimura assumes that

$$\lambda_{ij} = \begin{cases} \alpha & i \simeq j \\ \beta & i \not\simeq j. \end{cases}$$

These assumptions translate into the infinitesimal generator

$$
\Lambda = \begin{array}{c} \\ A \\ G \\ C \\ T \end{array}
\begin{array}{cccc}
A & G & C & T \\
\left(\begin{array}{cccc}
-(\alpha + 2\beta) & \alpha & \beta & \beta \\
\alpha & -(\alpha + 2\beta) & \beta & \beta \\
\beta & \beta & -(\alpha + 2\beta) & \alpha \\
\beta & \beta & \alpha & -(\alpha + 2\beta)
\end{array} \right)
\end{array}.
$$

Kimura's chain is reversible with the uniform distribution as its equilibrium distribution.   ∎

Example 8.5.3 *Circuit Theory*

Interesting but naive models of electrical circuits can be constructed using continuous-time Markov chains. Consider a model with $m + 1$ nodes labeled $0, \ldots, m$. Nodes 0 and 1 correspond to the terminals of a battery. Node 0 has potential 0 and node 1 has potential 1. Each of the nodes can be occupied by electrons. For the sake of simplicity in discussing potential differences, we assume that an electron carries a positive rather than a negative charge. Suppose node j has a capacity of e_j electrons. The number of electrons present at node j is given by a random variable $n_j(t)$ at time t. At nodes 0 and 1 we assume that $n_0(t) = 0$ and $n_1(t) = e_1$. Provided we define appropriate infinitesimal transition probabilities, the random count vector $\mathbf{n}(t) = [n_0(t), n_1(t), \ldots, n_m(t)]$ constitutes a continuous-time Markov chain [86].

Transitions of this chain correspond to the transfer of electrons between pairs of nodes from $2, \ldots, m$, the absorption of an electron at node 0 from one of the nodes $2, \ldots, m$, and the introduction of an electron into one of the nodes $2, \ldots, m$ from node 1. Electrons absorbed by node 0 are immediately passed to the battery so that the count $n_0(t)$ remains at 0 for all time.

Likewise, electrons leaving node 1 are immediately replaced by the battery so that $n_1(t)$ remains at e_1 for all time. Provided we refer to electron absorptions and introductions as transfer events, we can devise a useful model by giving the infinitesimal transfer rates between states. These will be phrased in terms of the conductance $c_{jk} = c_{kj}$ between two nodes j and k. The reciprocal of a conductance is a resistance. If we imagine the two nodes connected by wire, then conductance indicates the mobility of the electrons through the wire. In the absence of a wire between the nodes, the conductance is 0. The transfer rate λ_{jk} between nodes j and k should incorporate the conductance c_{jk}, the possible saturation of each node by electrons, and the fact that electrons repel. The particular choice

$$\lambda_{jk} = c_{jk}\frac{n_j}{e_j}\frac{e_k - n_k}{e_k}$$

succinctly captures these requirements.

To monitor the mean number of electrons at node j, we derive a differential equation involving the transfer of electrons during a short time interval of duration s. Conditioning on the electron counts at time t, we find that

$$
\begin{aligned}
&\mathrm{E}[n_j(t+s) - n_j(t)] \\
&= \mathrm{E}\{\mathrm{E}[n_j(t+s) - n_j(t) \mid \mathbf{n}(t)]\} \\
&= \sum_{k \neq j} \mathrm{E}\left\{ \mathrm{E}\left[c_{kj}\left(\frac{e_j - n_j(t)}{e_j}\frac{n_k(t)}{e_k} - \frac{n_j(t)}{e_j}\frac{e_k - n_k(t)}{e_k}\right)s \mid \mathbf{n}(t)\right]\right\} + o(s) \\
&= \sum_{k \neq j} \mathrm{E}\left[c_{kj}\left(\frac{n_k(t)}{e_k} - \frac{n_j(t)}{e_j}\right)\right]s + o(s).
\end{aligned}
$$

Forming the corresponding difference quotient and sending s to 0 give the differential equation

$$\frac{d}{dt}\mathrm{E}[n_j(t)] = \sum_{k \neq j} c_{kj}[p_k(t) - p_j(t)],$$

where $p_j(t) = \mathrm{E}[n_j(t)/e_j]$ and $p_k(t) = \mathrm{E}[n_k(t)/e_k]$ are the potentials at nodes j and k. The term $c_{kj}[p_k(t) - p_j(t)]$ represents the current flow from node j to node k and summarizes Ohm's law. At equilibrium, the mean number of electrons entering and leaving node j is 0. This translates into Kirchhoff's law

$$0 = \frac{d}{dt}\mathrm{E}[n_j(t)] = \sum_{k \neq j} c_{kj}(p_k - p_j),$$

where $p_k(t) = p_k$ and $p_j(t) = p_j$ are constant. An alternative Markov chain model for current flow is presented in reference [41]. ∎

8.6   Calculation of Matrix Exponentials

From the definition of the matrix exponential e^A, it is easy to deduce that it is continuous in A and satisfies $e^{A+B} = e^A e^B$ whenever $AB = BA$. It is also straightforward to check the differentiability condition

$$\frac{d}{dt}e^{tA} \;=\; Ae^{tA} \;=\; e^{tA}A.$$

Proofs of these facts depend on the introduction of vector and matrix norms. Of more practical importance is how one actually calculates e^{tA} [114]. In some cases it is possible to do so analytically. For instance, if u and v are column vectors with the same number of components, then

$$e^{suv^t} \;=\; \begin{cases} I + suv^t & \text{if } v^t u = 0 \\ I + \frac{e^{sv^t u}-1}{v^t u}uv^t & \text{otherwise.} \end{cases}$$

This follows from the formula $(uv^t)^i = (v^t u)^{i-1}uv^t$. The special case where $u = (-\alpha, \beta)^t$ and $v = (1,-1)^t$ permits explicit calculation of the finite-time transition matrix

$$P(s) \;=\; \exp\left[s\begin{pmatrix} -\alpha & \alpha \\ \beta & -\beta \end{pmatrix} \right]$$

for a two-state Markov chain.

If A is a diagonalizable $n \times n$ matrix, then we can write $A = TDT^{-1}$ for D a diagonal matrix with ith diagonal entry ρ_i. Here ρ_i is an eigenvalue of A with eigenvector equal to the ith column of T. Because $A = TDT^{-1}$, we find that $A^2 = TDT^{-1}TDT^{-1} = TD^2T^{-1}$ and in general $A^i = TD^iT^{-1}$. Hence,

$$\begin{aligned} e^{tA} &= \sum_{i=0}^{\infty} \frac{1}{i!}(tA)^i \\ &= \sum_{i=0}^{\infty} \frac{1}{i!}T(tD)^i T^{-1} \\ &= Te^{tD}T^{-1}, \end{aligned} \tag{8.11}$$

where

$$e^{tD} \;=\; \begin{pmatrix} e^{\rho_1 t} & \cdots & 0 \\ \vdots & \ddots & \vdots \\ 0 & \cdots & e^{\rho_n t} \end{pmatrix}.$$

Equation (8.11) suggests that the behavior of e^{tA} is determined by its dominant eigenvalue. This is the eigenvalue with largest real part. Suppose for the sake of argument that a dominant eigenvalue exists and is real. If the dominant eigenvalue is negative, then e^{tA} will tend to the zero matrix as t tends to ∞. If the dominant eigenvalue is positive, then usually the entries of e^{tA} will diverge. Finally, if the dominant eigenvalue is 0, then e^{tA} may converge to a constant matrix. As demonstrated in Section 8.4, the infinitesimal generator of an irreducible chain falls in this third category. Seneta [137] develops these ideas rigorously.

Even if we cannot calculate e^{tA} analytically, we can usually do so numerically. For instance when $t > 0$ is small, we can approximate e^{tA} by the truncated series $\sum_{i=0}^{n}(tA)^i/i!$ for n small. For larger t such truncation can lead to serious errors. If the truncated expansion is sufficiently accurate for all $t \le c$, then for arbitrary t one can exploit the property $e^{(s+t)A} = e^{sA}e^{tA}$ of the matrix exponential. Thus, if $t > c$, take the smallest positive integer k such that $2^{-k}t \le c$ and approximate $e^{2^{-k}tA}$ by the truncated series. Applying the multiplicative property, we can compute e^{tA} by squaring $e^{2^{-k}tA}$, squaring the result $e^{2^{-k+1}tA}$, squaring the result of this, and so forth, a total of k times.

Problem 2 features a method of computing matrix exponentials specifically tailored to infinitesimal generators. This uniformization technique is easy to implement numerically.

Example 8.6.1 *Application to Kimura's Model*

Because the infinitesimal generator Λ in Kimura's model is a symmetric matrix, it has only real eigenvalues. These are 0, -4β, $-2(\alpha + \beta)$, and $-2(\alpha + \beta)$. The reader can check that the four corresponding eigenvectors

$$\frac{1}{2}\begin{pmatrix} 1 \\ 1 \\ 1 \\ 1 \end{pmatrix}, \quad \frac{1}{2}\begin{pmatrix} -1 \\ -1 \\ 1 \\ 1 \end{pmatrix}, \quad \frac{1}{\sqrt{2}}\begin{pmatrix} 0 \\ 0 \\ -1 \\ 1 \end{pmatrix}, \quad \frac{1}{\sqrt{2}}\begin{pmatrix} -1 \\ 1 \\ 0 \\ 0 \end{pmatrix}$$

are orthogonal unit vectors. Therefore, the matrix T constructed by concatenating these vectors is orthogonal with inverse T^t. Equation (8.11) gives

$$e^{t\Lambda} = \begin{pmatrix} \frac{1}{2} & \frac{-1}{2} & 0 & \frac{-1}{\sqrt{2}} \\ \frac{1}{2} & \frac{-1}{2} & 0 & \frac{1}{\sqrt{2}} \\ \frac{1}{2} & \frac{1}{2} & \frac{-1}{\sqrt{2}} & 0 \\ \frac{1}{2} & \frac{1}{2} & \frac{1}{\sqrt{2}} & 0 \end{pmatrix} \begin{pmatrix} 1 & 0 & 0 & 0 \\ 0 & e^{-4\beta t} & 0 & 0 \\ 0 & 0 & e^{-2(\alpha+\beta)t} & 0 \\ 0 & 0 & 0 & e^{-2(\alpha+\beta)t} \end{pmatrix}$$

$$\times \begin{pmatrix} \frac{1}{2} & \frac{1}{2} & \frac{1}{2} & \frac{1}{2} \\ \frac{-1}{2} & \frac{-1}{2} & \frac{1}{2} & \frac{1}{2} \\ 0 & 0 & \frac{-1}{\sqrt{2}} & \frac{1}{\sqrt{2}} \\ \frac{-1}{\sqrt{2}} & \frac{1}{\sqrt{2}} & 0 & 0 \end{pmatrix}.$$

From this representation, we can calculate typical entries such as

$$p_{AA}(t) = \frac{1}{4} + \frac{1}{4}e^{-4\beta t} + \frac{1}{2}e^{-2(\alpha+\beta)t}$$

$$p_{AG}(t) = \frac{1}{4} + \frac{1}{4}e^{-4\beta t} - \frac{1}{2}e^{-2(\alpha+\beta)t}$$

$$p_{AC}(t) = \frac{1}{4} - \frac{1}{4}e^{-4\beta t}.$$

In fact, all entries of the finite-time transition matrix take one of these three forms. ∎

8.7 Kendall's Birth-Death-Immigration Process

In this section we tackle a continuous-time Markov chain important in biological and chemical applications. This chain has nonconstant transition intensities and an infinite number of states. As a preliminary to understanding Kendall's birth-death-immigration process, we derive nonrigorously the forward equations governing such a chain X_t. Let $p_{ij}(t)$ be the probability that the chain is in state j at time t given it was in state i at time 0. Also let $\lambda_{ij}(t)$ be the intensity of a transition at time t from state i to state j. Our point of departure is the Chapman-Kolmorgorov relation

$$\begin{aligned}
p_{ij}(t+h) &= p_{ij}(t)\Pr(X_{t+h} = j \mid X_t = j) \\
&\quad + \sum_{k \neq j} p_{ik}(t)\Pr(X_{t+h} = j \mid X_t = k).
\end{aligned} \quad (8.12)$$

We now assume that

$$\begin{aligned}
\Pr(X_{t+h} = j \mid X_t = j) &= 1 - \lambda_j(t)h + o(h) \\
\Pr(X_{t+h} = j \mid X_t = k) &= \lambda_{kj}(t)h + o(h)
\end{aligned}$$

for continuous functions $\lambda_j(t)$ and $\lambda_{kj}(t)$. Inserting these approximations into the time-inhomogeneous Chapman-Kolmorgorov relation (8.12) and rearranging terms yields the difference quotient

$$\frac{p_{ij}(t+h) - p_{ij}(t)}{h} = -p_{ij}(t)\lambda_j(t) + \sum_{k \neq j} p_{ik}(t)\lambda_{kj}(t) + \frac{o(h)}{h}.$$

Sending h to 0 therefore gives the system of forward equations

$$\frac{d}{dt}p_{ij}(t) = -p_{ij}(t)\lambda_j(t) + \sum_{k \neq j} p_{ik}(t)\lambda_{kj}(t)$$

with the obvious initial conditions $p_{ij}(0) = 1_{\{i=j\}}$.

In Kendall's birth-death-immigration process, X_t counts the number of particles at time t. Particles are of a single type and independently die and reproduce. The death rate per particle is $\mu(t)$, and the birth rate per particle is $\alpha(t)$. Reproduction occurs throughout the lifetime of a particle, not at its death. A Poisson process with intensity $\nu(t)$ feeds new particles into the process. Each new immigrant starts an independently evolving clan of particles. If there are initially i particles, then the forward equations can be summarized as

$$\frac{d}{dt}p_{i0}(t) = -\nu(t)p_{i0}(t) + \mu(t)p_{i1}(t)$$

and

$$\frac{d}{dt}p_{ij}(t) = -[\nu(t) + j\alpha(t) + j\mu(t)]p_{ij}(t) + [\nu(t) + (j-1)\alpha(t)]p_{i,j-1}(t)$$
$$+ (j+1)\mu(t)p_{i,j+1}(t)$$

for $j > 0$. Note here the assumption that each birth involves a single daughter particle.

To better understand this infinite system of coupled ordinary differential equations, we define the generating function

$$G(s,t) = \sum_{j=0}^{\infty} p_{ij}(t)s^j$$

for $s \in [0, 1]$. If we multiply the jth forward equation by s^j and sum on j, then we find that

$$\frac{\partial}{\partial t}G(s,t) = \sum_{j=0}^{\infty} \frac{d}{dt}p_{ij}(t)s^j$$

$$= -\nu(t)\sum_{j=0}^{\infty} p_{ij}(t)s^j - [\alpha(t) + \mu(t)]s\sum_{j=1}^{\infty} jp_{ij}(t)s^{j-1}$$

$$+ \nu(t)s\sum_{j=1}^{\infty} p_{i,j-1}(t)s^{j-1} + \alpha(t)s^2\sum_{j=1}^{\infty} (j-1)p_{i,j-1}(t)s^{j-2}$$

$$+ \mu(t)\sum_{j=0}^{\infty} (j+1)p_{i,j+1}(t)s^j$$

$$= -\nu(t)G(s,t) - [\alpha(t) + \mu(t)]s\frac{\partial}{\partial s}G(s,t) + \nu(t)sG(s,t)$$

$$+ \alpha(t)s^2\frac{\partial}{\partial s}G(s,t) + \mu(t)\frac{\partial}{\partial s}G(s,t).$$

Collecting terms yields the partial differential equation

$$\frac{\partial}{\partial t}G(s,t) = [\alpha(t)s - \mu(t)](s-1)\frac{\partial}{\partial s}G(s,t) + \nu(t)(s-1)G(s,t) \quad (8.13)$$

with the initial condition $G(s,0) = s^i$.

Before solving equation (8.13), it is worth solving for the mean number of particles $m_i(t)$ at time t. In view of the fact that $m_i(t) = \frac{\partial}{\partial s} G(s,t)|_{s=1}$, we can differentiate equation (8.13) and derive the ordinary differential equation

$$
\begin{aligned}
\frac{d}{dt} m_i(t) &= \frac{\partial^2}{\partial s \partial t} G(1,t) \\
&= [\alpha(t) - \mu(t)] \frac{\partial}{\partial s} G(1,t) + \nu(t) G(1,t) \\
&= [\alpha(t) - \mu(t)] m_i(t) + \nu(t)
\end{aligned}
$$

with the initial condition $m_i(0) = i$. To solve this first-order ordinary differential equation, we multiply both sides by $e^{w(t)}$, where

$$
w(t) = \int_0^t [\mu(\tau) - \alpha(\tau)] \, d\tau.
$$

This action produces

$$
\frac{d}{dt} \left[m_i(t) e^{w(t)} \right] = \nu(t) e^{w(t)}.
$$

Integrating and rearranging then yields the solution

$$
m_i(t) = i e^{-w(t)} + \int_0^t \nu(\tau) e^{w(\tau) - w(t)} \, d\tau.
$$

The special case where $\alpha(t)$, $\mu(t)$, and $\nu(t)$ are constant simplifies to

$$
m_i(t) = i e^{(\alpha - \mu)t} + \frac{\nu}{\alpha - \mu} \left[e^{(\alpha - \mu)t} - 1 \right].
$$

When $\alpha < \mu$, the forces of birth and immigration eventually balance the force of death, and the process reaches equilibrium. The equilibrium distribution has mean $\lim_{t \to \infty} m_i(t) = \nu/(\mu - \alpha)$.

Example 8.7.1 *Inhomogeneous Poisson Process*

If we take $i = 0$, $\alpha(t) = 0$, and $\mu(t) = 0$, then Kendall's process coincides with an inhomogeneous Poisson process. The reader can check that the partial differential equation (8.13) reduces to

$$
\frac{\partial}{\partial t} G(s,t) = \nu(t)(s - 1) G(s,t)
$$

with solution

$$
G(s,t) = e^{-(1-s) \int_0^t \nu(\tau) d\tau}.
$$

From $G(s,t)$ we reap the Poisson probabilities

$$p_{0j}(t) = \frac{\left(\int_0^t \nu(\tau)\,d\tau\right)^j}{j!} e^{-\int_0^t \nu(\tau)\,d\tau}.$$

■

Example 8.7.2 *Inhomogeneous Pure Death Process*

When $i > 0$, $\alpha(t) = 0$, and $\nu(t) = 0$, Kendall's process represents a pure death process. The partial differential equation (8.13) reduces to

$$\frac{\partial}{\partial t}G(s,t) = -\mu(t)(s-1)\frac{\partial}{\partial s}G(s,t)$$

with solution

$$G(s,t) = 1 - e^{-\int_0^t \mu(\tau)\,d\tau} + se^{-\int_0^t \mu(\tau)\,d\tau}$$

if $i = 1$. In other words, a single initial particle is still alive at time t with probability $\exp[-\int_0^t \mu(\tau)\,d\tau]$. Another interpretation of this process is possible if t is taken as the distance traveled by a particle through some attenuating medium with attenuation coefficient $\mu(t)$. Death corresponds to the particle being stopped. The solution for general $i > 0$ is given by the binomial generating function

$$G(s,t) = \left[1 - e^{-\int_0^t \mu(\tau)\,d\tau} + se^{-\int_0^t \mu(\tau)\,d\tau}\right]^i$$

because all i particles behave independently.   ■

Example 8.7.3 *Inhomogeneous Pure Birth Process*

When $i > 0$, $\mu(t) = 0$, and $\nu(t) = 0$, Kendall's process represents a pure birth process. The partial differential equation (8.13) becomes

$$\frac{\partial}{\partial t}G(s,t) = \alpha(t)s(s-1)\frac{\partial}{\partial s}G(s,t)$$

with solution

$$G(s,t) = \left\{ \frac{se^{-\int_0^t \alpha(\tau)\,d\tau}}{1 - s\left[1 - e^{-\int_0^t \alpha(\tau)\,d\tau}\right]} \right\}^i,$$

which is the generating function of a negative binomial distribution with success probability $\exp[-\int_0^t \alpha(\tau)\,d\tau]$ and required number of successes i. When $i = 1$,

$$p_{1,j}(t) = e^{-\int_0^t \alpha(\tau)\,d\tau}\left[1 - e^{-\int_0^t \alpha(\tau)\,d\tau}\right]^{j-1}$$

is just the probability that a single ancestral particle generates $j-1$ descendant particles. If we reverse time in this process, then births appear to be deaths, and $p_{1j}(t)$ coincides with the probability that all $j-1$ descendant particles present at time t die before time 0 and the ancestral particle lives throughout the time interval. ∎

8.8   Solution of Kendall's Equation

Remarkably enough, the dynamics of Kendall's birth-death-immigration process can be fully specified by solving equation (8.13). Let us begin by supposing that the immigration rate $\nu(t)$ is identically 0 and that $X_0 = 1$. In this situation, $G(s,t)$ is given by the formidable expression

$$
\begin{aligned}
G(s,t) &= \frac{e^{w(t)} - (s-1)[\int_0^t \alpha(\tau)e^{w(\tau)}d\tau - 1]}{e^{w(t)} - (s-1)\int_0^t \alpha(\tau)e^{w(\tau)}d\tau} \qquad (8.14) \\
&= 1 + \frac{1}{\frac{e^{w(t)}}{s-1} - \int_0^t \alpha(\tau)e^{w(\tau)}d\tau},
\end{aligned}
$$

where $w(t) = \int_0^t [\mu(\tau) - \alpha(\tau)]\, d\tau$. It follows from (8.14) that $G(s,0) = s$, consistent with starting with a single particle.

To find $G(s,t)$, we consider a curve $s(t)$ in (s,t) space parameterized by t and determined implicitly by the relation $G(s,t) = s_0$, where s_0 is some constant in $[0,1]$. Differentiating $G(s,t) = s_0$ with respect to t produces

$$
\frac{\partial}{\partial t}G(s,t) + \frac{\partial}{\partial s}G(s,t)\frac{ds}{dt} = 0.
$$

Comparing this equation to equation (8.13), we conclude that

$$
\frac{ds}{dt} = (\alpha s - \mu)(1-s) = [\alpha - \mu + \alpha(s-1)](1-s), \qquad (8.15)
$$

where we have omitted the dependence of the various functions on t.

If we let $u = 1 - s$ and $r = \mu - \alpha$, then the ordinary differential equation (8.15) is equivalent to

$$
\frac{du}{dt} = (r + \alpha u)u.
$$

The further transformation $v = \ln u$ gives

$$
\frac{dv}{dt} = r + \alpha e^v,
$$

which in turn yields

$$
\frac{d}{dt}(v - w) = \alpha e^v = \alpha e^{v-w}e^w
$$

for $w(t) = \int_0^t r(\tau) \, d\tau$. Once we write this last equation as

$$\frac{d}{dt} e^{-(v-w)} = -\alpha e^w,$$

the solution

$$e^{-v(t)+w(t)} - e^{-v(0)+w(0)} = -\int_0^t \alpha(\tau) e^{w(\tau)} d\tau$$

is obvious. When we impose the initial condition $s(0) = s_0$ and recall the definitions of v and u, it follows that

$$\frac{e^{w(t)}}{1-s} - \frac{1}{1-s_0} = -\int_0^t \alpha(\tau) e^{w(\tau)} d\tau.$$

A final rearrangement now gives

$$s_0 = 1 + \frac{1}{\frac{e^{w(t)}}{s-1} - \int_0^t \alpha(\tau) e^{w(\tau)} d\tau},$$

which validates the representation (8.14) of $G(s,t) = s_0$.

Before adding the complication of immigration, let us point out that the generating function (8.14) collapses to the appropriate expression for $G(s,t)$ when the process is a pure birth process or a pure death process. It is also noteworthy that $G(s,t)^i$ is the generating function of X_t when $X_0 = i$ rather than $X_0 = 1$. This is just a manifestation of the fact that particles behave independently in the model.

The key to including immigration is the observation that when an immigrant particle arrives, it generates a clan of particles similar to the clan of particles issuing from a particle initially present. However, the clan originating from the immigrant has less time to develop. This suggests that we consider the behavior of Kendall's birth-death process starting with a single particle at some later time $u > 0$. Denote the generating function associated with this delayed process by $G(s,t,u)$ for $t \geq u$. Our discussion above indicates that $G(s,t,u)$ is given by expression (8.14), provided we replace 0 by u in the lower limits of integration appearing in (8.14) and in the definition of the function $w(t)$.

If we now assume $X_0 = 0$, then the generating function $H(s,t)$ of X_t in the presence of immigration is given by

$$H(s,t) = \exp\left\{\int_0^t [G(s,t,u) - 1]\nu(u) \, du\right\}. \tag{8.16}$$

We will prove this by constructing an appropriate marked Poisson process and appealing to Campbell's formula (6.13). Recall that immigrant particles arrive according to a Poisson process with intensity $\nu(t)$. Now imagine marking a new immigrant particle at time u by the size of the clan y it generates at the subsequent time t. The random marked points (U, Y) constitute a marked Poisson process Π with intensity $p(y \mid u)\nu(u)$, where $p(y \mid u)$ is the conditional discrete density of Y given the immigration time u. Thus, formula (6.13) implies that $X_t = \sum_{(u,y)\in\Pi} y$ has generating function

$$H(s, t) \;=\; \exp\Big\{ \int \sum_y (s^y - 1)p(y \mid u)\nu(u)\, du \Big\}.$$

In view of the identities $\sum_y p(y \mid u) = 1$ and $\sum_y s^y p(y \mid u) = G(s, t, u)$, this proves equation (8.16). Because particles behave independently, if initially $X_0 = i$ instead of $X_0 = 0$, then the generating function of X_t is

$$E(s^{X_t}) \;=\; G(s, t, 0)^i \exp\Big\{ \int_0^t [G(s, t, u) - 1]\nu(u)\, du \Big\}. \qquad (8.17)$$

In the special case where birth, death, and immigration rates are constant, the above expression fortunately simplifies. Thus, the two explicit formulas

$$\int_u^t (\mu - \alpha)\, d\tau \;=\; (\mu - \alpha)(t - u)$$

$$\int_u^t \alpha e^{\int_u^\tau (\mu-\alpha) d\eta}\, d\tau \;=\; \frac{\alpha}{\mu - \alpha}[e^{(\mu-\alpha)(t-u)} - 1]$$

lead to

$$G(s, t, u) \;=\; 1 + \cfrac{1}{\cfrac{e^{(\mu-\alpha)(t-u)}}{s-1} - \cfrac{\alpha}{\mu-\alpha}[e^{(\mu-\alpha)(t-u)} - 1]}$$

$$=\; 1 + \frac{(s - 1)(\mu - \alpha)e^{(\alpha-\mu)(t-u)}}{\mu - \alpha s + \alpha(s - 1)e^{(\alpha-\mu)(t-u)}},$$

which in turn implies

$$\exp\Big\{ \int_0^t [G(s, t, u) - 1]\nu\, du \Big\} \;=\; e^{\frac{\nu}{\alpha} \int_0^t \frac{d}{du} \ln(\mu - \alpha s + \alpha(s-1)e^{(\alpha-\mu)(t-u)})\, du}$$

$$=\; \Big(\frac{\mu - \alpha}{\mu - \alpha s + \alpha(s - 1)e^{(\alpha-\mu)t}} \Big)^{\frac{\nu}{\alpha}}.$$

From these pieces the full generating function (8.17) can be assembled.

8.9 Problems

1. Let $\Lambda = (\lambda_{ij})$ be an $m \times m$ matrix and $\pi = (\pi_i)$ be a $1 \times m$ row vector. Show that the equality $\pi_i \lambda_{ij} = \pi_j \lambda_{ji}$ is true for all pairs

(i, j) if and only if $\mathrm{diag}(\pi)\Lambda = \Lambda^t \mathrm{diag}(\pi)$, where $\mathrm{diag}(\pi)$ is a diagonal matrix with ith diagonal entry π_i. Now suppose Λ is an infinitesimal generator with equilibrium distribution π. If $P(t) = e^{t\Lambda}$ is its finite-time transition matrix, then show that detailed balance $\pi_i \lambda_{ij} = \pi_j \lambda_{ji}$ for all pairs (i, j) is equivalent to finite-time detailed balance $\pi_i p_{ij}(t) = \pi_j p_{ji}(t)$ for all pairs (i, j) and times $t \geq 0$.

2. Suppose that Λ is the infinitesimal generator of a continuous-time finite-state Markov chain, and let $\mu \geq \max_i \lambda_i$. If $R = I + \mu^{-1}\Lambda$, then prove that R has nonnegative entries and that

$$S(t) \;=\; \sum_{i=0}^{\infty} e^{-\mu t} \frac{(\mu t)^i}{i!} R^i$$

coincides with $P(t)$. (Hint: Verify that $S(t)$ satisfies the same defining differential equation and the same initial condition as $P(t)$.)

3. Consider a continuous-time Markov chain with infinitesimal generator Λ and equilibrium distribution π. If the chain is at equilibrium at time 0, then show that it experiences $t \sum_i \pi_i \lambda_i$ transitions on average during the time interval $[0, t]$, where $\lambda_i = \sum_{j \neq i} \lambda_{ij}$ and λ_{ij} denotes a typical off-diagonal entry of Λ.

4. Let $P(t) = [p_{ij}(t)]$ be the finite-time transition matrix of a finite-state irreducible Markov chain. Show that $p_{ij}(t) > 0$ for all i, j, and $t > 0$. Thus, no state displays periodic behavior. (Hint: Use Problem 2.)

5. Let $-f_i(t) = \ln p_{ii}(t) = \ln \Pr(X_t = i \mid X_0 = i)$ for a continuous-time Markov chain X_t on a countably-infinite state space. Assuming the finite-time transition matrix $P(t)$ satisfies $P(0) = I$, $\lim_{t \downarrow 0} P(t) = I$, and $P(s + t) = P(s)P(t)$ for all $s, t > 0$, demonstrate that the limit

$$\omega_i \;=\; \lim_{t \downarrow 0} \frac{f_i(t)}{t}$$

exists for all i. Note that $\omega_i = \lambda_i = \sum_{j \neq i} \lambda_{ij}$ for a chain on a finite-state space. (Hint: Apply problem 10 of Chapter 5.)

6. Let A and B be the 2×2 real matrices

$$A \;=\; \begin{pmatrix} a & -b \\ b & a \end{pmatrix}, \qquad B \;=\; \begin{pmatrix} \lambda & 0 \\ 1 & \lambda \end{pmatrix}.$$

Show that

$$e^A \;=\; e^a \begin{pmatrix} \cos b & -\sin b \\ \sin b & \cos b \end{pmatrix}, \qquad e^B \;=\; e^\lambda \begin{pmatrix} 1 & 0 \\ 1 & 1 \end{pmatrix}.$$

(Hints: Note that 2×2 matrices of the form $\begin{pmatrix} a & -b \\ b & a \end{pmatrix}$ are isomorphic to the complex numbers under the correspondence $\begin{pmatrix} a & -b \\ b & a \end{pmatrix} \leftrightarrow a + bi$. For the second case write $B = \lambda I + C$.)

7. Define matrices

$$A = \begin{pmatrix} a & 0 \\ 1 & a \end{pmatrix}, \qquad B = \begin{pmatrix} b & 1 \\ 0 & b \end{pmatrix}.$$

Show that $AB \neq BA$ and that

$$e^A e^B = e^{a+b} \begin{pmatrix} 1 & 1 \\ 1 & 2 \end{pmatrix}$$

$$e^{A+B} = e^{a+b} \left[\cosh(1) \begin{pmatrix} 1 & 0 \\ 0 & 1 \end{pmatrix} + \sinh(1) \begin{pmatrix} 0 & 1 \\ 1 & 0 \end{pmatrix} \right].$$

Hence, $e^A e^B \neq e^{A+B}$. (Hint: Use Problem 6 to calculate e^A and e^B. For e^{A+B} write $A + B = (a + b)I + R$ with R satisfying $R^2 = I$.)

8. Prove that $\det(e^A) = e^{\operatorname{tr}(A)}$, where tr is the trace function. (Hint: Since the diagonalizable matrices are dense in the set of matrices [74], by continuity you may assume that A is diagonalizable.)

9. In our discussion of mean hitting times in Section 7.5, we derived the formula $t = (I - Q)^{-1}\mathbf{1}$ for the vector of mean times spent in the transient states $\{1, \ldots, m\}$ enroute to the absorbing states $\{m+1, \ldots, n\}$. If we pass to continuous time and replace transition probabilities p_{ij} by transition intensities λ_{ij}, then show that the mean time t_i spent in the transient states beginning at state i satisfies the equation

$$t_i = \frac{1}{\lambda_i} + \sum_{\{j : j \neq i, 1 \leq j \leq m\}} \frac{\lambda_{ij}}{\lambda_i} t_j.$$

Prove that the solution to this system can be expressed as

$$t = (I - Q)^{-1}\omega = -\Upsilon^{-1}\mathbf{1},$$

where ω is the $m \times 1$ column vector with ith entry λ_i^{-1} and Υ is the upper left $m \times m$ block of the infinitesimal generator $\Lambda = (\lambda_{ij})$.

10. In Kimura's model, suppose that two new species bifurcate at time 0 from an ancestral species and evolve independently thereafter. Show that the probability that the two species possess the same base at a given site at time t is

$$\frac{1}{4} + \frac{1}{4}e^{-8\beta t} + \frac{1}{2}e^{-4(\alpha+\beta)t}.$$

(Hint: By symmetry this formula holds regardless of what base was present at the site in the ancestral species.)

11. Ehrenfest's model of diffusion involves a box with n gas molecules. Suppose the box is divided in half by a rigid partition with a very small hole. Molecules drift aimlessly around and occasionally pass through the hole. During a short time interval h, a given molecule changes sides with probability $\lambda h + o(h)$. Show that a single molecule at time $t > 0$ is on the same side of the box as it started at time 0 with probability $\frac{1}{2}(1 + e^{-2\lambda t})$. Now consider the continuous-time Markov chain for the number of molecules in the left half of the box. Given that the n molecules behave independently, prove that finite-time transition probability $p_{ij}(t)$ amounts to

$$p_{ij}(t) = \left(\frac{1}{2}\right)^n \sum_{k=\max\{0,i+j-n\}}^{\min\{i,j\}} \binom{i}{k}\binom{n-i}{j-k}\left(1 + e^{-2\lambda t}\right)^{n-i-j+2k}$$
$$\times \left(1 - e^{-2\lambda t}\right)^{i+j-2k}.$$

(Hint: The summation index k is the number of molecules initially in the left half that end up in the left half at time t.)

12. A chemical solution initially contains $n/2$ molecules of each of the four types A, B, C, and D. Here n is a positive even integer. Each pair of A and B molecules collides at rate α to produce one C molecule and one D molecule. Likewise, each pair of C and D molecules collides at rate β to produce one A molecule and one B molecule. In this problem, we model the dynamics of these reactions as a continuous-time Markov chain X_t and seek the equilibrium distribution. The random variable X_t tracks the number of A molecules at time t [13].

(a) Argue that the infinitesimal transition rates of the chain amount to

$$\lambda_{i,i-1} = i^2 \alpha$$
$$\lambda_{i,i+1} = (n-i)^2 \beta.$$

What about the other rates?

(b) Show that the chain is irreducible and reversible.

(c) Use Kolmorgorov's formula and calculate the equilibrium distribution

$$\pi_k = \pi_0 \left(\frac{\beta}{\alpha}\right)^k \binom{n}{k}^2$$

for k between 0 and n.

(d) For the special case $\alpha = \beta$, demonstrate that

$$\pi_k = \frac{\binom{n}{k}^2}{\binom{2n}{n}}.$$

To do so first prove the identity

$$\sum_{k=0}^{n} \binom{n}{k}^2 = \binom{2n}{n}.$$

(e) To handle the case $\alpha \neq \beta$, we revert to the normal approximation to the binomial distribution. Argue that

$$\binom{n}{k} p^k q^{n-k} = q^n \binom{n}{k} \left(\frac{p}{q}\right)^k$$

$$\approx \frac{1}{\sqrt{2\pi npq}} e^{-\frac{(k-np)^2}{2npq}}$$

for $p + q = 1$. Show that this implies

$$\binom{n}{k}^2 \left(\frac{p^2}{q^2}\right)^k \approx \frac{1}{2\pi npq^{2n+1}} e^{-\frac{(k-np)^2}{npq}}.$$

Now choose p so that $p^2/q^2 = \beta/\alpha$ and prove that the equilibrium distribution is approximately normally distributed with mean and variance

$$E(X_\infty) = \frac{n\sqrt{\frac{\beta}{\alpha}}}{1 + \sqrt{\frac{\beta}{\alpha}}}$$

$$\text{Var}(X_\infty) = \frac{n\sqrt{\frac{\beta}{\alpha}}}{2\left(1 + \sqrt{\frac{\beta}{\alpha}}\right)^2}.$$

13. Let n indistinguishable particles independently execute the same continuous-time Markov chain with infinitesimal transition probabilities λ_{ij}. Define a new Markov chain called the composition chain for the particles by recording how many of the n total particles are in each of the s possible states. A state of the new chain is a sequence of nonnegative integers $(k_1, \ldots, k_s)$ such that $\sum_{i=1}^{s} k_i = n$. For instance, with $n = 3$ particles and $s = 2$ states, the composition chain has the four states $(3, 0)$, $(2, 1)$, $(1, 2)$, and $(0, 3)$. Find the infinitesimal transition probabilities of the composition chain. If the original chain is ergodic with equilibrium distribution $\pi = (\pi_1, \ldots, \pi_s)$, find the equilibrium distribution of the composition chain. Finally, show that the composition chain is reversible if the original chain is reversible.

14. Apply Problem 13 to the hemoglobin model in Example 8.5.1 with the understanding that the attachment sites operate independently. What are the particles? How many states can each particle occupy? Identify the infinitesimal transition probabilities and the equilibrium distribution based on the results of Problem 13.

15. Prove that $G(s, t)$ defined by equation (8.14) satisfies the partial differential equation (8.13) with initial condition $G(s, 0) = s$.

16. In the homogeneous version of Kendall's process, show that

$$\text{Var}(X_t) = \frac{\nu}{(\alpha - \mu)^2} \left[\alpha e^{(\alpha - \mu)t} - \mu \right] \left[e^{(\alpha - \mu)t} - 1 \right]$$
$$+ \frac{i(\alpha + \mu)e^{(\alpha - \mu)t}}{\alpha - \mu} \left[e^{(\alpha - \mu)t} - 1 \right]$$

when $X_0 = i$.

17. Continuing Problem 16, demonstrate that

$$\text{Cov}(X_{t_2}, X_{t_1}) = e^{(\alpha - \mu)(t_2 - t_1)} \text{Var}(X_{t_1})$$

for $0 \le t_1 \le t_2$. (Hints: First show that

$$\text{Cov}(X_{t_2}, X_{t_1}) = \text{Cov}[\text{E}(X_{t_2} \mid X_{t_1}), X_{t_1}].$$

Then apply Problem 16.)

18. In the homogeneous version of Kendall's process, show that the generating function $G(s, t)$ of X_t satisfies

$$\lim_{t \to \infty} G(s, t) = \frac{\left(1 - \frac{\alpha}{\mu}\right)^{\nu/\alpha}}{\left(1 - \frac{\alpha s}{\mu}\right)^{\nu/\alpha}} \tag{8.18}$$

when $\alpha < \mu$.

19. Continuing Problem 18, prove that the equilibrium distribution π has jth component

$$\pi_j = \left(1 - \frac{\alpha}{\mu}\right)^{\nu/\alpha} \left(-\frac{\alpha}{\mu}\right)^j \binom{-\frac{\nu}{\alpha}}{j}.$$

Do this by expansion of the generating function on the right-hand side of equation (8.18) and by applying Kolmorgorov's method to Kendall's process. Note that the process is reversible.

9

Branching Processes

9.1 Introduction

A branching process models the reproduction of particles such as human beings, cells, or neutrons. In the simplest branching processes, time is measured discretely in generations, and particles are of only one type. Each particle is viewed as living one generation; during this period it produces offspring contributing to the next generation. The key assumption that drives the theory is that particles reproduce independently according to the same probabilistic law. Interactions between particles are forbidden. Within this context one can ask and at least partially answer interesting questions concerning the random number X_n of particles at generation n. For instance, what are the mean $E(X_n)$ and the variance $Var(X_n)$? What is the extinction probability $Pr(X_n = 0)$ on or before generation n, and what is the ultimate extinction probability $\lim_{n\to\infty} Pr(X_n = 0)$?

Probabilists have studied many interesting elaborations of the simple branching process paradigm. For example, in some applications it is natural to include immigration of particles from outside the system and to investigate the stochastic balance between immigration and extinction. The fact that branching processes are Markov chains suggests the natural generalization to continuous time. Here each particle lives an exponentially distributed length of time. Reproduction comes as the particle dies. We have already met one such process in the guise of the time-homogeneous Kendall process. A final generalization is to processes with multiple particle types. In continuous time, each type has its own mean lifetime and own reproductive pattern. Particles of one type can produce both offspring of their own type and offspring of other types.

The theory of branching processes has a certain spin that sets it apart from the general theory of Markov chains. Our focus in this chapter is on elementary results and applications to biological models. We stress computational topics such as the finite Fourier transform and the matrix exponential function rather than asymptotic results. Readers interested in pursuing the theory of branching processes in more detail should consult the references [11, 12, 48, 64, 68, 79, 82, 130]. Statistical inference questions arising in branching processes are considered in reference [65].

9.2   Examples of Branching Processes

We commence our discuss of branching processes in discrete time by assuming that all particles are of the same type and that no immigration occurs. The reproductive behavior of the branching process is encapsulated in the progeny generating function $Q(s) = \sum_{k=0}^{\infty} q_k s^k$ for the number of progeny (equivalently, offspring or daughter particles) born to a single particle. If the initial number of particles $X_0 = 1$, then $Q(s)$ is the generating function of X_1. For the sake of brevity in this chapter, we refer to probability generating functions simply as generating functions. Before launching into a discussion of theory, it is useful to look at a few concrete models of branching processes.

Example 9.2.1 *Cell Division*

A cell eventually either dies with probability q_0 or divides with probability q_2. In a cell culture, cells can be made to reproduce synchronously at discrete generation times. Starting from a certain number of progenitor cells, the number of cells at successive generations forms a branching process. The progeny generating function of this process is $Q(s) = q_0 + q_2 s^2$.
■

Example 9.2.2 *Neutron Chain Reaction*

In a fission reactor, a free neutron starts a chain reaction by striking and splitting a nucleus. Typically, this generates a fixed number m of secondary neutrons. These secondary neutrons are either harmlessly absorbed or strike further nuclei and release tertiary neutrons, and so forth. The progeny generating function $Q(s) = q_0 + q_m s^m$ of the resulting branching process has mean $\mu = m q_m$. The chain reaction environment is said to be subcritical, critical, or supercritical depending on whether $\mu < 1$, $\mu = 1$, or $\mu > 1$. A nuclear reactor is carefully modulated by drawing off excess neutrons to maintain the critical state and avoid an explosion.
■

Example 9.2.3 *Survival of Family Names*

In most cultures, family names (surnames) are passed through the male line. The male descendants of a given man constitute a branching process to a good approximation. Bienaymé, Galton, and Watson introduced branching processes to study the phenomenon of extinction of family names. We will see later that extinction is certain for subcritical and critical processes. A supercritical process either goes extinct or eventually grows at a geometric rate. These remarks partially explain why some long established countries have relatively few family names. ∎

Example 9.2.4 *Epidemics*

The early stages of an epidemic are well modeled by a branching process. If the number of infected people is small, then they act approximately independently of each other. The coefficient q_k in the progeny generating function $Q(s) = \sum_{k=0}^{\infty} q_k s^k$ is the probability that an infected individual infects k further people before he or she dies or recovers from the infection. The extinction question is of paramount importance in assessing the efficacy of vaccines in preventing epidemics. ∎

Example 9.2.5 *Survival of Mutant Genes*

If a dominant deleterious (harmful) mutation occurs at an autosomal genetic locus, then the person affected by the mutation starts a branching process of mutant people. (An autosomal locus is a gene location on a non-sex chromosome.) For instance, a clan of related people afflicted with the neurological disorder Huntington's disease can be viewed from this perspective [125]. Instead of sons, we follow the descendants, male and female, carrying the mutation. Because a deleterious gene is rare, we can safely assume that carriers mate only with normal people. On average, half of the children born to a carrier will be normal, and half will be carriers. The fate of each child is determined independently. Usually the reproductive fitness of carriers is sufficiently reduced so that the branching process is subcritical. ∎

9.3   Elementary Theory

For the sake of simplicity, we now take $X_0 = 1$ unless noted to the contrary. To better understand the nature of a branching process, we divide the X_n descendants at generation n into X_1 clans. The kth clan consists of the descendants of the kth progeny of the founder of the process. Translating this verbal description into symbols gives the representation

$$X_n = \sum_{k=1}^{X_1} X_{nk}, \tag{9.1}$$

where X_{nk} is the number of particles at generation n in the kth clan. The assumptions defining a branching process imply that the X_{nk} are independent, probabilistic replicas of X_{n-1}. Our calculations in Example 2.4.4 consequently indicate that $\mathrm{E}(X_n) = \mathrm{E}(X_1)\,\mathrm{E}(X_{n-1})$. If we let μ be the mean number $\mathrm{E}(X_1) = Q'(1)$ of progeny per particle, then this recurrence relation entails $\mathrm{E}(X_n) = \mu^n$.

Calculation of the variance $\mathrm{Var}(X_n)$ also yields to the analysis of Example 2.4.4. If σ^2 is the variance of X_1, then the representation (9.1) implies

$$
\begin{aligned}
\mathrm{Var}(X_n) &= \mu\,\mathrm{Var}(X_{n-1}) + \sigma^2\,\mathrm{E}(X_{n-1})^2 \\
&= \mu\,\mathrm{Var}(X_{n-1}) + \sigma^2\mu^{2(n-1)}.
\end{aligned}
\tag{9.2}
$$

We claim that this recurrence relation has solution

$$
\mathrm{Var}(X_n) = \begin{cases} n\sigma^2, & \mu = 1 \\ \dfrac{\mu^{n-1}(1-\mu^n)\sigma^2}{1-\mu}, & \mu \neq 1 \end{cases}
\tag{9.3}
$$

subject to the initial condition $\mathrm{Var}(X_0) = 0$. The stated formula obviously satisfies the recurrence (9.2) for $\mu = 1$. When $\mu \neq 1$ and $n = 0$, the formula also holds. If we assume by induction that it is true for $n - 1$, then the calculation

$$
\begin{aligned}
\mathrm{Var}(X_n) &= \mu\frac{\mu^{n-2}(1-\mu^{n-1})\sigma^2}{1-\mu} + \sigma^2\mu^{2(n-1)} \\
&= \mu\sigma^2\frac{\mu^{n-2}(1-\mu^{n-1}) + \mu^{2n-3}(1-\mu)}{1-\mu} \\
&= \frac{\mu^{n-1}(1-\mu^n)\sigma^2}{1-\mu}
\end{aligned}
$$

combining equations (9.2) and (9.3) completes the proof for $\mu \neq 1$.

Finally, the representation (9.1) implies that X_n has generating function $Q_n(s) = Q(Q_{n-1}(s))$, which is clearly the n-fold functional composition of $Q(s)$ with itself. The next example furnishes one of the rare instances in which $Q_n(s)$ can be explicitly found.

Example 9.3.1 *Geometric Progeny Distribution*

In a sequence of Bernoulli trials with success probability p and failure probability $q = 1 - p$, the number Y of failures until the first success has a geometric distribution with generating function

$$
Q(s) = \sum_{k=0}^{\infty} pq^k s^k = \frac{p}{1-qs}.
$$

The mean and variance of Y are

$$
\mu = \left.\frac{pq}{(1-qs)^2}\right|_{s=1} = \frac{q}{p}
$$

and

$$\sigma^2 = \left.\frac{2pq^2}{(1-qs)^3}\right|_{s=1} + \mu - \mu^2 = \frac{q}{p^2}.$$

We can verify inductively that

$$Q_n(s) = \begin{cases} \frac{n-(n-1)s}{n+1-ns}, & p=q \\ p\frac{q^n-p^n-(q^{n-1}-p^{n-1})qs}{q^{n+1}-p^{n+1}-(q^n-p^n)qs}, & p \neq q. \end{cases} \tag{9.4}$$

When $n=1$, the second of these formulas holds because

$$\frac{p(q-p)}{q^2-p^2-(q-p)qs} = \frac{p(q-p)}{(q+p)(q-p)-(q-p)qs}$$
$$= \frac{p}{1-qs}.$$

Assuming that the second formula in (9.4) is true for $n-1$, we reason that

$$Q_n(s) = Q(Q_{n-1}(s))$$
$$= \frac{p}{1-qp\frac{q^{n-1}-p^{n-1}-(q^{n-2}-p^{n-2})qs}{q^n-p^n-(q^{n-1}-p^{n-1})qs}}$$
$$= p\frac{q^n-p^n-(q^{n-1}-p^{n-1})qs}{q^n(1-p)-p^n(1-q)-q^{n-1}(1-p)qs+p^{n-1}(1-q)qs}$$
$$= p\frac{q^n-p^n-(q^{n-1}-p^{n-1})qs}{q^{n+1}-p^{n+1}-(q^n-p^n)qs}.$$

Inductive proof of the first formula in (9.4) is left to the reader. ∎

9.4 Extinction

Starting with a single particle at generation 0, we now ask for the probability s_∞ that a branching process eventually goes extinct. To characterize s_∞, we condition on the number of progeny $X_1 = k$ born to the initial particle. If extinction is to occur, then each of the clans emanating from the k progeny must go extinct. By independence, this event occurs with probability s_∞^k. Thus, s_∞ satisfies the functional equation $s_\infty = \sum_{k=0}^\infty q_k s_\infty^k = Q(s_\infty)$, where $Q(s)$ is the progeny generating function.

One can find the extinction probability by functional iteration starting at $s = 0$. Let s_n be the probability that extinction occurs in the branching process at or before generation n. Then $s_0 = 0$, $s_1 = q_0 = Q(s_0)$, and, in general, $s_{n+1} = Q(s_n) = Q_{n+1}(0)$. This recurrence relation can be deduced by conditioning once again on the number of progeny in the first generation.

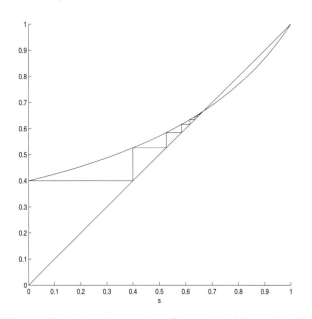

FIGURE 9.1. Extinction Iterates in a Supercritical Branching Process

If extinction is to occur at or before generation $n+1$, then extinction must occur in n additional generations or sooner for each clan emanating from a daughter particle of the founding particle.

On probabilistic grounds it is obvious that the sequence s_n increases monotonely to the extinction probability s_∞. To understand what is happening analytically, we need to know the number of roots of $s = Q(s)$ and which of these roots is s_∞. Because $Q''(s) = \sum_{k=2}^{\infty} k(k-1)q_k s^{k-2} \geq 0$, the curve $Q(s)$ is convex. It starts at $Q(0) = q_0 > 0$ above the diagonal line $t = s$ in Figure 9.1. (Note that if $q_0 = 0$, then the process can never go extinct.) On the interval $[0,1]$, the curve $Q(s)$ and the line $t = s$ intersect in either one or two points. The point $s = 1$ is certainly an intersection point because $Q(1) = \sum_{k=0}^{\infty} q_k = 1$. When the progeny mean $\mu = Q'(1) > 1$, the curve $Q(s)$ intersects $t = s$ at $s = 1$ from below, and there is a second intersection point to the left of $s = 1$. When $\mu < 1$, the second intersection point occurs to the right of $s = 1$.

The cobweb diagram in Figure 9.1 following the progress of the iterates s_n makes it clear that they not only monotonely increase, but they are also bounded above by the smaller of the two roots. Thus, the limit s_∞ of the s_n exists and is bounded above by the smaller root. Taking limits in the recurrence $s_{n+1} = Q(s_n)$ demonstrates that s_∞ coincides with the smaller root of $s = Q(s)$. In other words, extinction is certain for subcritical and critical processes and uncertain for supercritical processes.

TABLE 9.1. Functional Iteration for an Extinction Probability

Iteration n	Iterate s_n	Iteration n	Iterate s_n
0	.000	10	.847
1	.498	20	.873
2	.647	30	.878
3	.719	40	.879
4	.761	50	.880
5	.788		

Example 9.4.1 *Lotka's American Surname Data*

As a numerical example, consider the data of Lotka [109, 121] on the extinction of surnames among white males in the United States. Using 1920 census data, he computed the progeny generating function

$$Q(s) = .4982 + .2103s + .1270s^2 + .0730s^3 + .0418s^4 + .0241s^5$$
$$+ .0132s^6 + .0069s^7 + .0035s^8 + .0015s^9 + .0005s^{10}$$

for the number of sons of a random father. Because the average number of sons per father $\mu = Q'(1) > 1$, we anticipate an extinction probability $s_\infty < 1$. Table 9.1 lists some representative functional iterates. Convergence to the extinction probability .880 is slow but sure. ∎

Example 9.4.2 *Extinction Probability for the Geometric Distribution*

Consider once again the geometric distribution of Example 9.3.1. The two roots of the equation $s = \frac{p}{1-qs}$ are 1 and $\frac{p}{q} = \frac{1}{\mu}$. When $\mu > 1$, the extinction probability is $s_\infty = \frac{p}{q}$. When $\mu < 1$, the root $\frac{p}{q}$ lies outside the interval $[0, 1]$, and the extinction probability is $s_\infty = 1$. Finally, when $\mu = 1$, the two roots coincide. Thus, extinction is certain if $\mu \le 1$ and uncertain otherwise.

The same conclusions can be reached by considering the behavior of $s_n = Q_n(0)$ as suggested by formula (9.4). If $p < q$, then

$$\lim_{n\to\infty} Q_n(0) = \lim_{n\to\infty} \frac{p}{q} \frac{1 - \left(\frac{p}{q}\right)^n}{1 - \left(\frac{p}{q}\right)^{n+1}} = \frac{p}{q}.$$

If $p > q$, then

$$\lim_{n\to\infty} Q_n(0) = \lim_{n\to\infty} \frac{1 - \left(\frac{q}{p}\right)^n}{1 - \left(\frac{q}{p}\right)^{n+1}} = 1.$$

Finally, if $p = q$, then

$$\lim_{n \to \infty} Q_n(0) = \lim_{n \to \infty} \frac{n}{n+1} = 1.$$

∎

If T denotes the generation at which extinction occurs in a subcritical process, then $\Pr(T > n) = 1 - Q_n(0) = 1 - s_n$. The tail-probability method of Example 2.5.1 therefore implies

$$\mathrm{E}(T) = \sum_{n=0}^{\infty} (1 - s_n)$$

$$\mathrm{E}(T^2) = \sum_{n=0}^{\infty} (2n + 1)(1 - s_n). \tag{9.5}$$

Truncated versions of these sums permit one to approximate the first two moments of T. Problem 9 develops practical error bounds on this procedure. Finally, it is useful to contrast the critical case to the subcritical case. For instance, with a critical geometric generating function, we find that

$$\mathrm{E}(T) = \sum_{n=0}^{\infty} (1 - s_n) = \sum_{n=0}^{\infty} \left(1 - \frac{n}{n+1}\right) = \infty.$$

Now let $Y_n = 1 + \sum_{k=1}^{n} X_k$ be the total number of descendants up to generation n. The limit $Y_\infty = \lim_{n \to \infty} Y_n$ exists because the Y_n form an increasing sequence. The next proposition collects pertinent facts about this interesting random variable.

Proposition 9.4.1 *If $r_k = \Pr(Y_\infty = k)$ and $R(s) = \sum_{k=1}^{\infty} r_k s^k$, then the following hold:*

(a) *The extinction probability $s_\infty = \Pr(Y_\infty < \infty) = \sum_{k=1}^{\infty} r_k$. Therefore, in a supercritical process $Y_\infty = \infty$ occurs with positive probability.*

(b) *The generating function $R(s)$ satisfies $R(s) = sQ(R(s))$.*

(c) *For $\mu < 1$, the mean and variance of Y_∞ are $\mathrm{E}(Y_\infty) = \frac{1}{1-\mu}$ and $\mathrm{Var}(Y_\infty) = \frac{\sigma^2}{(1-\mu)^3}$.*

(d) *If the power $Q(s)^j$ has expansion $\sum_{k=0}^{\infty} q_{jk} s^k$, then $r_j = \frac{1}{j} q_{j,j-1}$.*

Proof: To prove part (a), note that there is at least one particle per generation if and only if the process does not go extinct. Hence, Y_∞ is finite if and only if the process goes extinct. For part (b), note that we have by analogy to equation (9.1) the representation

$$Y_\infty = 1 + \sum_{k=1}^{X_1} Y_{\infty k}, \tag{9.6}$$

where the $Y_{\infty k}$ are independent, probabilistic replicas of Y_∞. The generating function version of this representation is precisely $R(s) = sQ(R(s))$ as described in Example 2.4.4. To verify part (c), observe that equation (9.6) implies

$$
\begin{aligned}
\mathrm{E}(Y_\infty) &= 1 + \mu\, \mathrm{E}(Y_\infty) \\
\mathrm{Var}(Y_\infty) &= \mu\, \mathrm{Var}(Y_\infty) + \sigma^2\, \mathrm{E}(Y_\infty)^2.
\end{aligned}
$$

Assuming for the sake of simplicity that both $\mathrm{E}(Y_\infty)$ and $\mathrm{Var}(Y_\infty)$ are finite, we can solve the first of these equations to derive the stated expression for $\mathrm{E}(Y_\infty)$. Inserting this solution into the second equation and solving gives $\mathrm{Var}(Y_\infty)$. Proof of part (d) involves the tricky Lagrange inversion formula and is consequently omitted [33]. ∎

Example 9.4.3 *Total Descendants for the Geometric Distribution*

When $Q(s) = \frac{p}{1-qs}$, application of part (b) of Proposition 9.4.1 gives $R(s) = ps/[1 - qR(s)]$ or $qR(s)^2 - R(s) + ps = 0$. The smaller root

$$
R(s) = \frac{1 - \sqrt{1 - 4pqs}}{2q}
$$

of this quadratic is the relevant one. Indeed, based on the representation

$$
\begin{aligned}
\frac{1 + \sqrt{1 - 4pq}}{2q} &= \frac{p + q + \sqrt{(p+q)^2 - 4pq}}{2q} \\
&= \frac{p + q + |p - q|}{2q},
\end{aligned}
$$

taking the larger root produces the contradictory results $R(1) > 1$ in the subcritical case where $q < p$ and $R(1) = 1$ in the supercritical case where $q > p$. In the subcritical case, Proposition 9.4.1 or differentiation of $R(s)$ shows that $\mathrm{E}(Y_\infty) = \frac{p}{p-q}$ and $\mathrm{Var}(Y_\infty) = \frac{pq}{(p-q)^3}$. ∎

9.5 Immigration

We now modify the definition of a branching process to allow for immigration at each generation. In other words, we assume that the number of particles X_n at generation n is the sum $U_n + Z_n$ of the progeny U_n of generation $n-1$ plus a random number of immigrant particles Z_n independent of U_n. To make our theory as simple as possible, we take the Z_n to be independent and identically distributed with common mean α and variance β^2. If the process without immigration is subcritical ($\mu < 1$), then particle numbers eventually reach a stochastic equilibrium between extinction and immigration. The goal of this section is to characterize the equilibrium distribution.

Our point of departure in the subcritical case is the representation

$$U_n = \sum_{k=1}^{X_{n-1}} V_{n-1,k} \tag{9.7}$$

of the progeny of generation $n - 1$ partitioned by parent particle k. From this representation it follows by the usual conditioning arguments that

$$
\begin{aligned}
\mathrm{E}(X_n) &= \mathrm{E}(X_{n-1})\,\mathrm{E}(V_{n-1,1}) + \mathrm{E}(Z_n) \\
&= \mu\,\mathrm{E}(X_{n-1}) + \alpha \\
\mathrm{Var}(X_n) &= \mathrm{E}(X_{n-1})\,\mathrm{Var}(V_{n-1,1}) + \mathrm{Var}(X_{n-1})\,\mathrm{E}(V_{n-1,1})^2 + \mathrm{Var}(Z_n) \\
&= \sigma^2\,\mathrm{E}(X_{n-1}) + \mu^2\,\mathrm{Var}(X_{n-1}) + \beta^2.
\end{aligned}
\tag{9.8}
$$

If we assume $\lim_{n\to\infty}\mathrm{E}(X_n) = \mathrm{E}(X_\infty)$ and $\lim_{n\to\infty}\mathrm{Var}(X_n) = \mathrm{Var}(X_\infty)$ for some random variable X_∞, then taking limits on n in the two equations in (9.8) yields

$$
\begin{aligned}
\mathrm{E}(X_\infty) &= \mu\,\mathrm{E}(X_\infty) + \alpha \\
\mathrm{Var}(X_\infty) &= \sigma^2\,\mathrm{E}(X_\infty) + \mu^2\,\mathrm{Var}(X_\infty) + \beta^2.
\end{aligned}
$$

Solving these two equations in succession produces

$$
\begin{aligned}
\mathrm{E}(X_\infty) &= \frac{\alpha}{1-\mu} \\
\mathrm{Var}(X_\infty) &= \frac{\alpha\sigma^2 + \beta^2(1-\mu)}{(1-\mu)^2(1+\mu)}.
\end{aligned}
\tag{9.9}
$$

It is interesting that $\mathrm{E}(X_\infty)$ is the product of the average number of immigrants α per generation times the average clan size $\frac{1}{1-\mu}$ per particle identified in part (c) of Proposition 9.4.1.

Our next aim is to find the distribution of X_∞. Let $P_n(s)$ be the generating function of X_n and $R(s)$ be the common generating function of the Z_n. Then the decomposition $X_n = U_n + Z_n$ and equation (9.7) imply

$$P_n(s) = P_{n-1}(Q(s))R(s), \tag{9.10}$$

where $Q(s)$ is the progeny generating function. Iterating equation (9.10) yields

$$P_n(s) = P_0(Q_n(s)) \prod_{k=0}^{n-1} R(Q_k(s)), \tag{9.11}$$

where $Q_k(s)$ is again the k-fold functional composition of $Q(s)$ with itself. The finite product (9.11) tends to the infinite product

$$P_\infty(s) = \prod_{k=0}^{\infty} R(Q_k(s)) \tag{9.12}$$

representation of the generating function of X_∞. Observe here that the leading term $P_0(Q_n(s))$ on the right of equation (9.11) tends to 1 because $Q_n(0)$ tends to the extinction probability 1 and $Q_n(0) \leq Q_n(s) \leq 1$ for all $s \in [0, 1]$.

The problem now is to recover the coefficients p_k of $P_\infty(s) = \sum_{k=0}^\infty p_k s^k$. This is possible using the values of $P_\infty(s)$ for s on the boundary of the unit circle. Once we reparameterize by setting $s = e^{2\pi i t}$ for $t \in [0, 1]$, we recognize p_k as the kth Fourier series coefficient of the periodic function $P_\infty(e^{2\pi i t}) = \sum_{k=0}^\infty p_k e^{2\pi i k t}$. Obviously,

$$p_k = \int_0^1 P_\infty(e^{2\pi i t}) e^{-2\pi i k t} dt$$

can be approximated by the Riemann sum

$$\frac{1}{n} \sum_{j=0}^{n-1} P_\infty\left(e^{\frac{2\pi i j}{n}}\right) e^{-\frac{2\pi i j k}{n}} \tag{9.13}$$

for n large. The n sums in (9.13) for $0 \leq k \leq n - 1$ collectively define the finite Fourier transform of the sequence of n numbers $P_\infty(e^{2\pi i j/n})$ for $0 \leq j \leq n - 1$. To compute a finite Fourier transform, one can use an algorithm known as the fast Fourier transform [70, 92].

In summary, we can compute the p_k by

(a) choosing n so large that all p_k with $k \geq n$ can be ignored,

(b) approximating $P_\infty(e^{2\pi i j/n})$ by the finite product

$$\prod_{k=0}^m R(Q_k(e^{2\pi i j/n}))$$

with m large,

(c) taking finite Fourier transforms of the finite product.

To check the accuracy of these approximations, we numerically compute the moments of X_∞ from the resulting p_k for $0 \leq k \leq n - 1$ and compare the results to the theoretical moments [94].

Example 9.5.1 *Huntington's Disease*

Huntington's disease is caused by a deleterious dominant gene cloned in 1993 [145]. In the late 1950s, Reed and Neil estimated that carriers of the Huntington gene have a fitness $f \approx 0.81$ [125]. Here f is the ratio of the expected number of offspring of a carrier to the expected number of offspring of a normal person. In a stationary population, each person has on average one daughter and one son. Each carrier produces on average

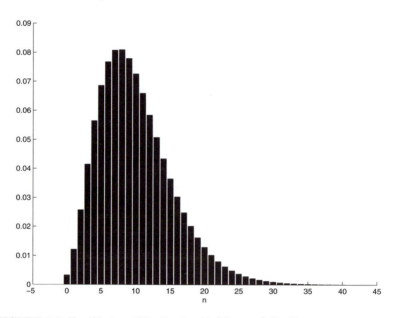

FIGURE 9.2. Equilibrium Distribution $\Pr(X_\infty = n)$ for Huntington's Disease

$0.81 \times 2 \times \frac{1}{2} = 0.81$ carrier children. If we let the progeny generating function $Q(s)$ be Poisson with mean $\mu = 0.81$, then the ultimate number of people Y_∞ affected starting from a single mutation has mean and variance

$$\mathrm{E}(Y_\infty) = \frac{1}{1-\mu} = 5.24$$

$$\mathrm{Var}(Y_\infty) = \frac{\sigma^2}{(1-\mu)^3} = 118.09$$

based on part (c) of Proposition 9.4.1 with $\sigma^2 = \mu$. The generation T at which extinction of the mutation occurs has mean $\mathrm{E}(T) = 3.01$ and variance $\mathrm{Var}(T) = \mathrm{E}(T^2) - \mathrm{E}(T)^2 = 11.9$ [100].

Let us also assume that the surrounding normal population exhibits a mutation rate of $\nu = 2 \times 10^{-6}$ and contains $r = 237,500$ females. Immigration into the branching process occurs whenever a parental gene mutates just prior to transmission to a child. By the "law of rare events" explained in Example 12.3.1, the number of new mutants at each generation is approximately Poisson with mean $\alpha = 2r2\nu = 4r\nu$. Here $2r$ is the total number of births per generation, and 2ν is the probability that either of the two parental genes contributed to a child mutate. According to equation (9.9), the mean and variance of the equilibrium distribution are

$$\mathrm{E}(X_\infty) = \frac{4r\nu}{1-\mu}$$
$$= 10$$

$$\text{Var}(X_\infty) = \frac{4rv\sigma^2 + 4rv(1-\mu)}{(1-\mu)^2(1+\mu)}$$
$$= 29.08.$$

Figure 9.2 depicts the equilibrium distribution as computed by the Fourier series method. ∎

9.6 Multitype Branching Processes

In a multitype branching process, one follows a finite number of independently acting particles that reproduce and die. Each particle is classified in one of n possible categories. In a continuous-time process, a type i particle lives an exponentially distributed length of time with death intensity λ_i. At the end of its life, a type i particle reproduces both particles of its own type and particles of other types. Suppose that on average it produces f_{ij} particles of type j.

We would like to calculate the average number of particles $m_{ij}(t)$ of type j at time $t \geq 0$ starting with a single particle of type i at time 0. Since particles of type j at time $t+s$ either arise from particles of type j at time t that do not die during $(t, t+s)$ or from particles of type k that die during $(t, t+s)$ and reproduce particles of type j, we find that

$$m_{ij}(t+s) = m_{ij}(t)(1-\lambda_j s) + \sum_k m_{ik}(t)\lambda_k f_{kj} s + o(s).$$

Forming the corresponding difference quotients and sending s to 0 yield the system of differential equations

$$m'_{ij}(t) = \sum_k m_{ik}(t)\lambda_k(f_{kj} - 1_{\{k=j\}}),$$

which we summarize as the matrix differential equation $M'(t) = M(t)\Omega$ for the $n \times n$ matrices $M(t) = [m_{ij}(t)]$ and $\Omega = [\lambda_i(f_{ij} - 1_{\{i=j\}})]$. The solution is provided by the matrix exponential $M(t) = e^{t\Omega}$ subject to the initial condition $M(0) = I$ [12].

The asymptotic behavior of $M(t)$ is determined qualitatively by the eigenvalue ρ of Ω with largest real part. Such a dominant eigenvalue exists and is real for an irreducible process [11, 12, 68, 82, 137]. In an irreducible process, a particle of one type can ultimately produce descendants of every other type. A descendant may be a granddaughter, great granddaughter, and so forth rather than a daughter directly. Irreducibility can be checked by examining the off-diagonal elements of Ω. When a process is irreducible, it is said to subcritical, critical, or supercritical according as $\rho < 0$, $\rho = 0$, or $\rho > 0$, respectively. The components of $M(t)$ tend to zero for a subcritical process and to infinity for a supercritical process.

To investigate the phenomenon of extinction, let $e_i(t)$ be the probability that the process is extinct at time t given that it begins with a single particle of type i at time 0. We can characterize the $e_i(t)$ by deriving a system of nonlinear ordinary differential equations. In almost all cases, this system must be solved numerically. In deriving the system of differential equations, suppose a type i particle produces d_1 type 1 daughter particles, d_2 type 2 daughter particles, and so on, to d_n type n daughter particles, with probability $p_{i,(d_1,\ldots,d_n)}$. To ease the notational burden, we write

$$\mathbf{d} = (d_1,\ldots,d_n)$$

$$p_{i\mathbf{d}} = p_{i,(d_1,\ldots,d_n)}$$

$$\mathbf{e}(t)^{\mathbf{d}} = \prod_{i=1}^{n} e_i(t)^{d_i}.$$

Given these conventions, we contend that

$$e_i(t+s) = (1-\lambda_i s)e_i(t) + \lambda_i s \sum_{\mathbf{d}} p_{i\mathbf{d}}\mathbf{e}(t)^{\mathbf{d}} + o(s). \qquad (9.14)$$

The logic behind this expression is straightforward. Either the original particle does not die during the time interval $(0, s)$, or it does and leaves behind a vector of $\mathbf{d}$ daughter particles with probability $p_{i\mathbf{d}}$. Each of the clans emanating from one of the daughter particles goes extinct independently of the remaining clans. Rearranging expression (9.14) into a difference quotient and sending s to 0 yield the differential equation

$$e_i'(t) = -\lambda_i e_i(t) + \lambda_i \sum_{\mathbf{d}} p_{i\mathbf{d}}\mathbf{e}(t)^{\mathbf{d}}.$$

The probabilities $e_i(t)$ are increasing because once extinction occurs, it is permanent. To find the limit of $e_i(t)$ as t tends to infinity, we set $e_i'(t)$ equal to 0. This action determines the ultimate extinction probabilities $\mathbf{e} = (e_1,\ldots,e_n)$ through the algebraic equations

$$e_i = \sum_{\mathbf{d}} p_{i\mathbf{d}}\mathbf{e}^{\mathbf{d}}. \qquad (9.15)$$

It is noteworthy that these equations do not depend on the average life expectancies of the different particle types but only on their reproductive patterns. For subcritical and critical irreducible processes, all extinction probabilities are 1. For supercritical irreducible processes, all extinction probabilities are strictly less than 1 [11, 12, 68, 82].

9.7   Viral Reproduction in HIV

As an illustration of the above theory, consider the following branching process model of how the HIV virus infects CD4 cells of the immune system in the first stage of an HIV infection [120]. Particles in this simplified model correspond to either virus particles (virions) in plasma or two types of infected CD4 cells. Virions are type 1 particles, latently infected CD4 cells are type 2 particles, and actively infected CD4 cells are type 3 particles. In actively infected CD4 cells, HIV is furiously replicating. As a first approximation, we will assume that replicated virions are released in a burst when the cell membrane ruptures and a large number of virions spill into the plasma. HIV does not replicate in latently infected cells. Therefore, a latently infected cell either converts to an actively infected cell or quietly dies without replenishing the plasma load of virus.

Let us now consider the fate of each type of particle. Type 1 particles or virions die or are eliminated from plasma at rate σ. A virion enters and infects an uninfected CD4 cell at rate βR, where β is the infection rate per CD4 cell, and where R is the fixed number of uninfected CD4 cells in plasma. (A defect of the model is that as time progresses R should decline as more and more CD4 cells are infected. In the earliest stage of an infection, we can ignore this effect.) Let θ be the probability that a CD4 cell commences its infection in a latent rather than in an active state. In the branching process paradigm, we interpret death broadly to mean either natural virion death, virion elimination, or virion removal by infection. Thus, our verbal description translates into the quantitative assumptions $\lambda_1 = \sigma + \beta R$ and

$$ f_{11} = 0, \qquad f_{12} = \frac{\theta \beta R}{\lambda_1}, \qquad f_{13} = \frac{(1-\theta)\beta R}{\lambda_1}. $$

Latently infected cells are eliminated at rate μ or convert to actively infected cells at rate α. Again broadly interpreting death in the branching process sense, we take $\lambda_2 = \mu + \alpha$ and

$$ f_{21} = 0, \qquad f_{22} = 0, \qquad f_{23} = \frac{\alpha}{\lambda_2}. $$

Finally, actively infected CD4 cells are eliminated at rate μ or burst due to viral infection at rate δ. When an actively infected CD4 cell bursts, it dumps an average π virions into the plasma. These new virions start the cycle anew. For type 3 particles we have $\lambda_3 = \mu + \delta$ and

$$ f_{31} = \frac{\delta \pi}{\lambda_3}, \qquad f_{32} = 0, \qquad f_{33} = 0. $$

Straightforward algebra shows that the branching process matrix

$$ \Omega = \begin{pmatrix} -\sigma - \beta R & \theta \beta R & (1-\theta)\beta R \\ 0 & -\mu - \alpha & \alpha \\ \delta \pi & 0 & -\mu - \delta \end{pmatrix} $$

and consequently that Ω has characteristic polynomial

$$
\det(\Omega - xI) = \det \begin{pmatrix} -\sigma - \beta R - x & \theta \beta R & (1 - \theta)\beta R \\ 0 & -\mu - \alpha - x & \alpha \\ \delta\pi & 0 & -\mu - \delta - x \end{pmatrix}
$$

$$
= -(\sigma + \beta R + x)(\mu + \alpha + x)(\mu + \delta + x)
$$
$$
+ \delta\pi[\theta\beta R\alpha + (\mu + \alpha + x)(1 - \theta)\beta R].
$$

Inspection of the cubic polynomial $p(x) = \det(\Omega - xI)$ demonstrates that the coefficients of x^3 and x^2 are negative. The sign of the coefficient of x is unclear, but the constant term of $p(x)$ is certainly positive if

$$
\frac{\delta\pi[\theta\beta R\alpha + (\mu + \alpha)(1 - \theta)\beta R]}{(\sigma + \beta R)(\mu + \alpha)(\mu + \delta)} = \frac{\delta\pi\beta R[\alpha + (1 - \theta)\mu]}{(\sigma + \beta R)(\mu + \alpha)(\mu + \delta)}
$$
$$
> 1. \tag{9.16}
$$

Regardless of the sign of the coefficient of x, the coefficients of $p(x)$ experience one change of sign. Thus, Descartes' rule of signs [149] implies that there is a positive root of the equation $p(x) = 0$ and therefore that the mean viral load in plasma grows explosively. It turns out that this root is the dominant eigenvalue of Ω. Conversely, one can prove that the mean viral load declines to 0 when the reproduction number on the right-hand side of inequality (9.16) satisfies

$$
\frac{\delta\pi\beta R[\alpha + (1 - \theta)\mu]}{(\sigma + \beta R)(\mu + \alpha)(\mu + \delta)} < 1.
$$

Problem 16 proves this more subtle assertion and gives a biological interpretation of the reproduction number.

It is impossible to calculate extinction probabilities in this model without specifying the model more fully. One possibility is to make the somewhat more realistic assumption that reproduction of virions by an actively infected cell occurs continuously rather than as a gigantic burst. If the cell sheds single virions with a steady intensity γ, then we need to adjust γ so that the expected number of virions produced continuously over the cell's lifetime matches the expected number of virions produced instantaneously by bursting. This leads to the condition $\gamma/\mu = \delta\pi/\lambda_3$ determining γ. We also must adjust our conception of death. The branching process model requires that reproduction occur simultaneously with death. Hence, in our revised model, an actively infected cell dies with rate $\mu + \gamma$. At its death, it leaves behind no particles with probability $\mu/(\mu+\gamma)$ or an actively infected cell and a virion with probability $\gamma/(\mu+\gamma)$.

With these amendments, the system of extinction equations (9.15) becomes

$$
e_1 = \frac{\sigma}{\lambda_1} + \frac{\theta\beta R}{\lambda_1}e_2 + \frac{(1 - \theta)\beta R}{\lambda_1}e_3
$$

$$e_2 \;=\; \frac{\mu}{\lambda_2} + \frac{\alpha}{\lambda_2} e_3$$

$$e_3 \;=\; \frac{\mu}{\mu+\gamma} + \frac{\gamma}{\mu+\gamma} e_1 e_3.$$

This system of equations reduces to a single quadratic equation for e_1 if we (a) substitute for e_2 in the first equation using the second equation, (b) solve the third equation for e_3 in terms of e_1, and (c) substitute the result into the modified first equation. We leave the messy details to the reader.

9.8   Problems

1. If p and α are constants in the open interval $(0,1)$, then show that $Q(s) = 1 - p(1-s)^\alpha$ is a generating function with nth functional iterate

$$Q_n(s) \;=\; 1 - p^{1+\alpha+\cdots+\alpha^{n-1}}(1-s)^{\alpha^n}.$$

Remember to check that the coefficients of $Q(s)$ are nonnegative and sum to 1.

2. Let X_n be the number of particles at generation n in a supercritical branching process with progeny mean μ and variance σ^2. If $X_0 = 1$ and $Z_n = X_n/\mu^n$, then find $\lim_{n\to\infty} \mathrm{E}(Z_n)$ and $\lim_{n\to\infty} \mathrm{Var}(Z_n)$. The fact that these limits exist and are finite when $\mu > 1$ correctly suggests that Z_n tends to a limiting random variable Z_∞.

3. Consider a supercritical branching process X_n with progeny generating function $Q(s)$ and extinction probability s_∞. Show that

$$\Pr(1 \leq X_n \leq k)s_\infty^k \;\leq\; Q_n'(s_\infty)$$

for all $k \geq 1$ and that

$$Q_n'(s_\infty) \;=\; Q'(s_\infty)^n.$$

Use these results and the Borel-Cantelli lemma to prove that

$$\Pr\left(\lim_{n\to\infty} X_n = \infty\right) \;=\; 1 - s_\infty.$$

4. Continuing Example 9.2.5, let $P(s)$ be the generating function for the total number of carrier and normal children born to a carrier of the mutant gene. Express the progeny generating function $Q(s)$ of the mutant-gene branching process in terms of $P(s)$. Find the mean μ and variance σ^2 of $Q(s)$ in terms of the mean α and variance β^2 of $P(s)$. How small must α be in order for extinction of the mutant-gene process to be certain?

5. The generating function $\frac{p}{1-qs}$ is an example of a fractional linear transformation $\frac{\alpha s + \beta}{\gamma s + \delta}$ [73]. To avoid trivial cases where the fractional linear transformation is undefined or constant, we impose the condition that $\alpha\delta - \beta\gamma \neq 0$. The restricted set of fractional linear transformations (or Möbius functions) forms a group under functional composition. This group is the homomorphic image of the group of invertible 2×2 matrices under the correspondence

$$\begin{pmatrix} \alpha & \beta \\ \gamma & \delta \end{pmatrix} \quad \longrightarrow \quad \frac{\alpha s + \beta}{\gamma s + \delta}. \tag{9.17}$$

A group homomorphism is a function between two groups that preserves the underlying algebraic operation. Show that the correspondence (9.17) qualifies as a group homomorphism in the sense that if $f_i(s) = \frac{\alpha_i s + \beta_i}{\gamma_i s + \delta_i}$ for $i = 1, 2$, then

$$\begin{pmatrix} \alpha_1 & \beta_1 \\ \gamma_1 & \delta_1 \end{pmatrix} \begin{pmatrix} \alpha_2 & \beta_2 \\ \gamma_2 & \delta_2 \end{pmatrix} \longrightarrow f_1(f_2(s)).$$

The homomorphism (9.17) correctly pairs the two identity elements $\begin{pmatrix} 1 & 0 \\ 0 & 1 \end{pmatrix}$ and $f(s) = s$ of the groups.

6. Suppose $Q(s)$ is a generating function with mean μ and variance σ^2. If $\mu - 1$ is small and positive, then verify that $Q(s)$ has approximate extinction probability $e^{-2(\mu-1)/\sigma^2}$. Show that this approximation equals 0.892 for Lotka's demographic data. (Hints: Put $t = \ln s$ and $L(t) = \ln Q(s)$. Expand $L(t)$ in a second order Taylor's series around $t = 0$.)

7. Let s_∞ be the extinction probability of a supercritical branching process with progeny generating function $Q(s) = \sum_{k=0}^{\infty} q_k s^k$. If the mean μ of $Q(s)$ is fixed, then one can construct counterexamples showing that s_∞ is not necessarily increasing as a function of q_0. As a case in point, let $Q(s) = P\left(\frac{1}{2} + \frac{1}{2}s\right)$ and consider

$$P(s) = \frac{1}{6} + \frac{5}{6}s^3$$

$$P(s) = \frac{3}{32} + \frac{15}{24}s^2 + \frac{5}{32}s^4 + \frac{3}{24}s^5.$$

Check numerically that these two choices lead to the extinction probabilities $s_\infty = 0.569$ and $s_\infty = 0.594$ and coefficients $q_0 = 0.271$ and $q_0 = 0.264$.

8. Newton's method offers an alternative method of finding the extinction probability s_∞ of a supercritical generating function $Q(s)$. Let

$s_0 = 0$ and $t_0 = 0$ be the initial values in the iteration schemes

$$s_{n+1} = Q(s_n), \qquad t_{n+1} = t_n + \frac{Q(t_n) - t_n}{1 - Q'(t_n)}.$$

Prove that t_n is an increasing sequence satisfying $s_n \leq t_n \leq s_\infty$ for all $n \geq 0$. It follows that $\lim_{n \to \infty} t_n = s_\infty$ and that Newton's method converges faster than functional iteration.

9. In a subcritical branching process, let T be the generation at which the process goes extinct starting from a single particle at generation 0. If $s_k = \Pr(T \leq k)$ and $Q(s)$ is the progeny generating function, then prove the error estimates

$$\frac{Q'(s_n)}{1 - Q'(s_n)}(1 - s_n) \leq E(T) - \sum_{k=0}^{n}(1 - s_k)$$

$$\leq \frac{Q'(1)}{1 - Q'(1)}(1 - s_n)$$

for $E(T)$. (Hint: Use the first equation in (9.5).)

10. In a branching process, let $R(s) = \sum_{k=0}^{\infty} r_k s^k$ be the generating function of the total number of particles Y_∞ over all generations starting from a single particle. If the progeny generating function is $Q(s)$, find r_k and the mean and variance of Y_∞ in each of the cases

$$Q(s) = e^{\lambda(s-1)}$$
$$Q(s) = (q + ps)^n$$
$$Q(s) = \left(\frac{p}{1 - qs}\right)^n,$$

where $q = 1 - p$. Furthermore, show that the Poisson, binomial, or negative binomial form of each of these progeny generating functions is preserved when we substitute $\frac{1}{2} + \frac{1}{2}s$ for s. Comment on the relevance of this last result to Problem 4.

11. Suppose X_n denotes the number of particles in a branching process with immigration. Let μ be the mean number of progeny per particle and α the mean number of new immigrants per generation. An ordinary branching process corresponds to the case $\alpha = 0$. For $k > j$ show that

$$E(X_k \mid X_j) = \begin{cases} (k - j)\alpha + X_j, & \mu = 1 \\ \frac{\alpha(1 - \mu^{k-j})}{1 - \mu} + \mu^{k-j} X_j, & \mu \neq 1 \end{cases}$$

$$\text{Cov}(X_k, X_j) = \mu^{k-j}\, \text{Var}(X_j).$$

12. In a subcritical branching process with immigration, let $Q(s)$ be the progeny generating function and $R(s)$ the generating function of the number of new immigrants at each generation. If the equilibrium distribution has generating function $P_\infty(s)$, then show that

$$P_\infty(s) \;=\; P_\infty(Q(s))R(s).$$

For the choices $Q(s) = 1 - p + ps$ and $R(s) = e^{-\lambda(1-s)}$, find $P_\infty(s)$. (Hint: Let $P_\infty(s)$ be a Poisson generating function.)

13. Branching processes can be used to model the formation of polymers [118]. Consider a large batch of identical subunits in solution. Each subunit has $m > 1$ reactive sites that can attach to similar reactive sites on other subunits. For the sake of simplicity, assume that a polymer starts from a fixed ancestral subunit and forms a tree structure with no cross linking of existing subunits. Also assume that each reactive site behaves independently and bonds to another site with probability p. Subunits attached to the ancestral subunit form the first generation of a branching process. Subunits attached to these subunits form the second generation and so forth. In this problem we investigate the possibility that polymers of infinite size form. In this case the solution turns into a gel. Show that the progeny distribution for the first generation is binomial with m trials and success probability p and that the progeny distribution for subsequent generations is binomial with $m-1$ trials and success probability p. Show that the extinction probability t_∞ satisfies

$$
\begin{aligned}
t_\infty &= (1 - p + ps_\infty)^m \\
s_\infty &= (1 - p + ps_\infty)^{m-1},
\end{aligned}
$$

where s_∞ is the extinction probability for a line of descent emanating from a first-generation subunit. Prove that polymers of infinite size occur if and only if $(m-1)p > 1$.

14. Yeast cells reproduce by budding. Suppose at each generation a yeast cell either dies with probability p, survives without budding with probability q, or survives with budding off a daughter cell with probability r. In the ordinary branching process paradigm, a surviving cell is considered a new cell. If we refuse to take this view, then what is the distribution of the number of daughter cells budded off by a single yeast cell before its death? Show that the extinction probability of a yeast cell line is 1 when $p \geq r$ and $\frac{p}{r}$ when $p < r$ [54].

15. At an X-linked recessive disease locus, there are two alleles, the normal allele (denoted $+$) and the disease allele (denoted $-$). Construct a two-type branching process for carrier females (genotype $+/-$) and affected males (genotype $-$). Calculate the expected numbers f_{ij} of

offspring of each type assuming that carrier females average 2 children, affected males average $2f$ children, all mates are $+/+$ or $+$, and children occur in a 1:1 sex ratio. Note that a branching process model assumes that all children are born simultaneously with the death of a parent. The sensible choice for the death rate λ in a continuous-time model of either type parent is the reciprocal of the generation time, say about $\frac{1}{25}$ per year in humans.

16. In the HIV branching process model, it is of interest to calculate the reproductive potential of a virion in plasma. Because virus reproduction takes place in CD4 cells, let r_i be the expected number of new virions that a particle of type i eventually generates. Show that these numbers obey the following equations:

$$
\begin{aligned}
r_1 &= \frac{\theta\beta R}{\sigma + \beta R} r_2 + \frac{(1-\theta)\beta R}{\sigma + \beta R} r_3 \\
r_2 &= \frac{\alpha}{\mu + \alpha} r_3 \\
r_3 &= \frac{\delta}{\mu + \delta} \pi.
\end{aligned}
$$

From these equations calculate

$$
r_1 = \frac{\delta\pi\beta R[\alpha + (1-\theta)\mu]}{(\sigma + \beta R)(\mu + \alpha)(\mu + \delta)}.
$$

If the reproduction number $r_1 < 1$, then virus numbers keep dropping until extinction. Conversely, virus numbers grow exponentially when $r_1 > 1$. The case $r_1 = 1$ is indeterminate, but a full stochastic analysis of the branching process model demonstrates that extinction is certain in this case as well [12].

17. Consider a multitype branching process with immigration. Suppose that each particle of type i has an exponential lifetime with death intensity λ_i and produces on average f_{ij} particles of type j at the moment of its death. Independently of death and reproduction, immigrants of type i enter the population according to a Poisson process with intensity α_i. If the Poisson immigration processes for different types are independent, then show that the mean number $m_i(t)$ of particles of type i satisfies the differential equation

$$
m_i'(t) = \alpha_i + \sum_j m_j(t)\lambda_j(f_{ji} - 1_{\{j=i\}}).
$$

Collecting the $m_i(t)$ and α_i into row vectors $m(t)$ and α, respectively, and the $\lambda_j(f_{ji} - 1_{\{j=i\}})$ into a matrix Ω, show that

$$
m(t) = m(0)e^{t\Omega} + \alpha\Omega^{-1}(e^{t\Omega} - I),
$$

assuming that Ω is invertible. If we replace the constant immigration intensity α_i by the exponentially decreasing immigration intensity $\alpha_i e^{-\mu t}$, then verify that

$$m(t) \;=\; m(0)e^{t\Omega} + \alpha(\Omega + \mu I)^{-1}(e^{t\Omega} - e^{-t\mu I}).$$

18. In a certain species, females die with intensity μ and males with intensity ν. All reproduction is through females at an intensity of λ per female. At each birth, the mother bears a daughter with probability p and a son with probability $1-p$. Interpret this model as a two-type, continuous-time branching process with X_t representing the number of females and Y_t representing the number of males, and show that

$$
\begin{aligned}
\mathrm{E}(X_t) \;&=\; \mathrm{E}(X_0)e^{(\lambda p - \mu)t} \\
\mathrm{E}(Y_t) \;&=\; \mathrm{E}(X_0)\frac{\lambda(1-p)}{\lambda p + \nu - \mu}e^{(\lambda p - \mu)t} \\
&\quad + \left[\mathrm{E}(Y_0) - \mathrm{E}(X_0)\frac{\lambda(1-p)}{\lambda p + \nu - \mu}\right]e^{-\nu t}.
\end{aligned}
$$

19. In some applications of continuous-time branching processes, it is awkward to model reproduction as occurring simultaneously with death. Birth-death processes offer an attractive alternative. In a birth-death process, a type i particle experiences death at rate μ_i and reproduction of daughter particles of type j at rate β_{ij}. Each reproduction event generates one and only one daughter particle. Thus, in a birth-death process each particle continually buds off daughter particles until it dies. In contrast, each particle of a multitype continuous-time branching process produces a burst of offspring at the moment of its death. This problem considers how we can reconcile these two modes of reproduction. There are two ways of doing this, one exact and one approximate.

   (a) Show that in a birth-death process, a particle of type i produces the count vector $\mathbf{d} = (d_1, \ldots, d_n)$ of daughter particles with probability

   $$p_{i\mathbf{d}} \;=\; \frac{\mu_i}{(\mu_i + \beta_i)^{|\mathbf{d}|+1}}\binom{|\mathbf{d}|}{d_1 \ldots d_n}\prod_{k=1}^{n}\beta_{ik}^{d_k},$$

   where $\beta_i = \sum_{j=1}^{n}\beta_{ij}$ and $|\mathbf{d}| = d_1 + \cdots + d_n$. (Hint: Condition on the time of death. The number of daughter particles of a given type produced up to this time follows a Poisson distribution.)

   (b) If we delay all offspring until the moment of death, then we get a branching process approximation to the birth-death process.

What is the death rate λ_i in the branching process approximation? Show that the approximate process has progeny generating
function

$$P_i(\mathbf{s}) = \sum_{\mathbf{d}} p_{i\mathbf{d}} \mathbf{s}^{\mathbf{d}}$$

$$= \sum_{m=0}^{\infty} \frac{\mu_i}{(\mu_i + \beta_i)^{m+1}} \left(\sum_{j=1}^{n} \beta_{ij} s_j \right)^m$$

for a type i particle.

(c) In the branching process approximation, demonstrate that a
type i particle averages $f_{ij} = \frac{\partial}{\partial s_j} P_i(\mathbf{1}) = \beta_{ij}/\mu_i$ type j daughter
particles. (Hint: Differentiate the identity $(1-x)^{-1} = \sum_{m=0}^{\infty} x^m$
for $|x| < 1$.)

(d) Explain in laymen's term the meaning of the ratio defining f_{ij}.

(e) Alternatively, we can view the mother particle as dying in one of
two ways. Either it dies in the ordinary way at rate μ_i, or it disappears at a reproduction event and is replaced by an identical
substitute and a single daughter particle. Eventually one of the
substitute particles dies in the ordinary way before reproduc-
ing, corresponding to death in the original birth-death process.
What is the death rate λ_i in this exact branching process analog
of the birth-death process? Justify in words the progeny generating function

$$P_i(\mathbf{s}) = \frac{\mu_i}{\mu_i + \beta_i} + \sum_{k=1}^{r} \frac{\beta_{ik}}{\mu_i + \beta_i} s_i s_k.$$

(f) We can turn the approximate correspondence discussed in parts
(b) and (c) around and seek to mimic a branching process by a
birth-death process. The most natural method involves matching the mean number of daughter particles f_{ij} in the branching
process to the mean number of daughter particles budded off in
the birth-death process. Given the λ_i and the f_{ij}, what are the
natural values for the death rates μ_i and the birth rates β_{ij} in
the birth-death approximation to the branching process?

10
Martingales

10.1 Introduction

Martingales generalize the notion of a fair game in gambling. Theory to the contrary, many gamblers still believe that they simply need to hone their strategies to beat the house. Probabilists know better. The real payoff with martingales is their practical value throughout probability theory. This chapter introduces martingales, develops some relevant theory, and delves into a few applications. As a prelude, readers are urged to review the material on conditional expectations in Chapter 1. In the current chapter we briefly touch on the convergence properties of martingales, the optional stopping theorem, and large deviation bounds via Azuma's inequality. More extensive treatments of martingale theory appear in the books [18, 19, 42, 64, 82, 93, 152]. Our other referenced sources either provide elementary accounts comparable in difficulty to the current material [102, 130] or interesting special applications [4, 106, 140, 147].

10.2 Definition and Examples

A sequence of integrable random variables X_n forms a martingale relative to a second sequence of random variables Y_n provided

$$\mathrm{E}(X_{n+1} \mid Y_1, \ldots, Y_n) \;=\; X_n \qquad\qquad (10.1)$$

for all $n \geq 1$. In many applications $X_n = Y_n$. It saves space and is often conceptually simpler to replace the collection of random variables $Y_1, \ldots, Y_n$ by the σ-algebra of events $\mathcal{F}_n$ that it generates. Note that these σ-algebras are increasing in the sense that any $A \in \mathcal{F}_n$ satisfies $A \in \mathcal{F}_{n+1}$. This relationship is written $\mathcal{F}_n \subset \mathcal{F}_{n+1}$, and the sequence $\mathcal{F}_n$ is said to be a filter. This somewhat odd terminology is motivated by the fact that the $\mathcal{F}_n$ contain more information or detail as n increases. In any case, we now rephrase the definition of a martingale to be a sequence of integrable random variables X_n satisfying

$$E(X_{n+1} \mid \mathcal{F}_n) \;=\; X_n \tag{10.2}$$

relative to a filter $\mathcal{F}_n$. Readers who find this definition unnecessarily abstract are urged to fall back on the original definition (10.1).

Before turning to specific examples, let us deduce a few elementary properties of martingales. First, equation (10.2) implies that the random variable X_n is measurable with respect to $\mathcal{F}_n$. When $\mathcal{F}_n$ is generated by $Y_1, \ldots, Y_n$, then X_n is expressible as a measurable function of $Y_1, \ldots, Y_n$. Second, iteration of the identity

$$
\begin{aligned}
E(X_{n+1}) \;&=\; E[E(X_{n+1} \mid \mathcal{F}_n)] \\
&=\; E(X_n)
\end{aligned}
$$

leads to $E(X_n) = E(X_1)$ for all $n > 1$. Third, an obvious inductive argument using the tower property (1.6) gives

$$
\begin{aligned}
E(X_{n+k} \mid \mathcal{F}_n) \;&=\; E[E(X_{n+k} \mid \mathcal{F}_{n+1}) \mid \mathcal{F}_n] \\
&=\; E(X_{n+1} \mid \mathcal{F}_n) \\
&=\; X_n
\end{aligned}
$$

for all $k > 0$. Fourth, if the X_n are square integrable and $i \leq j \leq k \leq l$, then

$$
\begin{aligned}
E[(X_j - X_i)(X_l - X_k)] \;&=\; E\{E[(X_j - X_i)(X_l - X_k) \mid \mathcal{F}_j]\} \\
&=\; E\{(X_j - X_i)\,E[(X_l - X_k) \mid \mathcal{F}_j]\} \\
&=\; E[(X_j - X_i)(X_j - X_j)] \\
&=\; 0.
\end{aligned}
$$

In other words, the increments $X_j - X_i$ and $X_l - X_k$ are orthogonal. This fact allows us to calculate the variance of

$$X_n \;=\; X_1 + \sum_{i=2}^{n} (X_i - X_{i-1})$$

via

$$\mathrm{Var}(X_n) \;=\; \mathrm{Var}(X_1) + \sum_{i=2}^{n} \mathrm{Var}(X_i - X_{i-1}), \tag{10.3}$$

exactly as if we were dealing with sums of independent random variables. The fact that $\mathrm{Var}(X_n)$ is increasing in n follows immediately from the decomposition (10.3). Finally, the special case

$$\mathrm{Var}(X_{m+n} - X_m) \;=\; \mathrm{Var}(X_{m+n}) - \mathrm{Var}(X_m) \qquad (10.4)$$

of the orthogonal increments property is worth highlighting.

Example 10.2.1 *Sums of Random Variables*

If Y_n is a sequence of independent random variables with common mean $\mu = \mathrm{E}(Y_1) = 0$, then the partial sums $S_n = Y_1 + \cdots + Y_n$ constitute a martingale relative to the filter $\mathcal{F}_n$ generated by $\{Y_1, \ldots, Y_n\}$. The martingale property (10.2) follows from the calculation

$$
\begin{aligned}
\mathrm{E}(S_{n+1} \mid \mathcal{F}_n) &= \mathrm{E}(Y_{n+1} \mid \mathcal{F}_n) + \mathrm{E}(S_n \mid \mathcal{F}_n) \\
&= \mathrm{E}(Y_{n+1}) + S_n \\
&= S_n.
\end{aligned}
$$

When $\mu \neq 0$, the modified sequence $S_n - n\mu$ forms a martingale. If the Y_n are dependent random variables, then the sequence

$$
\begin{aligned}
S_n &= \sum_{i=1}^{n} [Y_i - \mathrm{E}(Y_i \mid \mathcal{F}_{i-1})] \\
&= S_{n-1} + Y_n - \mathrm{E}(Y_n \mid \mathcal{F}_{n-1})
\end{aligned}
$$

provides a zero-mean martingale because

$$
\begin{aligned}
\mathrm{E}(S_{n+1} \mid \mathcal{F}_n) &= S_n + \mathrm{E}(Y_{n+1} \mid \mathcal{F}_n) - \mathrm{E}(Y_{n+1} \mid \mathcal{F}_n) \\
&= S_n.
\end{aligned}
$$

■

Example 10.2.2 *Products of Independent Random Variables*

If Y_n is a sequence of independent random variables with common mean $\mu = \mathrm{E}(Y_1) = 1$, then the partial products $X_n = \prod_{i=1}^{n} Y_i$ constitute a martingale relative to the filter $\mathcal{F}_n$ generated by $\{Y_1, \ldots, Y_n\}$. The martingale property (10.2) follows from the calculation

$$
\begin{aligned}
\mathrm{E}\Big(\prod_{i=1}^{n+1} Y_i \mid \mathcal{F}_n\Big) &= \mathrm{E}(Y_{n+1} \mid \mathcal{F}_n) \prod_{i=1}^{n} Y_i \\
&= \prod_{i=1}^{n} Y_i.
\end{aligned}
$$

When $\mu \neq 1$, the modified sequence $\mu^{-n} X_n$ is a martingale.

This martingale arises in Wald's theory of sequential testing in statistics. Let $Z_1, Z_2, \ldots$ be an i.i.d. sequential of random variables with density $f(z)$ under the simple null hypothesis H_o and density $g(z)$ under the simple alternative hypothesis H_a. Both of these densities are relative to a common measure μ such as Lebesgue measure or counting measure. If H_o is true, then

$$\int \frac{g(z)}{f(z)} f(z)\, d\mu(z) \;=\; \int g(z)\, d\mu(z) \;=\; 1.$$

It follows that the likelihood ratio statistics

$$X_n \;=\; \prod_{i=1}^{n} \frac{g(Z_i)}{f(Z_i)}$$

constitute a martingale. ∎

Example 10.2.3 *Martingale Differences*

If X_n is a martingale with respect to the filter $\mathcal{F}_n$, then the differences $\{X_{m+n} - X_m\}_{n\geq 1}$ are a martingale with respect to the filter $\{\mathcal{F}_{m+n}\}_{n\geq 1}$. This assertion is a consequence of the identity

$$\mathrm{E}(X_{m+n} - X_m \mid \mathcal{F}_{m+n-1}) \;=\; X_{m+n-1} - X_m.$$

∎

Example 10.2.4 *Doob's Martingale*

Let Z be an integrable random variable and $\mathcal{F}_n$ be a filter. Then the sequence $X_n = \mathrm{E}(Z \mid \mathcal{F}_n)$ is a martingale because

$$\mathrm{E}[\mathrm{E}(Z \mid \mathcal{F}_{n+1}) \mid \mathcal{F}_n] \;=\; \mathrm{E}(Z \mid \mathcal{F}_n).$$

In many examples, $\mathcal{F}_n$ is the σ-algebra defined by n independent random variables $Y_1, \ldots, Y_n$. If the random variable Z is a function of $Y_1, \ldots, Y_m$ alone, then $X_m = Z$, and for $1 \leq n < m$

$$X_n(y_1, \ldots, y_n) \;=\; \mathrm{E}(Z \mid Y_1 = y_1, \ldots, Y_n = y_n) \tag{10.5}$$

$$= \int \cdots \int Z(y_1, \ldots, y_m)\, dF_{n+1}(y_{n+1}) \cdots dF_m(y_m),$$

where $F_k(y_k)$ is the distribution function of Y_k. In other words, the conditional expectation X_n is generated by integrating over the last $m - n$ arguments of Z. This formula for X_n is correct because Fubini's theorem gives

$$\mathrm{E}[1_A(Y_1, \ldots, Y_n)Z]$$

$$= \int \cdots \int 1_A(y_1, \ldots, y_n) X_n(y_1, \ldots, y_n)\, dF_1(y_1) \cdots dF_n(y_n)$$

for any event A depending only on $Y_1, \ldots, Y_n$. ∎

Example 10.2.5 *Branching Processes*

Suppose Y_n counts the number of particles at generation n in a discrete-time branching process. (Here the index n starts at 0 rather than 1.) Let $\mathcal{F}_n$ be the σ-algebra generated by $Y_0, \ldots, Y_n$, and let μ be the mean of the progeny distribution. We argued in Chapter 9 that

$$\mathrm{E}(Y_{n+1} \mid Y_n) \;=\; \mathrm{E}(Y_{n+1} \mid \mathcal{F}_n) \;=\; \mu Y_n.$$

It follows that $X_n = \mu^{-n} Y_n$ is a martingale relative to the filter $\mathcal{F}_n$. ∎

Example 10.2.6 *Wright-Fisher Model of Genetic Drift*

In the Wright-Fisher model of Example 7.3.2, the proportion $X_n = \frac{1}{2m} Y_n$ of a_1 alleles at generation n provides a martingale relative to the filter $\mathcal{F}_n$ determined by the random counts $Y_1, \ldots, Y_n$ of a_1 alleles at generations 1 through n. Indeed,

$$\begin{aligned}
\mathrm{E}(X_{n+1} \mid \mathcal{F}_n) &= \frac{1}{2m} \mathrm{E}(Y_{n+1} \mid Y_n) \\
&= X_n
\end{aligned}$$

is obvious from the nature of the binomial sampling with success probability X_n in forming the population at generation $n + 1$. ∎

10.3   Martingale Convergence

Our purpose in this section is to inquire when a martingale X_n possesses a limit. The theory is much simpler if we stipulate that the X_n have finite second moments. This condition is not sufficient for convergence as Example 10.2.1 shows. However, if the second moments $\mathrm{E}(X_n^2)$ are uniformly bounded, then in general we get a convergent sequence. The next proposition paves the way by generalizing Chebyshev's inequality.

Proposition 10.3.1 (Kolmogorov) *Suppose that relative to the filter $\mathcal{F}_n$ the square-integrable martingale X_n has mean $\mathrm{E}(X_n) = 0$. The inequality*

$$\Pr(\max\{|X_1|, \ldots, |X_n|\} > \epsilon) \;\le\; \frac{\mathrm{Var}(X_n)}{\epsilon^2} \tag{10.6}$$

then holds for all $\epsilon > 0$ and $n \ge 1$.

Proof: Let $M = m$ be the smallest subscript between 1 and n such that $|X_m| > \epsilon$. If no such subscript exists, then set $M = 0$. The calculation

$$\begin{aligned}
\mathrm{E}(X_n^2) &= \mathrm{E}(X_n^2 1_{\{M=0\}}) + \sum_{m=1}^{n} \mathrm{E}(X_n^2 1_{\{M=m\}}) \\
&\ge \sum_{m=1}^{n} \mathrm{E}(X_n^2 1_{\{M=m\}}) \tag{10.7}
\end{aligned}$$

follows because the events $\{M = 0\}, \ldots, \{M = n\}$ are mutually exclusive and exhaustive. In view of the fact that the event $\{M = m\} \in \mathcal{F}_m$, we have

$$
\begin{aligned}
\mathrm{E}(X_n^2 1_{\{M=m\}}) &= \mathrm{E}[1_{\{M=m\}} \mathrm{E}(X_n^2 \mid \mathcal{F}_m)] \\
&= \mathrm{E}\{1_{\{M=m\}} \mathrm{E}[(X_n - X_m)^2 + X_m^2 \mid \mathcal{F}_m]\} \\
&\geq \mathrm{E}[1_{\{M=m\}} \mathrm{E}(X_m^2 \mid \mathcal{F}_m)] \quad (10.8) \\
&= \mathrm{E}(1_{\{M=m\}} X_m^2) \\
&\geq \epsilon^2 \Pr(M = m).
\end{aligned}
$$

In this derivation, we have employed the identity

$$
\mathrm{E}[X_m(X_n - X_m) \mid \mathcal{F}_m] = X_m(X_m - X_m) = 0.
$$

Combining inequality (10.8) with inequality (10.7) yields

$$
\mathrm{E}(X_n^2) \geq \epsilon^2 \sum_{m=1}^{n} \Pr(M = m) = \epsilon^2 \Pr(M > 0),
$$

which is clearly equivalent to inequality (10.6).   ∎

Proposition 10.3.2 *Suppose the martingale X_n relative to the filter $\mathcal{F}_n$ has uniformly bounded second moments. Then $X_\infty = \lim_{n \to \infty} X_n$ exists almost surely and*

$$
\lim_{n \to \infty} \mathrm{E}(|X_\infty - X_n|^2) = 0. \quad (10.9)
$$

Proof: To prove that X_∞ exists almost surely, it suffices to show that the sequence X_n is almost surely a Cauchy sequence. Let A_{km} be the event $\{|X_{m+n} - X_m| > \frac{1}{k}$ for some $n \geq 1\}$. On the complement of the event $A = \bigcup_{k \geq 1} \bigcap_{m \geq 0} A_{km}$, the sequence X_n is Cauchy. The inequality

$$
\Pr(A) \leq \sum_{k \geq 1} \liminf_{m \to \infty} \Pr(A_{km})
$$

suggests that we verify $\lim_{m \to \infty} \Pr(A_{km}) = 0$. To achieve this goal, we apply Proposition 10.3.1 to the difference $X_{m+n} - X_m$, taking into account Example 10.2.3 and equality (10.4). It follows that

$$
\begin{aligned}
\Pr(A_{km}) &= \lim_{l \to \infty} \Pr\left(\max_{1 \leq n \leq l} |X_{m+n} - X_m| > \frac{1}{k} \right) \\
&\leq \lim_{l \to \infty} k^2 \mathrm{Var}(X_{m+l} - X_m) \\
&= \lim_{l \to \infty} k^2 [\mathrm{Var}(X_{m+l}) - \mathrm{Var}(X_m)].
\end{aligned}
$$

Because the sequence $\mathrm{Var}(X_n)$ is increasing and uniformly bounded, this last inequality yields

$$
\begin{aligned}
0 &\leq \lim_{m\to\infty} \mathrm{Pr}(A_{km}) \\
&\leq k^2 \lim_{m\to\infty} \lim_{l\to\infty} [\mathrm{Var}(X_{m+l}) - \mathrm{Var}(X_m)] \\
&= 0,
\end{aligned}
$$

which finishes the proof that X_n converges almost surely.

To establish the limit (10.9), we note that Fatou's lemma and equality (10.4) imply

$$
\begin{aligned}
0 &\leq \lim_{m\to\infty} \mathrm{E}(|X_\infty - X_m|^2) \\
&= \lim_{m\to\infty} \mathrm{E}(\lim_{n\to\infty} |X_{m+n} - X_m|^2) \\
&\leq \lim_{m\to\infty} \liminf_{n\to\infty} \mathrm{E}(|X_{m+n} - X_m|^2) \\
&= \lim_{m\to\infty} \liminf_{n\to\infty} [\mathrm{Var}(X_{m+n}) - \mathrm{Var}(X_m)] \\
&= 0.
\end{aligned}
$$

This completes the proof. ∎

Example 10.3.1 *Strong Law of Large Numbers*

Let Y_n be a sequence of independent random variables with common mean μ and variance σ^2. According to the analysis of Example 10.2.1, the sums $X_n = \sum_{i=1}^n i^{-1}(Y_i - \mu)$ constitute a martingale. Because $\mathrm{E}(X_n) = 0$ and $\mathrm{Var}(X_n) = \sigma^2 \sum_{i=1}^n i^{-2}$, this martingale has uniformly bounded second moments and therefore converges almost surely by Proposition 10.3.2. In view of the identity $Y_i - \mu = i(X_i - X_{i-1})$, we can write

$$
\frac{1}{n}\sum_{i=1}^n (Y_i - \mu) = \frac{1}{n}\left(\sum_{i=1}^n iX_i - \sum_{i=1}^n iX_{i-1}\right) = X_n - \frac{1}{n}\sum_{i=1}^n X_{i-1} \quad (10.10)
$$

with the convention $X_0 = 0$. Because $\lim_{n\to\infty} X_n = X_\infty$ exists, it is easy to show that

$$
\lim_{n\to\infty} \frac{1}{n}\sum_{i=1}^n X_{i-1} = X_\infty
$$

as well. In conjunction with equation (10.10), this yields

$$
\lim_{n\to\infty} \frac{1}{n}\sum_{i=1}^n (Y_i - \mu) = X_\infty - X_\infty = 0 \quad (10.11)
$$

and proves the strong law of large numbers. Other proofs exist that do not require the Y_n to have finite variance. ∎

Example 10.3.2 *Convergence of a Supercritical Branching Process*

Returning to Example 10.2.5, suppose that the process is supercritical. Taking into account that $E(X_n) = 1$ by design, we can invoke Proposition 10.3.2 provided the variances $Var(X_n) = \mu^{-2n} Var(Y_n)$ are uniformly bounded. If σ^2 is the variance of the progeny distribution, then equation (9.3) indicates that

$$Var(X_n) = \frac{\sigma^2(1 - \frac{1}{\mu^n})}{\mu(\mu - 1)}$$

$$\leq \frac{\sigma^2}{\mu(\mu - 1)}$$

for $\mu > 1$. Thus, Proposition 10.3.2 implies that $\lim_{n\to\infty} X_n = X_\infty$ exists.

Finding the distribution of X_∞ is difficult. For a geometric progeny generating function $Q(s) = \frac{p}{1-qs}$, progress can be made by deriving a functional equation characterizing the Laplace transform $L_\infty(t) = E(e^{-tX_\infty})$. If $Q_n(s)$ is the probability generating function of Y_n, then the Laplace transform of X_n can be expressed as

$$L_n(t) = E(e^{-\frac{tY_n}{\mu^n}}) = Q_n(e^{-\frac{t}{\mu^n}}).$$

In view of the defining equation $Q_{n+1}(s) = Q(Q_n(s))$, this leads directly to $L_{n+1}(\mu t) = Q(L_n(t))$, which produces the functional equation

$$L_\infty(\mu t) = Q(L_\infty(t)) \tag{10.12}$$

after taking limits on n. Straightforward algebra shows that the fractional linear transformation

$$L_\infty(t) = \frac{pt - p + q}{qt - p + q}$$

solves equation (10.12) when $Q(s) = \frac{p}{1-qs}$ and $\mu = \frac{q}{p}$. To identify the distribution with Laplace transform $L_\infty(t)$, note that

$$\frac{pt - p + q}{qt - p + q} = re^0 + (1 - r)\int_0^\infty e^{-tx}(1 - r)e^{-(1-r)x}dx \tag{10.13}$$

for $r = \frac{p}{q}$. In other words, X_∞ is a mixture of a point mass at 0 and an exponential distribution with intensity $1 - r$. The total mass r at 0 is just the extinction probability in this case.   ∎

10.4 Optional Stopping

Many applications of martingales involve stopping times. A stopping time T is a random variable that is adapted to a filter $\mathcal{F}_n$. The possible values of T are ∞ and the nonnegative integers. The word "adapted" here is a technical term meaning that the event $\{T = n\}$ is in $\mathcal{F}_n$ for all n. In less precise language, a stopping time can only depend on the past and present and cannot anticipate the future. A typical stopping time associated with a martingale X_n is $T_A = \min\{n : X_n \in A\}$, the first entry into a Borel set A of real numbers. The next proposition gives sufficient conditions for the mean value of the stopped process X_T to equal the martingale mean $\mathrm{E}(X_n)$.

Proposition 10.4.1 (Optional Stopping Theorem) *Let X_n be a martingale with common mean μ relative to the filter $\mathcal{F}_n$. If T is a stopping time for X_n satisfying*

(a) $\Pr(T < \infty) = 1$,

(b) $\mathrm{E}(|X_T|) < \infty$

(c) $\lim_{n \to \infty} \mathrm{E}(X_n 1_{\{T>n\}}) = 0$,

then $\mathrm{E}(X_T) = \mu$. Condition (c) holds if the second moments $\mathrm{E}(X_n^2)$ are uniformly bounded.

Proof: In view of the fact that the event $\{T = i\} \in \mathcal{F}_i$, we have

$$
\begin{aligned}
\mathrm{E}(X_T) &= \mathrm{E}(X_T 1_{\{T>n\}}) + \sum_{i=1}^{n} \mathrm{E}(X_T 1_{\{T=i\}}) \\
&= \mathrm{E}(X_T 1_{\{T>n\}}) + \sum_{i=1}^{n} \mathrm{E}(X_i 1_{\{T=i\}}) \\
&= \mathrm{E}(X_T 1_{\{T>n\}}) + \sum_{i=1}^{n} \mathrm{E}[\mathrm{E}(X_n \mid \mathcal{F}_i) 1_{\{T=i\}}] \\
&= \mathrm{E}(X_T 1_{\{T>n\}}) + \sum_{i=1}^{n} \mathrm{E}(X_n 1_{\{T=i\}}) \\
&= \mathrm{E}(X_T 1_{\{T>n\}}) + \mathrm{E}(X_n 1_{\{T\leq n\}}).
\end{aligned}
$$

Subtracting $\mu = \mathrm{E}(X_n 1_{\{T>n\}}) + \mathrm{E}(X_n 1_{\{T\leq n\}})$ from this identity yields

$$
\mathrm{E}(X_T) - \mu = \mathrm{E}(X_T 1_{\{T>n\}}) - \mathrm{E}(X_n 1_{\{T>n\}}). \tag{10.14}
$$

The dominated convergence theorem implies $\lim_{n \to \infty} \mathrm{E}(X_T 1_{\{T>n\}}) = 0$. This takes care of the first term on the right-hand side of equation (10.14). The second term tends to 0 by condition (c). Owing to the Cauchy-Schwarz inequality

$$
\mathrm{E}(X_n 1_{\{T>n\}})^2 \leq \mathrm{E}(X_n^2) \Pr(T > n), \tag{10.15}
$$

condition (c) holds whenever the second moments $\mathrm{E}(X_n^2)$ are uniformly bounded. ∎

Example 10.4.1 *Wald's Identity and the Sex Ratio*

Consider Example 10.2.1 with the understanding that $\sigma^2 = \text{Var}(Y_n)$ is finite, $\mu = \text{E}(Y_n)$ is not necessarily 0, and the Y_n are independent. If T is a stopping time relative to the filter $\mathcal{F}_n$ generated by $Y_1, \ldots, Y_n$, then Wald's identity says that the stopped sum S_T has mean $\text{E}(S_T) = \mu \, \text{E}(T)$ provided $\text{E}(T) < \infty$. We have already visited this problem when T is independent of the Y_n. To prove Wald's identity in general, we apply Proposition 10.4.1 to the martingale $R_n = S_n - n\mu$. Part (b) of the proposition requires checking that $\text{E}(|R_T|) < \infty$. This inequality follows from the calculation

$$
\begin{aligned}
\text{E}(|R_T|) \;&\leq\; \text{E}\left(\sum_{i=1}^{T} |Y_i - \mu| \right) \\
&=\; \text{E}\left(\sum_{n=1}^{\infty} \sum_{i=1}^{n} |Y_i - \mu| 1_{\{T=n\}} \right) \\
&=\; \text{E}\left(\sum_{i=1}^{\infty} |Y_i - \mu| 1_{\{T \geq i\}} \right) \\
&=\; \sum_{i=1}^{\infty} \text{E}(|Y_i - \mu|) \Pr(T \geq i) \\
&\leq\; \sigma \, \text{E}(T).
\end{aligned}
$$

Here we have used Schlömilch's Inequality $\text{E}(|Y_i - \mu|) \leq \sigma$ and the fact that $1_{\{T \geq i\}} = 1_{\{T > i-1\}}$ depends only on $Y_1, \ldots, Y_{i-1}$ and consequently is independent of Y_i. Part (c) of Proposition 10.4.1 is validated by noting that inequality (10.15) can be continued to

$$
\begin{aligned}
\text{E}(R_n 1_{\{T>n\}})^2 \;&\leq\; \text{E}(R_n^2) \Pr(T > n) \\
&=\; n\sigma^2 \Pr(T > n) \\
&\leq\; \sigma^2 \sum_{i=n+1}^{\infty} i \Pr(T = i).
\end{aligned}
$$

Because $\text{E}(T) < \infty$ by assumption, $\lim_{n \to \infty} \sum_{i=n+1}^{\infty} i \Pr(T = i) = 0$. This completes the proof.

The family planning model discussed in Examples 2.3.3 and 6.6.2 involves stopping times $T = N_{sd}$ that could conceivably change the sex ratio of females to males. However, Wald's identity rules this out. If Y_i is the indicator random variable recording whether the ith birth is female, then S_T is the number of daughters born to a couple with stopping time T. Wald's identity $\text{E}(S_T) = q \, \text{E}(T)$ implies that the proportion of daughters

$$
\frac{\text{E}(S_T)}{\text{E}(T)} \;=\; q
$$

over a large number of such families does not deviate from q.   ∎

Example 10.4.2 *Successive Random Permutations*

Consider the following recursive construction. Let Y_1 be the number of matches in a random permutation π_1 of the set $\{1, \ldots, n\}$. Throw out each integer i for which $\pi_1(i) = i$, and relabel the remaining integers $1, \ldots, n - Y_1$. Let Y_2 be the number of matches in an independent random permutation π_2 of the set $\{1, \ldots, n - Y_1\}$. Throw out each integer i for which $\pi_2(i) = i$, and relabel the remaining integers $1, \ldots, n - Y_1 - Y_2$. Continue this process N times until $\sum_{i=1}^{N} Y_i = n$.

We now calculate the mean of the random variable N by exploiting the martingale

$$
\begin{aligned}
X_n &= \sum_{i=1}^{n} [Y_i - \mathrm{E}(Y_i \mid \mathcal{F}_{i-1})] \\
&= \sum_{i=1}^{n} (Y_i - 1)
\end{aligned}
$$

relative to the filter $\mathcal{F}_n$ generated by $Y_1, \ldots, Y_n$. Example 10.2.1 establishes the martingale property, and Example 2.2.1 proves the identity

$$
\mathrm{E}(Y_i \mid \mathcal{F}_{i-1}) = 1.
$$

Because N is a stopping time for the bounded martingale sequence X_n, Proposition 10.4.1 implies that

$$
\begin{aligned}
0 &= \mathrm{E}(X_N) \\
&= \mathrm{E}\left(\sum_{i=1}^{N} Y_i\right) - \mathrm{E}(N) \\
&= n - \mathrm{E}(N),
\end{aligned}
$$

which obviously entails $\mathrm{E}(N) = n$. ∎

Example 10.4.3 *Sequential Testing in Statistics*

Consider Wald's likelihood ratio martingale of Example 10.2.2. Suppose we quit sampling at the first epoch T with either $X_T \geq \alpha^{-1}$ or $X_T \leq \beta$. In the former case, we decide in favor of the alternative hypothesis H_a, and in the latter case, we decide in favor of the null hypothesis H_o. We claim that the type I and type II errors satisfy

$$
\begin{aligned}
\Pr(\text{reject } H_o \mid H_o) &\leq \alpha \quad\quad\quad\quad (10.16) \\
\Pr(\text{reject } H_a \mid H_a) &\leq \beta.
\end{aligned}
$$

To prove this claim, we follow the arguments Williams [153], taking for granted that $\Pr(T < \infty) = 1$. If we set $T \wedge n = \min\{T, n\}$ and assume

H_o, then applying Proposition 10.4.1 to $T \wedge n$ and invoking Fatou's lemma yield

$$
\begin{aligned}
\alpha^{-1} \Pr(X_T \geq \alpha^{-1}) \;&\leq\; \mathrm{E}(X_T) \\
&=\; \mathrm{E}(\lim_{n \to \infty} X_{T \wedge n}) \\
&\leq\; \liminf_{n \to \infty} \mathrm{E}(X_{T \wedge n}) \\
&\leq\; \mathrm{E}(X_1) \\
&=\; 1.
\end{aligned}
$$

The extremes of this last string of inequalities produce inequality (10.16). In general, sequential testing is more efficient in reaching a decision than testing with fixed sample sizes. ∎

Example 10.4.4 *Hitting Probabilities in the Wright-Fisher Model*

Because the proportion $X_n = \frac{1}{2m} Y_n$ of a_1 alleles at generation n is a martingale, Proposition 10.4.1 can be employed to calculate the probability of eventual fixation of the a_1 allele. Condition (a) of the proposition holds because the underlying Markov chain must reach one of the two absorbing states 0 or $2m$. Conditions (b) and (c) are trivial to verify in light of the inequalities $0 \leq X_n \leq 1$. If T is the time of absorption at 0 or 1 and $Y_1 = i$, then Proposition 10.4.1 implies

$$
\begin{aligned}
\frac{i}{2m} \;&=\; \mathrm{E}(X_1) \\
&=\; \mathrm{E}(X_T) \\
&=\; 0 \cdot \Pr(X_T = 0) + 1 \cdot \Pr(X_T = 1).
\end{aligned}
$$

Thus, the a_1 allele is eventually fixed with probability $\frac{i}{2m}$. ∎

10.5   Large Deviation Bounds

In Chapter 1 we investigated various classical inequalities such as Chebyshev's inequality. For well-behaved random variables, much sharper results are possible. The next proposition gives Azuma's tail-probability bound for martingales with bounded differences. Hoeffding originally established the bound for sums of independent random variables.

Proposition 10.5.1 (Azuma-Hoeffding) *Suppose the sequence of random variables X_n forms a martingale with mean 0 relative to the filter $\mathcal{F}_n$. If under the convention $X_0 = 0$ there exists a sequence of constants c_n such that $\Pr(|X_n - X_{n-1}| \leq c_n) = 1$, then*

$$
\mathrm{E}\left(e^{\beta X_n}\right) \;\leq\; e^{(\beta^2/2) \sum_{k=1}^{n} c_k^2} \tag{10.17}
$$

for all $\beta > 0$. Inequality (10.17) entails the further inequalities

$$\Pr(X_n \geq \lambda) \leq e^{-\lambda^2/(2\sum_{k=1}^n c_k^2)} \qquad (10.18)$$

and

$$\Pr(|X_n| \geq \lambda) \leq 2e^{-\lambda^2/(2\sum_{k=1}^n c_k^2)} \qquad (10.19)$$

for all $\lambda > 0$.

Proof: Because the function e^u is convex, $e^{\alpha u + (1-\alpha)v} \leq \alpha e^u + (1-\alpha)e^v$ for any $\alpha \in [0,1]$ and pair u and v. Putting $u = -\beta c$, $v = \beta c$, and $\alpha = \frac{c-x}{2c}$ for $x \in [-c, c]$ therefore yields

$$e^{\beta x} \leq \frac{c-x}{2c}e^{-\beta c} + \frac{c+x}{2c}e^{\beta c}.$$

If we substitute $X_n - X_{n-1}$ for x and c_n for c in this inequality and take conditional expectations, then the hypothesis $|X_n - X_{n-1}| \leq c_n$ and the fact $\mathrm{E}(X_n - X_{n-1} \mid \mathcal{F}_{n-1}) = 0$ imply

$$\begin{aligned}
\mathrm{E}\left(e^{\beta X_n}\right) &= \mathrm{E}\left[\mathrm{E}\left(e^{\beta X_n} \mid \mathcal{F}_{n-1}\right)\right] \\
&= \mathrm{E}\left\{e^{\beta X_{n-1}} \mathrm{E}\left[e^{\beta(X_n - X_{n-1})} \mid \mathcal{F}_{n-1}\right]\right\} \\
&\leq \mathrm{E}\left(e^{\beta X_{n-1}}\right)\left(\frac{c_n - 0}{2c_n}e^{-\beta c_n} + \frac{c_n + 0}{2c_n}e^{\beta c_n}\right) \\
&\leq \mathrm{E}\left(e^{\beta X_{n-1}}\right)\frac{e^{-\beta c_n} + e^{\beta c_n}}{2}.
\end{aligned}$$

Induction on n and the convention $X_0 = 0$ now prove inequality (10.17) provided we can show that

$$\frac{e^u + e^{-u}}{2} \leq e^{\frac{u^2}{2}}. \qquad (10.20)$$

However, inequality (10.20) follows by expanding its right and left sides in Taylor's series and noting that the corresponding coefficients of u^{2n} satisfy $\frac{1}{(2n)!} \leq \frac{1}{2^n n!}$. Of course, the coefficients of the odd powers u^{2n+1} on both sides of inequality (10.20) vanish.

To prove inequality (10.18), we apply Markov's inequality in the form

$$\begin{aligned}
\Pr(X_n \geq \lambda) &\leq \mathrm{E}\left(e^{\beta X_n}\right)e^{-\beta\lambda} \\
&\leq e^{(\beta^2/2)\sum_{k=1}^n c_k^2 - \beta\lambda} \qquad (10.21)
\end{aligned}$$

for an arbitrary $\beta > 0$. The particular β minimizing the exponent on the right-hand side of (10.21) is clearly $\beta = \lambda/(\sum_{k=1}^n c_k^2)$. Substituting this choice in inequality (10.21) yields inequality (10.18). Because $-X_n$ also fulfills the hypotheses of the proposition, inequality (10.19) follows from inequality (10.18).   ∎

In applying Proposition 10.5.1 to a martingale X_n with nonzero mean μ, we replace X_n by the recentered martingale $X_n - \mu$. Recentering has no impact on the differences $(X_n - \mu) - (X_{n-1} - \mu) = X_n - X_{n-1}$, so the above proof holds without change.

Example 10.5.1 *Tail Bound for the Binomial Distribution*

Suppose Y_n is a sequence of independent random variables with common mean $\mu = \mathrm{E}(Y_n)$. If $|Y_n - \mu| \leq c_n$ with probability 1, then Proposition 10.5.1 applies to the martingale $S_n - n\mu = \sum_{i=1}^n Y_i - n\mu$. In particular, when each Y_n is a Bernoulli random variable with success probability μ, then $c_n = 1$ and

$$\Pr(|S_n - n\mu| \geq \lambda) \leq 2e^{-\frac{\lambda^2}{2n}}. \tag{10.22}$$

Inequality (10.22) has an interesting implication for the sequence of Bernstein polynomials that approximate a continuous function $f(x)$ satisfying a Lipschitz condition $|f(u) - f(v)| \leq d|u - v|$ for all $u, v \in [0, 1]$ [66]. As in Example 3.5.1, we put $\mu = x$ and argue that

$$\left| \mathrm{E}\left[f\left(\frac{S_n}{n} \right) \right] - f(x) \right|$$
$$\leq d\frac{\lambda}{n}\Pr\left(\left| \frac{S_n}{n} - x \right| < \frac{\lambda}{n} \right) + 2\|f\|_\infty \Pr\left(\left| \frac{S_n}{n} - x \right| \geq \frac{\lambda}{n} \right).$$

Invoking the large deviation bound (10.22) for the second probability gives the inequality

$$\left| \mathrm{E}\left[f\left(\frac{S_n}{n} \right) \right] - f(x) \right| \leq d\frac{\lambda}{n} + 4\|f\|_\infty e^{-\frac{\lambda^2}{2n}}.$$

For the choice $\lambda = \sqrt{2n \ln n}$, this yields the uniform bound

$$\left| \mathrm{E}\left[f\left(\frac{S_n}{n} \right) \right] - f(x) \right| \leq d\sqrt{\frac{2 \ln n}{n}} + \frac{4\|f\|_\infty}{n}$$
$$= O\left(\sqrt{\frac{\ln n}{n}} \right).$$

This is an improvement over the uniform bound $O(n^{-1/3})$ for continuously differentiable functions noted in Problem 20 of Chapter 3. It is worse than the best available uniform bound $O(n^{-1})$ for functions for that are twice continuously differentiable rather than merely Lipschitz [122].   ∎

Many applications of Proposition 10.5.1 involve a more complicated combination of independent random variables $Y_1, \ldots, Y_n$ than a sum. If Z is such a combination and $\mathcal{F}_i$ is the σ-algebra generated by $Y_1, \ldots, Y_i$, then Example 10.2.4 implies that the conditional expectations $X_i = \mathrm{E}(Z \mid \mathcal{F}_i)$ constitute a martingale with $X_n = Z$. To apply Proposition 10.5.1, we observe that equation (10.5) implies

$$
\begin{aligned}
& X_i(y_1, \ldots, y_i) \\
=\ & \int \cdots \int Z(y_1, \ldots, y_{i-1}, y_i, v_{i+1}, \ldots, v_n)\, d\omega_{i+1}(v_{i+1}) \cdots d\omega_n(v_n) \\
& X_{i-1}(y_1, \ldots, y_{i-1}) \hspace{5cm} (10.23) \\
=\ & \int \cdots \int Z(y_1, \ldots, y_{i-1}, v_i, v_{i+1}, \ldots, v_n)\, d\omega_i(v_i) \cdots d\omega_n(v_n),
\end{aligned}
$$

where ω_k is the multivariate distribution of Y_k. Obviously, the extra integration on ω_i introduced in the multiple integral defining X_i has no effect. Taking differences in (10.23) produces an integral expression for $X_i - X_{i-1}$ whose integrand is

$$
\begin{aligned}
& Z_i - Z_{i-1} \hspace{6cm} (10.24) \\
=\ & Z(y_1, \ldots, y_{i-1}, y_i, v_{i+1}, \ldots, v_n) - Z(y_1, \ldots, y_{i-1}, v_i, v_{i+1}, \ldots, v_n).
\end{aligned}
$$

If $|Z_i - Z_{i-1}| \leq c_i$ for some constant c_i, then the inequality $|X_i - X_{i-1}| \leq c_i$ follows after integration against the independent probability distributions $\omega_i, \ldots, \omega_n$. These considerations set the stage for applying Proposition 10.5.1 to $Z = X_n$.

Example 10.5.2 *Longest Common Subsequence*

In the longest common subsequence problem considered in Example 5.7.1, we can derive tail-probability bounds for M_n. Let Y_i be the random pair of letters chosen for position i of the two strings. Conditioning on the σ-algebra $\mathcal{F}_i$ generated by $Y_1, \ldots, Y_i$ creates a martingale $X_i = \mathrm{E}(M_n \mid \mathcal{F}_i)$ with $X_n = M_n$. We now bound $|X_i - X_{i-1}|$ by considering what happens when exactly one argument of $M_n(y_1, \ldots, y_n)$ changes. If this is argument i, then the pair $y_i = (u_i, v_i)$ becomes the pair $y_i^* = (u_i^*, v_i^*)$. In a longest common subsequence of $y = (y_1, \ldots, y_n)$, changing u_i to u_i^* creates or destroys at most one match. Likewise, changing v_i to v_i^* creates or destroys at most one match. Thus, the revised string y^* has a common subsequence whose length differs from $M_n(y)$ by at most 2. The same argument shows that y has a common subsequence whose length differs from $M(y^*)$ by at most 2. It follows that $|M_n(y) - M_n(y^*)| \leq 2$, and Proposition 10.5.1 yields

$$
\Pr[|M_n - \mathrm{E}(M_n)| \geq \lambda\sqrt{n}] \ \leq \ 2e^{-\frac{\lambda^2}{8}}.
$$

Further problems of this sort are treated in the references [140, 147].   ∎

Example 10.5.3 *Euclidean Traveling Salesman Problem*

In some cases it is possible to calculate a more subtle bound on the martingale differences $X_{i+1} - X_i$. The Euclidean traveling salesman problem is typical in this regard. In applying Proposition 10.5.1, we take Z to be $D_n(\{Y_1, \ldots, Y_n\})$ in the notation of Example 5.7.2. Consider the integrand (10.24) determining the martingale difference $X_i - X_{i-1}$. If S denotes the set $S = \{y_1, \ldots, y_{i-1}, v_{i+1}, \ldots, v_n\}$, then reasoning as in Example 5.7.2 leads to the inequalities

$$D_{n-1}(S) \leq D_n(S \cup \{y_i\}) \leq D_{n-1}(S) + 2 \min_{j>i} \|v_j - y_i\|$$

$$D_{n-1}(S) \leq D_n(S \cup \{v_i\}) \leq D_{n-1}(S) + 2 \min_{j>i} \|v_j - v_i\|$$

involving $Z_i = D_n(S \cup \{y_i\})$ and $Z_{i-1} = D_n(S \cup \{v_i\})$. It follows that

$$|Z_i - Z_{i-1}| \leq 2 \min_{j>i} \|v_j - y_i\| + 2 \min_{j>i} \|v_j - v_i\|$$

and consequently that

$$|X_i - X_{i-1}| \leq 2 \operatorname{E}(\min_{j>i} \|Y_j - y_i\|) + 2 \operatorname{E}(\min_{j>i} \|Y_j - Y_i\|).$$

We now estimate the right-tail probability $\Pr(\min_{j>i} \|Y_j - y\| \geq r)$ for any point y. The smallest area of the unit square at a distance of r or less from y is a quarter-circle of radius r. This extreme case occurs when y occupies a corner of the square. In view of this result,

$$\Pr\left(\min_{j>i} \|Y_j - y\| \geq r\right) \leq \left(1 - \frac{\pi r^2}{4}\right)^{n-i} \leq e^{-\frac{(n-i)\pi r^2}{4}}.$$

Application of Example 2.5.1 therefore yields

$$\operatorname{E}\left(\min_{j>i} \|Y_j - y\|\right) \leq \int_0^\infty e^{-\frac{(n-i)\pi r^2}{4}}\, dr$$

$$= \frac{1}{2} \int_{-\infty}^\infty e^{-\frac{(n-i)\pi r^2}{4}}\, dr$$

$$= \frac{1}{\sqrt{n-i}}$$

and

$$|X_i - X_{i-1}| \leq \frac{4}{\sqrt{n-i}}.$$

The case $i = n$ must be considered separately. If we use the crude inequality

$$|X_n - X_{n-1}| = |D_n - X_{n-1}| \leq 2\sqrt{2},$$

then the sum $\sum_{i=1}^n c_i^2$ figuring in Proposition 10.5.1 can be bounded by

$$\sum_{i=1}^n c_i^2 \;\leq\; (2\sqrt{2})^2 + 4^2 \sum_{i=1}^{n-1} \frac{1}{n-i} \;\leq\; 8 + 16(\ln n + 1).$$

This in turn translates into the Azuma-Hoeffding bound

$$\Pr[|D_n - E(D_n)| \geq \lambda] \;\leq\; 2e^{-\frac{\lambda^2}{64+32\ln n}}.$$

$\blacksquare$

10.6 Problems

1. Define the random variables Y_n inductively taking by $Y_0 = 1$ and Y_{n+1} to be uniformly distributed on the interval $(0, Y_n)$. Show that the sequence $X_n = 2^n Y_n$ is a martingale.

2. An urn contains b black balls and w white balls. Each time we randomly withdraw a ball, we replace it by $c+1$ balls of the same color. Let X_n be the fraction of white balls after n draws. Demonstrate that X_n is a martingale.

3. Let $Y_1, Y_2, \ldots$ be a sequence of independent random variables with zero means and common variance σ^2. If $X_n = Y_1 + \cdots + Y_n$, then show that $X_n^2 - n\sigma^2$ is a martingale.

4. Let $Y_1, Y_2, \ldots$ be a sequence of i.i.d. random variables with common generating function $M(t) = E(e^{tY_1})$. Prove that

$$X_n \;=\; M(t)^{-n} e^{t(Y_1 + \cdots + Y_n)}$$

is a martingale whenever $M(t) < \infty$.

5. Let Y_n be a finite-state, discrete-time Markov chain with transition matrix $P = (p_{ij})$. If v is a column eigenvector for P with nonzero eigenvalue λ, then verify that $X_n = \lambda^{-n} v_{Y_n}$ is a martingale, where v_{Y_n} is coordinate Y_n of v.

6. Suppose Y_n is the number of particles at the nth generation of a branching process. If s_∞ is the extinction probability, prove that $X_n = s_\infty^{Y_n}$ is a martingale. (Hint: If $Q(s)$ is the progeny generating function, then $Q(s_\infty) = s_\infty$.)

7. In Example 10.3.2, show that $\mathrm{Var}(X_\infty) = \frac{\sigma^2}{\mu(\mu-1)}$ by differentiating equation (10.12) twice. This result is consistent with the mean square convergence displayed in equation (10.9).

8. In Example 10.3.2, show that the fractional linear transformation

$$L_\infty(t) = \frac{pt - p + q}{qt - p + q}$$

solves equation (10.12) when $Q(s) = \frac{p}{1-qs}$ and $\mu = \frac{q}{p}$. Also verify equation (10.13).

9. In Example 10.2.2, suppose that each Y_n is equally likely to assume the values $\frac{1}{2}$ and $\frac{3}{2}$. Show that $\prod_{i=1}^\infty Y_i \equiv 0$, but $\prod_{i=1}^\infty E(Y_i) = 1$ [19]. (Hint: Apply the strong law of large numbers to the sequence $\ln Y_n$.)

10. Given $X_0 = \mu \in (0, 1)$, define X_n inductively by

$$X_{n+1} = \begin{cases} \alpha + \beta X_n, & \text{with probability } X_n \\ \beta X_n, & \text{with probability } 1 - X_n. \end{cases}$$

where $\alpha, \beta > 0$ and $\alpha + \beta = 1$. Prove that X_n is a martingale with (a) $X_n \in (0, 1)$, (b) $E(X_n) = \mu$, and (c) $\text{Var}(X_n) = [1-(1-\alpha^2)^n]\mu(1-\mu)$. Also prove that Proposition 10.3.2 implies that $\lim_{n\to\infty} X_n = X_\infty$ exists with $E(X_\infty) = \mu$ and $\text{Var}(X_\infty) = \mu(1 - \mu)$. (Hint: Derive a recurrence relation for $\text{Var}(X_{n+1})$ by conditioning on X_n.)

11. Let $S_n = X_1 + \cdots + X_n$ be a symmetric random walk on the integers $\{-a, \ldots, b\}$ starting at $S_0 = 0$. For the stopping time

$$T = \min\{n : S_n = -a \text{ or } S_n = b\},$$

prove that $\Pr(S_T = b) = a/(a + b)$ by considering the martingale S_n and that $E(T) = ab$ by considering the martingale $S_n^2 - n$. (Hints: Apply Proposition 10.4.1 and Problem 3.)

12. In the Wright-Fisher model of Example 10.2.6, show that

$$Z_n = \frac{X_n(1 - X_n)}{\left(1 - \frac{1}{2m}\right)^n}$$

is a martingale with values on $[0, 1]$. In view of Proposition 10.3.2, $\lim_{n\to\infty} Z_n = Z_\infty$ exists. Thus, $X_n(1 - X_n) \approx \left(1 - \frac{1}{2m}\right)^n Z_\infty$ for n large. In other words, X_n approaches either 0 or 1 at rate $1 - \frac{1}{2m}$.

13. In Proposition 10.4.1, suppose that $E(T) < \infty$ and that the differences $X_{n+1} - X_n$ satisfy $|X_{n+1} - X_n| \le d$ for some constant d and all n. Prove that these conditions imply conditions (b) and (c) of the proposition. (Hints: For condition (b), show that

$$|X_T| \le |X_1| + (T - 1)d.$$

For condition (c), use $\text{Var}(X_n) \le \text{Var}(X_1) + (n - 1)d^2$ and inequality (10.15).)

14. Let $Y_1, \ldots, Y_n$ be independent Bernoulli random variables with success probability μ. Graphically compare the large deviation bound (10.22) to Chebyshev's bound

$$\Pr(|S_n - n\mu| \geq \lambda) \leq \frac{n\mu(1-\mu)}{\lambda^2}$$

when $\mu = 1/2$. Which bound is better? If neither is uniformly better than the other, determine which combinations of values of n and λ favor Chebyshev's bound.

15. Suppose that $v_1, \ldots, v_n \in \mathrm{R}^m$ have Euclidean norms $\|v_i\|_2 \leq 1$. Let $Y_1, \ldots, Y_n$ be independent random variables uniformly distributed on the two-point set $\{-1, 1\}$. If $Z = \|Y_1 v_1 + \cdots + Y_n v_n\|_2$, then prove that

$$\Pr[Z - \mathrm{E}(Z) \geq \lambda\sqrt{n}] \leq e^{-\frac{\lambda^2}{8}}.$$

11
Diffusion Processes

11.1 Introduction

Despite their reputations for sophistication, diffusion processes are widely applied throughout science and engineering. Here we survey the theory at an elementary level, stressing intuition rather than rigor. Readers with the time and mathematical inclination should follow up this brief account by delving into serious presentations of the mathematics [64, 83]. A good grounding in measure theory is indispensable in understanding the theory. At the highest level of abstraction, diffusion processes can be treated via the Ito stochastic integral [24, 30].

Because of the hard work involved in mastering the abstract theory, there is a great deal to be said for starting with specific applications and heuristic arguments. The current chapter follows a few selected applications from population biology, neurophysiology, and population genetics. These serve to illustrate concrete techniques for calculating moments, first passage times, and equilibrium distributions. Ordinary and partial differential equations play a prominent role in the chapter. The final third of the chapter introduces a numerical method for approximating diffusion densities. This method is crafted to take advantage of the Gaussian nature of the increments to a diffusion process.

11.2 Basic Definitions and Properties

A diffusion process X_t is a continuous-time Markov process with approximately Gaussian increments over small time intervals. Its sample paths $t \mapsto X_t$ are continuous functions confined to an interval I, either finite or infinite. The process X_t is determined by the Markovian assumption and the distribution of its increments. For small s and $X_t = x$, the increment $X_{t+s} - X_t$ is nearly Gaussian (normal) with mean and variance

$$\mathrm{E}(X_{t+s} - X_t \mid X_t = x) \quad = \quad \mu(t, x)s + o(s) \qquad (11.1)$$
$$\mathrm{E}[(X_{t+s} - X_t)^2 \mid X_t = x] \quad = \quad \sigma^2(t, x)s + o(s). \qquad (11.2)$$

The functions $\mu(t, x)$ and $\sigma^2(t, x) \geq 0$ are called the infinitesimal mean and variance, respectively. Here the term "infinitesimal variance" is used rather than "infinitesimal second moment" because the approximation

$$\mathrm{Var}(X_{t+s} - X_t \mid X_t = x) \quad = \quad \mathrm{E}[(X_{t+s} - X_t)^2 \mid X_t = x] + o(s)$$

follows directly from approximation (11.1). If the infinitesimal mean and variance do not depend on time t, then the process is time homogeneous. The choices $\mu(t, x) = 0$, $\sigma^2(t, x) = 1$, and $X_0 = 0$ characterize standard Brownian motion.

To begin our nonrigorous, intuitive discussion of diffusion processes, we note that the normality assumption implies

$$\mathrm{E}(|X_{t+s} - X_t|^m \mid X_t = x) \quad = \quad \mathrm{E}\left(\left| \frac{X_{t+s} - X_t}{\sigma(t, x)\sqrt{s}} \right|^m \, \Big| \, X_t = x \right) \left[\sigma(t, x)\sqrt{s} \right]^m$$
$$= \quad o(s) \qquad (11.3)$$

for $m > 2$. This insight is crucial in various arguments involving Taylor series expansions. For instance, it allows us to deduce how X_t behaves under a smooth, invertible transformation. If $Y_t = g(t, X_t)$ denotes the transformed process, then

$$\begin{aligned}
Y_{t+s} - y \quad = \quad & \frac{\partial}{\partial t}g(t, x)s + \frac{\partial}{\partial x}g(t, x)(X_{t+s} - x) + \frac{1}{2}\frac{\partial^2}{\partial t^2}g(t, x)s^2 \\
& + \frac{\partial^2}{\partial t \partial x}g(t, x)s(X_{t+s} - x) + \frac{1}{2}\frac{\partial^2}{\partial x^2}g(t, x)(X_{t+s} - x)^2 \\
& + O[(|X_{t+s} - x| + s)^3]
\end{aligned}$$

for $X_t = x$ and $y = g(t, x)$. Taking conditional expectations and invoking equation (11.3) produce

$$\begin{aligned}
\mathrm{E}(Y_{t+s} - Y_t \mid Y_t = y) \quad = \quad & \frac{\partial}{\partial t}g(t, x)s + \frac{\partial}{\partial x}g(t, x)\mu(t, x)s \\
& + \frac{1}{2}\frac{\partial^2}{\partial x^2}g(t, x)\sigma^2(t, x)s + o(s).
\end{aligned}$$

Similarly,

$$\mathrm{Var}(Y_{t+s} - Y_t \mid Y_t = y) = \left[\frac{\partial}{\partial x}g(t,x)\right]^2 \sigma^2(t,x)s + o(s).$$

It follows that the transformed diffusion process Y_t has infinitesimal mean and variance

$$\mu_Y(t,y) = \frac{\partial}{\partial t}g(t,x) + \frac{\partial}{\partial x}g(t,x)\mu(t,x) + \frac{1}{2}\frac{\partial^2}{\partial x^2}g(t,x)\sigma^2(t,x)$$

$$\sigma_Y^2(t,y) = \left[\frac{\partial}{\partial x}g(t,x)\right]^2 \sigma^2(t,x), \qquad\qquad (11.4)$$

at $y = g(t,x)$.

In many cases of interest, the random variable X_t has a density function $f(t,x)$ that depends on the initial point $X_0 = x_0$. To characterize $f(t,x)$, we now give a heuristic derivation of Kolmogorov's forward partial differential equation. Our approach exploits the notion of probability flux. Here it helps to imagine a large ensemble of diffusing particles, each independently executing the same process. We position ourselves at some point x and record the rate at which particles pass through x from left to right minus the rate at which they pass from right to left. This rate, normalized by the total number of particles, is the probability flux at x. We can express the flux more formally as the negative derivative $-\frac{\partial}{\partial t}\mathrm{Pr}(X_t \leq x)$.

To calculate this time derivative, we rewrite the difference

$$\begin{aligned}
&\mathrm{Pr}(X_t \leq x) - \mathrm{Pr}(X_{t+s} \leq x)\\
={}&\mathrm{Pr}(X_t \leq x, X_{t+s} > x) + \mathrm{Pr}(X_t \leq x, X_{t+s} \leq x)\\
&- \mathrm{Pr}(X_t \leq x, X_{t+s} \leq x) - \mathrm{Pr}(X_t > x, X_{t+s} \leq x)\\
={}&\mathrm{Pr}(X_t \leq x, X_{t+s} > x) - \mathrm{Pr}(X_t > x, X_{t+s} \leq x).
\end{aligned}$$

The first of the resulting probabilities, $\mathrm{Pr}(X_t \leq x, X_{t+s} > x)$, can be expressed as

$$\mathrm{Pr}(X_t \leq x, X_{t+s} > x) = \int_0^\infty \int_{x-z}^x f(t,y)\phi_s(y,z)\,dy\,dz,$$

where the increment $Z = X_{t+s} - X_t$ has density $\phi_s(y,z)$ when $X_t = y$. The limits on the inner integral reflect the inequalities $x \geq y$ and $y + z > x$. In similar fashion, the second probability becomes

$$\mathrm{Pr}(X_t > x, X_{t+s} \leq x) = \int_{-\infty}^0 \int_x^{x-z} f(t,y)\phi_s(y,z)\,dy\,dz,$$

producing overall

$$\mathrm{Pr}(X_t \leq x) - \mathrm{Pr}(X_{t+s} \leq x) = \int_{-\infty}^\infty \int_{x-z}^x f(t,y)\phi_s(y,z)\,dy\,dz. \quad (11.5)$$

For small values of s, only values of y near x should contribute to the flux. Therefore, we can safely substitute the first-order expansion

$$f(t, y)\phi_s(y, z) \approx f(t, x)\phi_s(x, z) + \frac{\partial}{\partial x}\left[f(t, x)\phi_s(x, z)\right](y - x)$$

in equation (11.5). In light of equations (11.1) and (11.2), this yields

$$\Pr(X_t \le x) - \Pr(X_{t+s} \le x)$$

$$\approx \int_{-\infty}^{\infty}\int_{x-z}^{x}\left\{f(t, x)\phi_s(x, z) + \frac{\partial}{\partial x}\left[f(t, x)\phi_s(x, z)\right](y - x)\right\} dy\, dz$$

$$= \int_{-\infty}^{\infty}\left\{zf(t, x)\phi_s(x, z) - \frac{z^2}{2}\frac{\partial}{\partial x}\left[f(t, x)\phi_s(x, z)\right]\right\} dz$$

$$\approx \mu(t, x)f(t, x)s - \frac{1}{2}\frac{\partial}{\partial x}\left[\int_{-\infty}^{\infty}z^2\phi_s(x, z)\,dz f(t, x)\right]$$

$$\approx \mu(t, x)f(t, x)s - \frac{1}{2}\frac{\partial}{\partial x}\left[\sigma^2(t, x)f(t, x)\right]s.$$

Using equation (11.3), one can show that these approximations are good to order $o(s)$. Dividing by s and sending s to 0 give the flux

$$-\frac{\partial}{\partial t}\Pr(X_t \le x) = \mu(t, x)f(t, x) - \frac{1}{2}\frac{\partial}{\partial x}\left[\sigma^2(t, x)f(t, x)\right].$$

A final differentiation with respect to x now produces the Kolmogorov forward equation

$$\frac{\partial}{\partial t}f(t, x) = -\frac{\partial}{\partial x}\left[\mu(t, x)f(t, x)\right] + \frac{1}{2}\frac{\partial^2}{\partial x^2}\left[\sigma^2(t, x)f(t, x)\right]. \quad (11.6)$$

As t tends to 0, the density $f(t, x)$ concentrates all of its mass around the initial point x_0.

11.3 Examples of Diffusion Processes

Example 11.3.1 *Standard Brownian Motion*

If $\mu(t, x) = 0$ and $\sigma^2(t, x) = 1$, then the forward equation (11.6) becomes

$$\frac{\partial}{\partial t}f(t, x) = \frac{1}{2}\frac{\partial^2}{\partial x^2}f(t, x).$$

For $X_0 = 0$ one can check the solution

$$f(t, x) = \frac{1}{\sqrt{2\pi t}}e^{-\frac{x^2}{2t}}$$

by straightforward differentiation. Thus, X_t has a Gaussian density with mean 0 and variance t. As t tends to 0, X_t becomes progressively more concentrated around its starting point 0. Because X_t and the increment $X_{t+s} - X_t$ are effectively independent for $s > 0$ small, and because the sum of independent Gaussian random variables is Gaussian, X_t and $X_{t+s} - X_t$ are Gaussian and independent for large s as well as for small s. Of course, rigorous proof of this fact is more subtle. In general, we cannot expect a diffusion process X_t to be normally distributed just because its short-time increments are approximately normal.

The independent increments property of standard Brownian motion facilitates calculation of covariances. For instance, writing

$$X_{t+s} \quad = \quad X_{t+s} - X_t + X_t$$

makes it clear that $\text{Cov}(X_{t+s}, X_t) = \text{Var}(X_t) = t$. We will use this formula in finding the infinitesimal mean and variance of the Brownian bridge diffusion process described in Example 11.3.3. Although the sample paths of Brownian motion are continuous, they are extremely rough and nowhere differentiable [82]. ∎

Example 11.3.2 *Transformations of Standard Brownian Motion*

The transformed Brownian process $Y_t = \sigma X_t + \alpha t + x_0$ has infinitesimal mean and variance $\mu_Y(t, x) = \alpha$ and $\sigma_Y^2(t, x) = \sigma^2$. It is clear that Y_t is normally distributed with mean $\alpha t + x_0$ and variance $\sigma^2 t$. The further transformation $Z_t = e^{Y_t}$ leads to a process with infinitesimal mean and variance $\mu_Z(t, z) = z\alpha + \frac{1}{2}z\sigma^2$ and $\sigma_Z^2(t, z) = z^2\sigma^2$. Because Y_t is normally distributed, Z_t is lognormally distributed. ∎

Example 11.3.3 *Brownian Bridge*

To construct the Brownian bridge Y_t from standard Brownian motion X_t, we restrict t to the interval $[0, 1]$ and conditional on the event $X_1 = 0$. This ties down X_t at the two time points 0 and 1. The Brownian bridge diffusion process is important in evaluating the asymptotic distribution of the Kolmogorov-Smirnov statistic in nonparametric statistics [18].

To find the infinitesimal mean and variance of the Brownian bridge, we note that the vector (X_t, X_1, X_{t+s}) follows a multivariate normal distribution with mean vector $\mathbf{0} = (0, 0, 0)$ and covariance matrix

$$\begin{pmatrix} t & t & t \\ t & 1 & t+s \\ t & t+s & t+s \end{pmatrix}.$$

If Y_t equals X_t conditional on the event $X_1 = 0$, then Y_{t+s} conditional on the event $Y_t = y$ is just X_{t+s} conditional on the joint event $X_t = y$ and

$X_1 = 0$. It follows from the conditioning formulas in Section 1.8 that Y_{t+s} given $Y_t = y$ is normally distributed with mean and variance

$$\mathrm{E}(Y_{t+s} \mid Y_t = y) \;=\; (t, t+s) \begin{pmatrix} t & t \\ t & 1 \end{pmatrix}^{-1} \begin{pmatrix} y - 0 \\ 0 - 0 \end{pmatrix} + 0$$

$$\mathrm{Var}(Y_{t+s} \mid Y_t = y) \;=\; t + s - (t, t+s) \begin{pmatrix} t & t \\ t & 1 \end{pmatrix}^{-1} \begin{pmatrix} t \\ t+s \end{pmatrix}.$$

In view of the matrix inverse

$$\begin{pmatrix} t & t \\ t & 1 \end{pmatrix}^{-1} \;=\; \frac{1}{t(1-t)} \begin{pmatrix} 1 & -t \\ -t & t \end{pmatrix}^{-1},$$

straightforward algebra demonstrates that

$$\mathrm{E}(Y_{t+s} \mid Y_t = y) \;=\; y - \frac{ys}{1-t}$$

$$\mathrm{Var}(Y_{t+s} \mid Y_t = y) \;=\; s - \frac{s^2}{1-t}. \tag{11.7}$$

It follows that the Brownian bridge has infinitesimal mean and variance $\mu(t, y) = -y/(1-t)$ and $\sigma^2(t, y) = 1$. ∎

Example 11.3.4 *Bessel Process*

Consider a random process $X_t = (X_{1t}, \ldots, X_{nt})$ in R^n whose n components are independent standard Brownian motions. Let

$$Y_t \;=\; \sum_{i=1}^{n} X_{it}^2$$

be the squared distance from the origin. To calculate the infinitesimal mean and variance of this diffusion process, we write

$$Y_{t+s} - Y_t \;=\; \sum_{i=1}^{n} \left[2 X_{it}(X_{i,t+s} - X_{it}) + (X_{i,t+s} - X_{it})^2 \right].$$

It follows from this representation, independence, and equation (11.3) that

$$\mathrm{E}(Y_{t+s} - Y_t \mid X_t = x) \;=\; ns$$

$$\mathrm{E}[(Y_{t+s} - Y_t)^2 \mid X_t = x] \;=\; \sum_{i=1}^{n} 4 x_i^2 \, \mathrm{E}[(X_{i,t+s} - X_{it})^2 \mid X_t = x] + o(s)$$

$$\;=\; 4ys + o(s)$$

for $y = \sum_{i=1}^{n} x_i^2$. Because

$$\mathrm{E}(Y_{t+s} - Y_t \mid Y_t = y) \;=\; \mathrm{E}[\mathrm{E}(Y_{t+s} - Y_t \mid X_t = x) \mid Y_t = y]$$

$$\mathrm{Var}(Y_{t+s} - Y_t \mid Y_t = y) \;=\; \mathrm{E}[(Y_{t+s} - Y_t)^2 \mid Y_t = y] + o(s)$$

$$\;=\; \mathrm{E}\{\mathrm{E}[(Y_{t+s} - Y_t)^2 \mid X_t = x] \mid Y_t = y\} + o(s),$$

we conclude that Y_t has infinitesimal mean and variance

$$\begin{aligned} \mu_Y(t,y) &= n \\ \sigma_Y^2(t,y) &= 4y. \end{aligned}$$

The random distance $R_t = \sqrt{Y_t}$ is known as the Bessel process. Its infinitesimal mean and variance

$$\begin{aligned} \mu_R(t,r) &= \frac{n-1}{2r} \\ \sigma_R^2(t,r) &= 1 \end{aligned}$$

are immediate consequences of formula (11.4).   ∎

Example 11.3.5 *Diffusion Approximation to Kendall's Process*

Suppose in Kendall's model, the birth, death, and immigration rates α, δ, and ν are constant. (Here we have substituted δ for μ to avoid a collision with the symbol for the infinitesimal mean.) Given x particles at time t, there is one birth with probability αxs during the very short time interval $(t, t+s)$, one death with probability δxs, and one immigrant with probability νs. The probability of more than one event is $o(s)$. These considerations imply that

$$\begin{aligned} \mu(t,x) &= (\alpha - \delta)x + \nu \\ \sigma^2(t,x) &= (\alpha + \delta)x + \nu. \end{aligned}$$

This diffusion approximation is apt to be good for a moderate number of particles and poor for a small number of particles.   ∎

Example 11.3.6 *Neuron Firing and the Ornstein-Uhlenbeck Process*

Physiologists have long been interested in understanding how neurons fire [84, 146]. Firing involves the electrical potentials across the cell membranes of these basic cells of the nervous system. The typical neuron is composed of a compact cell body or soma into which thousands of small dendrites feed incoming signals. The soma integrates these signals and occasionally fires. When the neuron fires, a pulse, or action potential, is sent down the axon connected to the soma. The axon makes contact with other nerve cells or with muscle cells through junctions know as synapses. An action potential is a transient electrical depolarization of the cell membrane that propagates from the soma along the axon to the synapses. In a neuron's resting state, there is a potential difference of about -70 mV (millivolts) across the soma membrane. This is measured by inserting a microelectrode through the membrane. A cell is said to excited, or depolarized, if the soma potential exceeds the resting potential; it is inhibited, or hyperpolarized, in the reverse case. When the soma potential reaches a threshold of from

10 to 40 mV above the resting potential, the neuron fires. After the axon fires, the potential is reset to a level below the resting potential. An all or nothing action potential converts an analog potential difference into a digital pulse of information.

To model the firing of a neuron, we let X_t be the soma membrane potential at the time t. Two independent processes, one excitatory and one inhibitory, drive X_t. The soma receives excitatory pulses of magnitude ϵ according to a Poisson process with intensity α and inhibitory pulses of magnitude $-\delta$ according an independent Poisson process with intensity β. The potential is subject to exponential decay with time constant γ to the resting value x_r. When X_t reaches the fixed threshold s, the neuron fires. Afterwards, X_t is reset to the level x_0.

Because typical values of ϵ and δ range from 0.5 to 1 mV, a diffusion approximation is appropriate. In view of the Poisson nature of the inputs, we have

$$
\begin{aligned}
\mu(t, x) &= \alpha\epsilon - \beta\delta - \gamma(x - x_r) \\
\sigma^2(t, x) &= \alpha\epsilon^2 + \beta\delta^2.
\end{aligned}
$$

The transformed process $Y_t = X_t - \eta/\gamma$ with $\eta = \alpha\epsilon - \beta\delta + \gamma x_r$ is known as the Ornstein-Uhlenbeck process. It has infinitesimal mean and variance

$$
\begin{aligned}
\mu_Y(t, y) &= -\gamma\left(y + \frac{\eta}{\gamma}\right) + \eta \\
&= -\gamma y \\
\sigma_Y^2(t, y) &= \alpha\epsilon^2 + \beta\delta^2 \\
&= \sigma^2
\end{aligned}
$$

and is somewhat easier to study.

Fortunately, it is possible to relate the Ornstein-Uhlenbeck process to Brownian motion. Consider the complicated transformation

$$
Y_t = \alpha e^{-\gamma t}\left(W_{e^{\beta t}} - W_1 + \alpha^{-1}y_0\right)
$$

of standard Brownian motion W_t. If we write

$$
\begin{aligned}
Y_{t+s} - Y_t &= \alpha e^{-\gamma(t+s)}\left[W_{e^{\beta(t+s)}} - W_{e^{\beta t}}\right] \\
&\quad + \left(e^{-\gamma s} - 1\right)\alpha e^{-\gamma t}\left(W_{e^{\beta t}} - W_1 + \alpha^{-1}y_0\right),
\end{aligned}
$$

then is clear that this increment is Gaussian. Furthermore, setting $\beta = 2\gamma$ gives

$$
\begin{aligned}
\mathrm{E}\left[Y_{t+s} - Y_t \mid Y_t = y\right] &= \left(e^{-\gamma s} - 1\right)y \\
&= -\gamma y s + o(s) \\
\mathrm{Var}\left[Y_{t+s} - Y_t \mid Y_t = y\right] &= \alpha^2 e^{-2\gamma(t+s)}\left[e^{\beta(t+s)} - e^{\beta t}\right] \\
&= \alpha^2\left[1 - e^{-2\gamma s}\right] \\
&= 2\alpha^2\gamma s + o(s).
\end{aligned}
$$

Therefore, if we define α so that $\sigma^2 = 2\alpha^2\gamma$, then the infinitesimal mean $-\gamma y$ and variance σ^2 of Y_t coincide with those of the Ornstein-Uhlenbeck process. One interesting dividend of this approach is that we can immediately conclude that Y_t is Gaussian with mean and variance

$$
\begin{aligned}
\mathrm{E}(Y_t) &= y_0 e^{-\gamma t} \\
\mathrm{Var}(Y_t) &= \alpha^2 e^{-2\gamma t}\left(e^{2\gamma t} - 1\right) \\
&= \frac{\sigma^2}{2\gamma}\left(1 - e^{-2\gamma t}\right).
\end{aligned}
\tag{11.8}
$$

These are clearly compatible with the initial value $Y_0 = y_0$. ∎

Example 11.3.7 *Wright-Fisher Model with Mutation and Selection*

The Wright-Fisher model for the evolution of a deleterious or neutral gene postulates (a) discrete generations, (b) finite population size, (c) no immigration, and (d) random sampling from a gamete pool. In assumption (d), each current population member contributes to the infinite pool of potential gametes (sperm and eggs) in proportion to his or her fitness.

Mutation from the normal allele A_2 to the deleterious allele A_1 takes place at this stage with mutation rate η; backmutation is not permitted. Once the pool of potential gametes is formed, actual gametes are sampled randomly. In the neutral model introduced in Examples 7.3.2, 10.2.6, and 10.4.4, we neglect mutation and selection and treat the two alleles symmetrically.

The population frequencies (proportions) p and q of the two alleles A_1 and A_2 are the primary focus of the Wright-Fisher model [35, 45, 83]. These frequencies change over time in response to the forces of mutation, selection, and genetic drift (random sampling of gametes). Selection operates through fitness differences. Denote the average fitnesses of the genotypes A_1/A_1, A_1/A_2, and A_2/A_2 by w_{A_1/A_1}, w_{A_1/A_2}, and w_{A_2/A_2}, respectively. Because only relative fitness is important in formulating the dynamics of the Wright-Fisher model, we set $w_{A_1/A_1} = f < 1$ and $w_{A_1/A_2} = w_{A_2/A_2} = 1$ for a recessive disease. For a dominant disease, we set $w_{A_1/A_1} = w_{A_1/A_2} = f < 1$ and $w_{A_2/A_2} = 1$.

In a purely deterministic model, the frequency p_n of the disease allele A_1 for a dominant disease satisfies the recurrence

$$
p_{n+1} = \frac{fp_n^2 + (1+\eta)fp_nq_n + \eta q_n^2}{fp_n^2 + f2p_nq_n + q_n^2}.
\tag{11.9}
$$

Here individuals bearing the three genotypes contribute gametes in proportion to the product of their population frequencies and relative fitnesses. Because gametes are drawn independently from the pool at generation n, the three genotypes A_1/A_1, A_1/A_2, and A_2/A_2 occur in the Hardy-Weinberg proportions p_n^2, $2p_nq_n$, and q_n^2, respectively.

Given that we expect p_n to be of order η, equation (11.9) radically simplifies if we expand and drop all terms of order η^2 and higher. The resulting linear recurrence

$$p_{n+1} \;=\; \eta + f p_n \tag{11.10}$$

has fixed point $p_\infty = \frac{\eta}{1-f}$. Furthermore, because $p_{n+1} - p_\infty = f(p_n - p_\infty)$, the iterates p_n converge to p_∞ at linear rate f.

In contrast, the frequency p_n of the disease allele A_1 for a recessive disease satisfies the deterministic recurrence

$$p_{n+1} \;=\; \frac{f p_n^2 + (1+\eta)p_n q_n + \eta q_n^2}{f p_n^2 + 2 p_n q_n + q_n^2}.$$

Now we expect p_n to be of order $\sqrt{\eta}$. Expanding the recurrence and dropping all terms of order $\eta^{3/2}$ and higher yields the quadratic recurrence

$$p_{n+1} \;=\; \eta + p_n q_n + f p_n^2 \tag{11.11}$$

with fixed point $p_\infty = \sqrt{\eta/(1-f)}$. To determine the rate of convergence, we rewrite equation (11.11) as

$$p_{n+1} - p_\infty \;=\; [1 - (1-f)(p_n + p_\infty)](p_n - p_\infty).$$

This makes it clear that p_n converges to p_∞ at linear rate

$$1 - (1-f)2p_\infty \;=\; 1 - 2\sqrt{\eta(1-f)}.$$

These two special cases both entail a disease prevalence on the order of η, which typically falls in the range 10^{-7} to 10^{-5}. The two cases differ markedly in their rates of convergence to equilibrium. A dominant disease reaches equilibrium quickly unless the average fitness f of affecteds is very close to 1. By comparison, a recessive disease reaches equilibrium extremely slowly.

Random fluctuations in the frequency of the disease allele can be large in small populations. Let N_m be the size of the surrounding population at generation m. In some cases we will take N_m constant to simplify our mathematical development. In the stochastic theory, the deterministic frequency p_m of allele A_1 at generation m is replaced by the random frequency X_m of A_1. This frequency is the ratio of the total number Y_m of A_1 alleles present to the total number of genes $2N_m$. The Wright-Fisher model specifies that Y_m is binomially distributed with $2N_m$ trials and success probability $p(X_{m-1})$ determined by the proportion $p(X_{m-1})$ of A_1 alleles in the pool of potential gametes for generation m. In passing to a diffusion approximation, we take one generation as the unit of time and substitute

$$\mu(m, x_m) \;=\; \mathrm{E}(X_{m+1} - X_m \mid X_m = x_m) \tag{11.12}$$

$$\begin{aligned} &= p(x_m) - x_m \\ \sigma^2(m, x_m) &= \mathrm{Var}(X_{m+1} - X_m \mid X_m = x_m) \\ &= \frac{p(x_m)[1 - p(x_m)]}{2N_{m+1}} \end{aligned} \qquad (11.13)$$

for the infinitesimal mean $\mu(t, x)$ and variance $\sigma^2(t, x)$ of the diffusion process evaluated at time $t = m$ and position $x = x_m$.

Under neutral evolution, the gamete pool probability $p(x) = x$. This formula for $p(x)$ entails no systematic tendency for either allele to expand at the expense of the other allele. For a dominant disease, $p(x) = \eta + fx$, while for a recessive disease, $p(x) = \eta + x - (1 - f)x^2$. Most population geneticists substitute $p(x) = x$ in formula (11.13) defining the infinitesimal variance $\sigma^2(t, x)$. This action is justified for neutral and recessive inheritance, but less so for dominant inheritance where the allele frequency x is typically on the order of magnitude of the mutation rate η. It is also fair to point out that in the presence of inbreeding or incomplete mixing of a population, the effective population size is less than the actual population size [35]. For the sake of simplicity, we will ignore this evolutionary fact. ∎

11.4   Process Moments

Taking unconditional expectations in expression (11.1) gives

$$\mathrm{E}(X_{t+s}) = \mathrm{E}(X_t) + \mathrm{E}[\mu(t, X_t)]s + o(s).$$

Forming the obvious difference quotient and sending s to 0 therefore provide the ordinary differential equation

$$\frac{d}{dt}\mathrm{E}(X_t) = \mathrm{E}[\mu(t, X_t)] \qquad (11.14)$$

characterizing $\mathrm{E}(X_t)$. In the special case $\mu(t, x) = \alpha x + \beta$, it is easy to check that equation (11.14) has solution

$$\mathrm{E}(X_t) = \mathrm{E}(X_0)e^{\alpha t} + \frac{\beta}{\alpha}\left(e^{\alpha t} - 1\right), \qquad (11.15)$$

unless $\alpha = 0$, in which case $\mathrm{E}(X_t) = \mathrm{E}(X_0) + \beta t$.

Taking unconditional variances in expression (11.2) yields in a similar manner

$$\begin{aligned} \mathrm{Var}(X_{t+s}) &= \mathrm{E}[\mathrm{Var}(X_t + \Delta X_t \mid X_t)] + \mathrm{Var}[\mathrm{E}(X_t + \Delta X_t \mid X_t)] \\ &= \mathrm{E}[\sigma^2(t, X_t)s + o(s)] + \mathrm{Var}[X_t + \mu(t, X_t)s + o(s)] \\ &= \mathrm{E}[\sigma^2(t, X_t)]s + \mathrm{Var}(X_t) + 2\,\mathrm{Cov}[X_t, \mu(t, X_t)]s + o(s) \end{aligned}$$

for $\Delta X_t = X_{t+s} - X_t$. In this case taking the difference quotient and sending s to 0 give the ordinary differential equation

$$\frac{d}{dt}\operatorname{Var}(X_t) \;=\; \operatorname{E}[\sigma^2(t, X_t)] + 2\operatorname{Cov}[X_t, \mu(t, X_t)] \qquad (11.16)$$

rigorously derived in reference [47].

Example 11.4.1 *Moments of the Wright-Fisher Diffusion Process*

In the diffusion approximation to the Wright-Fisher model for a dominant disease, equation (11.12) implies

$$\mu(t, x) \;=\; \eta - (1 - f)x.$$

It follows from equation (11.15) that

$$\operatorname{E}(X_t) \;=\; \left[x_0 - \frac{\eta}{1-f}\right]e^{-(1-f)t} + \frac{\eta}{1-f}$$

for $X_0 = x_0$. The limiting value of $\eta/(1-f)$ is the same as the deterministic equilibrium. In the case of neutral evolution with $f = 1$ and $\eta = 0$, the mean $\operatorname{E}(X_t) = x_0$ is constant. With constant population size N, equations (11.13) and (11.16) therefore yield the differential equation

$$\begin{aligned}
\frac{d}{dt}\operatorname{Var}(X_t) &= \frac{\operatorname{E}(X_t) - \operatorname{E}(X_t^2)}{2N} \\
&= \frac{x_0 - x_0^2 - \operatorname{Var}(X_t)}{2N},
\end{aligned}$$

with solution

$$\operatorname{Var}(X_t) \;=\; x_0(1 - x_0)\left[1 - e^{-\frac{t}{2N}}\right].$$

This expression for $\operatorname{Var}(X_t)$ tends to $x_0(1 - x_0)$ as t tends to ∞, which is the variance of the limiting random variable

$$X_\infty \;=\; \begin{cases} 1 & \text{with probability } x_0 \\ 0 & \text{with probability } 1 - x_0. \end{cases}$$

Fan and Lange [46] calculate $\operatorname{Var}(X_t)$ for the dominant case. In the recessive case, this approach to $\operatorname{E}(X_t)$ and $\operatorname{Var}(X_t)$ breaks down because $\mu(t, x)$ is quadratic rather than linear in x. ∎

Example 11.4.2 *Moments of the Brownian Bridge*

For the Brownian bridge, equation (11.14) becomes

$$\frac{d}{dt}\operatorname{E}(X_t) \;=\; \operatorname{E}(X_0) - \frac{1}{1-t}\operatorname{E}(X_t).$$

The unique solution with $E(X_0) = 0$ is $E(X_t) = 0$ for all t. Equation (11.16) becomes

$$\frac{d}{dt}\,\text{Var}(X_t) \;=\; 1 - \frac{2}{1-t}\,\text{Var}(X_t).$$

This has solution

$$\text{Var}(X_t) \;=\; t(1-t)$$

subject to the initial value $\text{Var}(X_0) = 0$. These results match the results in formula (11.7) if we replace y by 0, s by t, and t by 0. ∎

11.5 First Passage Problems

Let $c < d$ be two points in the interior of the range I of a diffusion process X_t. Define T_c to be the first time t that $X_t = c$ starting from $X_0 \geq c$. If eventually $X_t < c$, then the continuity of the sample paths guarantees that $X_t = c$ at some first time T_c. It may be that $T_c = \infty$ with positive probability. Similar considerations apply to T_d, the first time t that $X_t = d$ starting from $X_0 \leq d$. The process X_t exits (c, d) at the time $T = \min\{T_c, T_d\}$. We consider two related problems involving these first passage times. One problem is to calculate the probability $u(x) = \Pr(T_d < T_c \mid X_0 = x)$ that the process exits via d starting from $x \in [c, d]$. It is straightforward to derive a differential equation determining $u(x)$ given the boundary conditions $u(c) = 0$ and $u(d) = 1$. With this end in mind, we assume that X_t is time homogeneous.

For $s > 0$ small and $x \in (c, d)$, the probability that X_t reaches either c or d during the time interval $[0, s]$ is $o(s)$. Thus,

$$u(x) \;=\; E[u(X_s) \mid X_0 = x] + o(s).$$

If we let $\Delta X_s = X_s - X_0$ and expand $u(X_s)$ in a second-order Taylor series, then we find that

$$
\begin{aligned}
u(X_s) \;&=\; u(x + \Delta X_s) \\
&=\; u(x) + u'(x)\Delta X_s + \frac{1}{2}\Big[u''(x) + r(\Delta X_s)\Big]\Delta X_s^2, \quad (11.17)
\end{aligned}
$$

where the relative error $r(\Delta X_s)$ tends to 0 as ΔX_s tends to 0. Invoking equations (11.1), (11.2), and (11.17) therefore yields

$$
\begin{aligned}
u(x) \;&=\; E[u(X_s)] + o(s) \\
&=\; u(x) + \mu(x)u'(x)s + \frac{1}{2}\sigma^2(x)u''(x)s + o(s),
\end{aligned}
$$

which, upon rearrangement and sending s to 0, gives the differential equation

$$0 = \mu(x)u'(x) + \frac{1}{2}\sigma^2(x)u''(x). \tag{11.18}$$

It is a simple matter to check that equation (11.18) can be solved explicitly by defining

$$v(x) = \int_l^x e^{-\int_l^y \frac{2\mu(z)}{\sigma^2(z)}dz} dy$$

and setting

$$u(x) = \frac{v(x) - v(c)}{v(d) - v(c)}. \tag{11.19}$$

Here the lower limit of integration l can be any point in the interval $[c, d]$. This particular solution also satisfies the boundary conditions.

Example 11.5.1 *Exit Probabilities in the Wright-Fisher Model*

In the diffusion approximation to the neutral Wright-Fisher model with constant population size N, we calculate

$$v(x) = \int_l^x e^{-\int_l^y 0\,dz} dy = x - l.$$

Thus, starting at a frequency of x for allele A_1, allele A_2 goes extinct before allele A_1 with probability

$$u(x) = \lim_{c\to 0,\, d\to 1} \frac{x - l - (c - l)}{d - l - (c - l)} = x.$$

This example is typical in the sense that $u(x) = (x - c)/(d - c)$ for any diffusion process with $\mu(x) = 0$. ∎

Example 11.5.2 *Exit Probabilities in the Bessel Process*

Consider two fixed radii $0 < r_0 < r_1$ in the Bessel process. It is straightforward to calculate

$$v(x) = \int_l^x e^{-\int_l^y \frac{n-1}{z} dz} dy$$

$$= \int_l^x \frac{l^{n-1}}{y^{n-1}} dy$$

$$= -\frac{l^{n-1}}{(n-2)x^{n-2}} + \frac{l^{n-1}}{(n-2)l^{n-2}}$$

for $n > 2$. This yields

$$u(x) \quad = \quad \frac{r_0^{-n+2} - x^{-n+2}}{r_0^{-n+2} - r_1^{-n+2}},$$

from which it follows that $u(x)$ tends to 1 as r_0 tends to 0. This fact is intuitively obvious because the Brownian path X_t is much more likely to hit the surface of a large outer sphere of radius r_1 before it hits the surface of a small inner sphere of radius r_0. In R^2 we have

$$u(x) \quad = \quad \frac{\ln x - \ln r_0}{\ln r_1 - \ln r_0},$$

and the same qualitative comments apply. ∎

Another important problem is to calculate the expectation

$$w(x) \quad = \quad E[g(T) \mid X_0 = x]$$

of a function of the exit time T from $[c, d]$. For instance, $g(t) = t^n$ gives the nth moment of T, and $g(t) = e^{-\theta t}$ gives the Laplace transform of T. We again derive an ordinary differential equation determining $w(x)$, but now the pertinent boundary conditions are $w(c) = w(d) = g(0)$. To emphasize the dependence of T on the initial position x, let us write T_x in place of T.

We commence our derivation with the expansion

$$
\begin{aligned}
w(x) \quad &= \quad E[g(T_x) \mid X_0 = x] \\
&= \quad E[g(T_{X_s} + s) \mid X_0 = x] + o(s) \\
&= \quad E[g(T_{X_s}) + g'(T_{X_s})s \mid X_0 = x] + o(s) \\
&= \quad E\{E[g(T_{X_s}) \mid X_s] \mid X_0 = x\} + E[g'(T_{X_s}) \mid X_0 = x]s + o(s) \\
&= \quad E[w(X_s) \mid X_0 = x] + E[g'(T_x) \mid X_0 = x]s + o(s).
\end{aligned}
$$

Employing the same reasoning used in deriving the differential equation (11.18) for $u(x)$, we deduce that

$$E[w(X_s) \mid X_0 = x] \quad = \quad w(x) + \mu(x)w'(x)s + \frac{1}{2}\sigma^2(x)w''(x)s + o(s).$$

It follows that

$$
\begin{aligned}
w(x) \quad = \quad &w(x) + \mu(x)w'(x)s + \frac{1}{2}\sigma^2(x)w''(x)s \\
&+ E[g'(T_x) \mid X_0 = x]s + o(s).
\end{aligned}
$$

Rearranging this and sending s to 0 produce the differential equation

$$0 \quad = \quad \mu(x)w'(x) + \frac{1}{2}\sigma^2(x)w''(x) + E[g'(T_x) \mid X_0 = x].$$

The special cases $g(t) = t$ and $g(t) = e^{-\theta t}$ correspond to the differential equations

$$0 = \mu(x)w'(x) + \frac{1}{2}\sigma^2(x)w''(x) + 1 \tag{11.20}$$

$$0 = \mu(x)w'(x) + \frac{1}{2}\sigma^2(x)w''(x) - \theta w(x), \tag{11.21}$$

respectively. The pertinent boundary conditions are $w(c) = w(d) = 0$ and $w(c) = w(d) = 1$.

Example 11.5.3 *Mean Exit Times in the Wright-Fisher Model*

In the diffusion approximation to the neutral Wright-Fisher model with constant population size N, equation (11.20) becomes

$$0 = \frac{x(1-x)}{4N}w''(x) + 1. \tag{11.22}$$

If we take $c = 0$ and $d = 1$, then $w(x)$ represents the expected time until fixation of one of the two alleles. To solve equation (11.22), observe that

$$\begin{aligned}
w'(x) &= -4N \int_{\frac{1}{2}}^{x} \frac{1}{y(1-y)} \, dy + k_1 \\
&= -4N \int_{\frac{1}{2}}^{x} \left[\frac{1}{y} + \frac{1}{(1-y)} \right] dy + k_1 \\
&= -4N \left[\ln x - \ln(1-x) \right] + k_1
\end{aligned}$$

for some constant k_1. Integrating again yields

$$\begin{aligned}
w(x) &= -4N \int_{\frac{1}{2}}^{x} \left[\ln y - \ln(1-y) \right] dy + k_1 x + k_2 \\
&= -4N \left[x \ln x + (1-x) \ln(1-x) \right] + k_1 x + k_2
\end{aligned}$$

for some constant k_2. The boundary condition $w(0) = 0$ implies $k_2 = 0$, and the boundary condition $w(1) = 0$ implies $k_1 = 0$. It follows that

$$w(x) = -4N \left[x \ln x + (1-x) \ln(1-x) \right].$$

This is proportional to N and attains a maximum of $4N \ln 2$ at $x = 1/2$. ∎

Example 11.5.4 *Mean Exit Times in the Bessel Process*

Again fix two radii $0 < r_0 < r_1$. If we let $v(x) = w'(x)$, then equation (11.20) becomes

$$0 = \frac{n-1}{2x}v(x) + \frac{1}{2}v'(x) + 1$$

for the Bessel process. This inhomogeneous differential equation has the particular solution $v(x) = -2x/n$. The general solution is the sum of this particular solution and an arbitrary multiple of a solution to the homogeneous differential equation

$$0 = \frac{n-1}{2x}v(x) + \frac{1}{2}v'(x).$$

This makes it clear that the general solution is

$$v(x) = -\frac{2x}{n} + ax^{-n+1}$$

for an arbitrary constant a. Integrating the general solution for $v(x)$ produces the general solution

$$w(x) = -\frac{x^2}{n} - \frac{ax^{-n+2}}{n-2} + b$$

for $w(x)$ when $n > 2$.

The arbitrary constants a and b are determined by the boundary conditions $w(r_0) = w(r_1) = 0$. Thus,

$$\begin{aligned} 0 &= w(r_1) - w(r_0) \\ &= \frac{r_0^2 - r_1^2}{n} + \frac{a(r_0^{-n+2} - r_1^{-n+2})}{n-2} \end{aligned}$$

gives

$$a = \frac{(n-2)(r_0^2 - r_1^2)}{n(r_1^{-n+2} - r_0^{-n+2})},$$

which tends to 0 as r_0 tends to 0. For x fixed, it follows that b tends to

$$b = \frac{r_1^2}{n}.$$

This gives the expected time

$$w(x) = \frac{r_1^2 - x^2}{n} \tag{11.23}$$

to reach the surface of the sphere of radius r_1 starting from the surface of the sphere of radius x. This formula for $w(x)$ also holds in $\mathbf{R}^2$, but the derivation is slightly different. ∎

11.6   Equilibrium Distribution

In certain situations, a time-homogeneous diffusion process will tend to equilibrium. To find the equilibrium distribution, we set the left-hand side of Kolmogorov's equation (11.6) equal to 0 and solve for the equilibrium distribution $f(x) = \lim_{t \to \infty} f(t, x)$. Integrating the equation

$$0 = -\frac{d}{dx}\left[\mu(x)f(x)\right] + \frac{1}{2}\frac{d^2}{dx^2}\left[\sigma^2(x)f(x)\right] \tag{11.24}$$

once gives

$$k_1 = -\mu(x)f(x) + \frac{1}{2}\frac{d}{dx}\left[\sigma^2(x)f(x)\right]$$

for some constant k_1. The choice $k_1 = 0$ corresponds to the intuitively reasonable condition of no probability flux at equilibrium. Dividing the no flux equation by $\sigma^2(x)f(x)$ yields

$$\frac{d}{dx}\ln[\sigma^2(x)f(x)] = \frac{2\mu(x)}{\sigma^2(x)}.$$

If we now choose l in the interior of the range I of X_t and integrate a second time, then we deduce that

$$\ln[\sigma^2(x)f(x)] = k_2 + \int_l^x \frac{2\mu(y)}{\sigma^2(y)}\,dy,$$

from which Wright's formula

$$f(x) = \frac{k_3 e^{\int_l^x \frac{2\mu(y)}{\sigma^2(y)}\,dy}}{\sigma^2(x)} \tag{11.25}$$

for the equilibrium distribution follows. An appropriate choice of the constant $k_3 = e^{k_2}$ serves to make $\int_I f(x)\,dx = 1$ when the equilibrium distribution exists and is unique.

Example 11.6.1 *Equilibrium for the Ornstein-Uhlenbeck Process*

Wright's formula (11.25) gives

$$f(y) = \frac{k_3 e^{-\int_0^y \frac{2\gamma z}{\sigma^2}\,dz}}{\sigma^2}$$

$$= \sqrt{\frac{\gamma}{\pi\sigma^2}}e^{-\gamma y^2/\sigma^2}$$

for the Ornstein-Uhlenbeck process. This is exactly the normal density one would predict by sending t to ∞ in the moment equations (11.8). The neuron firing process $X_t = Y_t + \eta/\gamma$ has the same equilibrium distribution shifted by the amount η/γ.   ∎

Example 11.6.2 *Equilibrium for a Recessive Disease Gene*

Equilibrium for a disease gene is maintained by the balance between selection and mutation. To avoid fixation of the deleterious allele and to ensure existence of the equilibrium distribution, backmutation of the deleterious allele to the normal allele must be incorporated into the model. In reality, the chance of fixation is so remote that backmutation does not enter into the following approximation of the equilibrium distribution $f(x)$. Because only small values of the disease gene frequency are likely, $f(x)$ is concentrated near 0. In the vicinity of 0, the approximation $x(1 - x) \approx x$ holds. For a recessive disease, these facts suggest that we use

$$\frac{2\mu(y)}{\sigma^2(y)} = \frac{2[\eta - (1-f)y^2]}{\frac{y(1-y)}{2N}}$$

$$\approx 4N\left[\frac{\eta}{y} - (1-f)y\right]$$

in Wright's formula (11.25) when the surrounding population size N is constant.

With this understanding,

$$f(x) \approx \frac{2Nk_3}{x}e^{4N\eta \ln(x/l)-2N(1-f)(x^2-l^2)}$$

$$= k_4 x^{4N\eta-1}e^{-2N(1-f)x^2}$$

for some constant $k_4 > 0$. The change of variables $z = 2N(1 - f)x^2$ shows that the mth moment of $f(x)$ is

$$\int_I x^m f(x)\, dx \approx k_4 \int_0^1 x^{m+4N\eta-1}e^{-2N(1-f)x^2}\, dx$$

$$= \frac{k_4}{4N(1-f)} \int_0^1 x^{m+4N\eta-2}e^{-2N(1-f)x^2}4N(1-f)x\, dx$$

$$= \frac{k_4}{4N(1-f)} \int_0^{2N(1-f)} \left[\frac{z}{2N(1-f)}\right]^{\frac{m+4N\eta-2}{2}} e^{-z}\, dz$$

$$\approx \frac{k_4}{2[2N(1-f)]^{\frac{m}{2}+2N\eta}} \int_0^\infty z^{\frac{m}{2}+2N\eta-1}e^{-z}\, dz$$

$$= \frac{k_4\Gamma(\frac{m}{2}+2N\eta)}{2[2N(1-f)]^{\frac{m}{2}+2N\eta}}.$$

Taking $m = 0$ identifies the normalizing constant

$$k_4 = \frac{2[2N(1-f)]^{2N\eta}}{\Gamma(2N\eta)}.$$

With this value of k_4 in hand, the mean of $f(x)$ is

$$\int_I xf(x)\, dx \approx \frac{\Gamma(2N\eta+\frac{1}{2})}{\sqrt{2N(1-f)}\Gamma(2N\eta)}.$$

When $N\eta$ is large, application of Stirling's formula implies that the mean is close to the deterministic equilibrium value $\sqrt{\eta/(1-f)}$. In practice, one should be wary of applying the equilibrium theory because the approach to equilibrium is so slow.   ∎

11.7   A Numerical Method for Diffusion Processes

It is straightforward to simulate a diffusion process X_t. The definition tells us to extend X_t to X_{t+s} by setting the increment $X_{t+s} - X_t$ equal to a normal deviate with mean $\mu(t,x)s$ and variance $\sigma^2(t,x)s$. The time increment s should be small, and each sampled normal variate should be independent. Techniques for generating random normal deviates are covered in standard texts on computational statistics and will not be discussed here [87, 95]. Of more concern is how to cope with a diffusion process with finite range I. Because a normally distributed random variable has infinite range, it is possible in principle to generate an increment that takes the simulated process outside I. One remedy for this problem is to take s extremely small. It also helps if the infinitesimal variance $\sigma^2(t,x)$ tends to 0 as x approaches the boundary of I. This is the case with the neutral Wright-Fisher process.

   Simulation offers a crude method of finding the distribution of X_t. Simply conduct multiple independent simulations and compute a histogram of the recorded values of X_t. Although this method is neither particularly accurate nor efficient, it has the virtue of yielding simultaneously the distributions of all of the X_t involved in the simulation process. Thus, if 1000 times are sampled per simulation, then the method yields all 1000 distributions, assuming that enough computer memory is available. Much greater accuracy can be achieved by solving Kolmogorov's forward equation. The ideal of an exact solution is seldom attained in practice, even for time-homogeneous problems. However, Kolmogorov's forward equation can be solved numerically by standard techniques for partial differential equations. Here we would like to discuss a nonstandard method for finding the distribution of X_t that directly exploits the definition of a diffusion process.

   This method recursively computes the distribution of X_{t_i} at n times points labeled $0 < t_1 < \cdots < t_n = t$. In the diffusion approximation to Markov chain models such as the Wright-Fisher model, it is reasonable to let $\delta t_i = t_{i+1} - t_i$ be one generation. It is also convenient to supplement these points with the initial point $t_0 = 0$. For each t_i, we would like to compute the probability that $X_{t_i} \in [a_{ij}, a_{i,j+1}]$ for r_i+1 points $a_{i0} < \cdots < a_{i,r_i}$. We will say more about these mesh points later. In the meanwhile, let p_{ij} denote the probability $\Pr(X_{t_i} \in [a_{ij}, a_{i,j+1}])$ and c_{ij} the center of probability $\mathrm{E}(X_{t_i} \mid X_{t_i} \in [a_{ij}, a_{i,j+1}])$. Our method carries forward approximations to both of these sequences starting from an arbitrary initial distribution for X_0.

In passing from time t_i to time t_{i+1}, the diffusion process redistributes a certain amount of probability mass from the interval $[a_{ij}, a_{i,j+1}]$ to the interval $[a_{i+1,k}, a_{i+1,k+1}]$. Given the definition of a diffusion process and the notation $m(i, x) = x + \mu(t_i, x)\delta t_i$ and $s^2(i, x) = \sigma^2(t_i, x)\delta t_i$, the amount redistributed is approximately

$$
\begin{aligned}
&p_{ij \to i+1,k} \\
&= \int_{a_{ij}}^{a_{i,j+1}} \frac{1}{\sqrt{2\pi s^2(i,x)}} \int_{a_{i+1,k}}^{a_{i+1,k+1}} e^{-\frac{[y-m(i,x)]^2}{2s^2(i,x)}} \, dy \, f(t_i, x) \, dx \\
&= \int_{a_{ij}}^{a_{i,j+1}} \frac{1}{\sqrt{2\pi}} \int_{\frac{a_{i+1,k}-m(i,x)}{s(i,x)}}^{\frac{a_{i+1,k+1}-m(i,x)}{s(i,x)}} e^{-\frac{z^2}{2}} \, dz \, f(t_i, x) \, dx. \qquad (11.26)
\end{aligned}
$$

(Here and in the remainder of this section, the equality sign indicates approximate equality.) Similarly the center of probability $c_{ij \to i+1,k}$ of the redistributed probability approximately satisfies

$$
\begin{aligned}
&c_{ij \to i+1,k} p_{ij \to i+1,k} \\
&= \int_{a_{ij}}^{a_{i,j+1}} \frac{1}{\sqrt{2\pi s^2(i,x)}} \int_{a_{i+1,k}}^{a_{i+1,k+1}} y e^{-\frac{[y-m(i,x)]^2}{2s^2(i,x)}} \, dy \, f(t_i, x) \, dx \qquad (11.27) \\
&= \int_{a_{ij}}^{a_{i,j+1}} \frac{1}{\sqrt{2\pi}} \int_{\frac{a_{i+1,k}-m(i,x)}{s(i,x)}}^{\frac{a_{i+1,k+1}-m(i,x)}{s(i,x)}} [m(i,x) + s(i,x)z] e^{-\frac{z^2}{2}} \, dz \, f(t_i, x) \, dx.
\end{aligned}
$$

Given these quantities, we calculate

$$
p_{i+1,k} = \sum_{j=0}^{r_i-1} p_{ij \to i+1,k}
$$

$$
c_{i+1,k} = \frac{1}{p_{i+1,k}} \sum_{j=0}^{r_i-1} c_{ij \to i+1,k} p_{ij \to i+1,k}, \qquad (11.28)
$$

assuming that X_{t_i} is certain to belong to one of the intervals $[a_{ij}, a_{i,j+1}]$.

To carry out this updating scheme, we must approximate the integrals $p_{ij \to i+1,k}$ and $c_{ij \to i+1,k} p_{ij \to i+1,k}$. If the interval $[a_{ij}, a_{i,j+1}]$ is fairly narrow, then the linear approximations

$$
\begin{aligned}
m(i, x) &= \mu_{ij0} + \mu_{ij1} x \\
s^2(i, x) &= \sigma_{ij}^2 \qquad (11.29) \\
f(t_i, x) &= f_{ij0} + f_{ij1} x
\end{aligned}
$$

should suffice for all x in the interval. The first two of these linear approximations follow directly from the diffusion model. The constants involved in the third approximation are determined by the equations

$$
\begin{aligned}
p_{ij} &= \int_{a_{ij}}^{a_{i,j+1}} (f_{ij0} + f_{ij1}x)\, dx \\
&= f_{ij0}(a_{i,j+1} - a_{ij}) + \frac{1}{2}f_{ij1}(a_{i,j+1} + a_{ij})(a_{i,j+1} - a_{ij}) \\
c_{ij}p_{ij} &= \int_{a_{ij}}^{a_{i,j+1}} x(f_{ij0} + f_{ij1}x)\, dx \\
&= \frac{1}{2}f_{ij0}(a_{i,j+1} + a_{ij})(a_{i,j+1} - a_{ij}) \\
&\quad + \frac{1}{3}f_{ij1}(a_{i,j+1}^2 + a_{i,j+1}a_{ij} + a_{ij}^2)(a_{i,j+1} - a_{ij})
\end{aligned}
$$

with inverses

$$
\begin{aligned}
f_{ij0} &= \frac{2p_{ij}(2a_{ij}^2 + 2a_{ij}a_{i,j+1} + 2a_{i,j+1}^2 - 3a_{ij}c_{ij} - 3a_{i,j+1}c_{ij})}{(a_{i,j+1} - a_{ij})^3} \\
f_{ij1} &= \frac{6p_{ij}(2c_{ij} - a_{ij} - a_{i,j+1})}{(a_{i,j+1} - a_{ij})^3}. \tag{11.30}
\end{aligned}
$$

Problem 13 asks the reader to check that the linear density $f_{ij0} + f_{ij1}x$ is nonnegative throughout the interval $(a_{ij}, a_{i,j+1})$ if and only if its center of mass c_{ij} lies in the middle third of the interval.

Given the linear approximations (11.29), we now show that the double integrals (11.26) and (11.27) reduce to expressions involving elementary functions and the standard normal distribution function

$$
\Phi(x) = \frac{1}{\sqrt{2\pi}} \int_{-\infty}^{x} e^{-y^2/2}\, dy.
$$

The latter can be rapidly evaluated by either a power series or a continued fraction expansion [95]. It also furnishes the key to evaluating the hierarchy of special functions

$$
\Phi_k(x) = \frac{1}{\sqrt{2\pi}} \int_{-\infty}^{x} y^k e^{-y^2/2}\, dy
$$

through the integration-by-parts recurrence

$$
\Phi_k(x) = -\frac{1}{\sqrt{2\pi}} x^{k-1} e^{-x^2/2} + (k-1)\Phi_{k-2}(x) \tag{11.31}
$$

beginning with $\Phi_0(x) = \Phi(x)$. We can likewise evaluate the related integrals

$$
\Psi_{jk}(x) = \int_{-\infty}^{x} y^j \Phi_k(y)\, dy
$$

via the integration-by-parts reduction

$$\Psi_{jk}(x) \;=\; \frac{1}{j+1}x^{j+1}\Phi_k(x) - \frac{1}{j+1}\Phi_{j+k+1}(y). \qquad (11.32)$$

Based on the definition of $\Phi(x)$, the integral (11.26) becomes

$$p_{ij\to i+1,k} \;=\; \int_{a_{ij}}^{a_{i,j+1}} \Phi\!\left(\frac{z-\mu_{ij0}-\mu_{ij1}x}{\sigma_{ij}}\right)\bigg|_{a_{i+1,k}}^{a_{i+1,k+1}} (f_{ij0}+f_{ij1}x)\,dx,$$

and based on the recurrence (11.31), the integral (11.27) becomes

$$c_{ij\to i+1,k}p_{ij\to i+1,k}$$

$$= \int_{a_{ij}}^{a_{i,j+1}} (\mu_{ij0}+\mu_{ij1}x)\Phi\!\left(\frac{z-\mu_{ij0}-\mu_{ij1}x}{\sigma_{ij}}\right)\bigg|_{a_{i+1,k}}^{a_{i+1,k+1}} (f_{ij0}+f_{ij1}x)\,dx$$

$$\quad - \frac{\sigma_{ij}}{\sqrt{2\pi}} \int_{a_{ij}}^{a_{i,j+1}} e^{-(z-\mu_{ij0}-\mu_{ij1}x)^2/(2\sigma_{ij}^2)}\bigg|_{a_{i+1,k}}^{a_{i+1,k+1}} (f_{ij0}+f_{ij1}x)\,dx.$$

To evaluate the one-dimensional integrals in these expressions for $p_{ij\to i+1,k}$ and $c_{ij\to i+1,k}p_{ij\to i+1,k}$, we make appropriate linear changes of variables so that $e^{-x^2/2}$ and $\Phi(x)$ appear in the integrands and then apply formulas (11.31) and (11.32) as needed. Although the details are messy, it is clear that these maneuvers reduce everything to combinations of elementary functions and the standard normal distribution function.

To summarize, the algorithm presented approximates the probability p_{ij} and center of probability c_{ij} of each interval $[a_{ij}, a_{i,j+1}]$ of a subdivision of the range I of X_t. Equation (11.30) converts these parameters into a piecewise-linear approximation to the density of the process in preparation for propagation to the next subdivision. The actual propagation of probability from an interval of the current subdivision to another interval of the next subdivision is accomplished by computing $p_{ij\to i+1,k}$ and $c_{ij\to i+1,k}p_{ij\to i+1,k}$ based on elementary functions and the standard normal distribution function. The pieces $p_{ij\to i+1,k}$ and $c_{ij\to i+1,k}$ are then reassembled into probabilities and centers of probabilities using equations (11.28).

Choice of the mesh points $a_{i0} < \cdots < a_{i,r_i}$ at time t_i is governed by several considerations. First, the probability $\Pr(X_{t_i} \notin [a_{i0}, a_{i,r_i}])$ should be negligible. Second, $\sigma^2(t_i, x)$ should be well approximated by a constant and $\mu(t_i, x)$ by a linear function on each interval $[a_{ij}, a_{i,j+1}]$. Third, the density $f(t_i, x)$ should be well approximated by a linear function on $[a_{ij}, a_{i,j+1}]$ as well. This last requirement is the hardest to satisfy in advance, but nothing prevents one from choosing the next subdivision adaptively based on the distribution of probability within the current subdivision. Adding more mesh points will improve accuracy at the expense of efficiency. Mesh points need not be uniformly spaced. It makes sense to cluster them in regions of high probability and rapid fluctuations of $f(t, x)$. Given the smoothness expected of $f(t, x)$, rapid fluctuations are unlikely.

Many of the probabilities $p_{ij \to i+1,k}$ are negligible. We can accelerate the algorithm by computing $p_{ij \to i+1,k}$ and $c_{ij \to i+1,k}$ only for $[a_{i+1,k}, a_{i+1,k+1}]$ close to $[a_{ij}, a_{i,j+1}]$. Because the conditional increment $X_{t_{i+1}} - X_{t_i}$ is normally distributed, it is very unlikely to extend beyond a few standard deviations σ_{ij} given X_{t_i} is in $[a_{ij}, a_{i,j+1}]$. Thus, the most sensible strategy is to visit each interval $[a_{ij}, a_{i,j+1}]$ in turn and propagate probability only to those intervals $[a_{i+1,k}, a_{i+1,k+1}]$ that lie a few standard deviations to the left or right of $[a_{ij}, a_{i,j+1}]$.

11.8   Application to the Wright-Fisher Process

We now apply the numerical method just described to the Wright-Fisher process. Our application confronts the general issue of how to deal with a diffusion approximation when it breaks down. In the case of the Wright-Fisher Markov chain, the diffusion approximation degrades for very low allele frequencies. Because of the interest in gene extinction, this is regrettable. However in the regime of low allele frequencies, we can always fall back on the Wright-Fisher Markov chain. As population size grows, the Markov chain updates become more computationally demanding. The most pressing concern thus becomes how to merge the Markov chain and diffusion approaches seamlessly into a single algorithm for following the evolution of an allele. Here we present one possible algorithm and apply it to understanding disease-gene dynamics in a population isolate.

The algorithm outlined in the previous section has the virtue of being posed in terms of distribution functions rather than density functions. For low allele frequencies, discreteness is inevitable, and density functions are unrealistic. In adapting the algorithm to the regime of low allele frequencies, it is useful to let

$$a_{ij} = \frac{j - \frac{1}{2}}{2N_i}$$

for $0 \le j \le q$ and some positive integer q. The remaining a_{ij} are distributed over the interval $[a_{iq}, 1]$ less uniformly. This tactic separates the possibility of exactly j alleles at time t_i, $0 \le j \le q$, from other possibilities. For $0 \le j \le q$, binomial sampling dictates that

$$p_{ij \to i+1,k} = \sum_l \binom{2N_{i+1}}{l} p^l (1-p)^{2N_{i+1}-l}$$

$$c_{ij \to i+1,k} = \frac{1}{p_{ij \to i+1,k}} \sum_l \binom{2N_{i+1}}{l} \frac{l}{2N_{i+1}} p^l (1-p)^{2N_{i+1}-l}$$

where $p = m(i, x)$ is the gamete pool probability at frequency $x = j/(2N_i)$ and the sums occur over all l such that $l/(2N_{i+1}) \in [a_{i+1,k}, a_{i+1,k+1})$. When

$0 \leq k \leq q$, it is sensible to set $c_{ij\rightarrow i+1,k} = k/(2N_{i+1})$. For $j > q$, we revert to the updates based on the normal approximation.

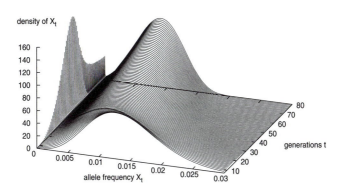

FIGURE 11.1. Density of the Frequency of a Recessive Gene

To illustrate this strategy for a recessive disease, we turn to Finland, a relatively isolated population of northern Europe. We assume that the Finnish population has grown exponentially from 1000 founders to $5,000,000$ contemporary people over a span of 80 generations. Our hypothetical recessive disease has mutation rate $\eta = 10^{-6}$, fitness $f = 0.5$, and a high initial gene frequency of $X_0 = 0.015$. The slow deterministic decay to the equilibrium gene frequency of $\sqrt{\eta/(1-f)} = 0.0014$ extends well beyond the present. Figure 11.1 plots the density of the frequency of the recessive gene from generation 7 to generation 80. The figure omits the first seven generations because the densities in that time range are too concentrated for the remaining densities to scale well. The left ridge of the gene density surface represents a moderate probability mass collecting in the narrow region where the gene is either extinct or in danger of going extinct.

As a technical aside, it is interesting to compare two versions of the algorithm. Version one carries forward probabilities but not centers of probabilities. Version two carries both forward. Version one is about twice as fast as version two, given the same mesh points at each generation. In Figure 11.1, version two relies on 175 intervals in the continuous region. With 2000 intervals in the continuous region, version one takes 25 times more computing cpu time and still fails to achieve the same accuracy at generation 80 as version two. Needless to say, version one is not recommended.

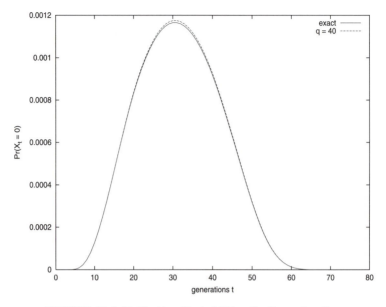

FIGURE 11.2. Extinction Probability of a Recessive Gene

Gene extinction is naturally of great interest. Figure 11.2 depicts the probability that the recessive gene is entirely absent from the population. This focuses our attention squarely on the discrete domain where we would expect the diffusion approximation to deteriorate. The solid curve of the graph shows the outcome of computing directly with the exact Wright-Fisher chain. At about generation 60, the matrix times vector multiplications implicit in the Markov chain updates start to slow the computations drastically. In this example, it took 14 minutes of computing time on a desktop PC to reach 80 generations. The hybrid algorithm with $q = 40$ intervals covering the discrete region and 500 intervals covering the continuous region takes only 11 seconds to reach generation 80. The resulting dashed curve is quite close to the solid curve in Figure 11.2, and setting $q = 50$ makes it practically identical.

11.9 Problems

1. Consider a diffusion process X_t with infinitesimal mean $\mu(t, x)$ and infinitesimal variance $\sigma^2(t, x)$. If the function $f(t)$ is strictly increasing and continuously differentiable, then argue that $Y_t = X_{f(t)}$ is a diffusion process with infinitesimal mean and variance

$$
\begin{aligned}
\mu_Y(t, y) &= \mu[f(t), y]f'(t) \\
\sigma_Y^2(t, y) &= \sigma^2[f(t), y]f'(t).
\end{aligned}
$$

Apply this result to the situation where Y_t equals y_0 at $t = 0$ and has $\mu_Y(t, y) = 0$ and $\sigma_Y^2(t, y) = \sigma^2(t)$. Show that Y_t is normally distributed with mean and variance

$$E(Y_t) = y_0$$

$$\text{Var}(Y_t) = \int_0^t \sigma^2(s)\, ds.$$

(Hint: Let X_t be standard Brownian motion.)

2. Show that

$$\text{Cov}(Y_{t+s}, Y_t) = \frac{\sigma^2 e^{-\gamma s}(1 - e^{-2\gamma t})}{2\gamma}$$

in the Ornstein-Uhlenbeck process when s and t are nonnegative.

3. Consider a diffusion process X_t with infinitesimal mean

$$\mu(t, x) = \begin{cases} 1, & x < 0 \\ 0, & x = 0 \\ -1, & x > 0 \end{cases}$$

and infinitesimal variance 1. Find the equilibrium distribution $f(x)$ of X_t.

4. In the diffusion approximation to a branching process with immigration, we set $\mu(t, x) = (\alpha - \delta)x + \nu$ and $\sigma^2(t, x) = (\alpha + \delta)x + \nu$, where α and δ are the birth and death rates per particle and ν is the immigration rate. Demonstrate that

$$E(X_t) = x_0 e^{\beta t} + \frac{\nu}{\beta}\left[e^{\beta t} - 1\right]$$

$$\text{Var}(X_t) = \frac{\gamma x_0 (e^{2\beta t} - e^{\beta t})}{\beta} + \frac{\gamma \nu (e^{2\beta t} - e^{\beta t})}{\beta^2}$$
$$- \frac{\gamma \nu (e^{2\beta t} - 1)}{2\beta^2} + \frac{\nu (e^{2\beta t} - 1)}{2\beta}$$

for $\beta = \alpha - \delta$, $\gamma = \alpha + \delta$, and $X_0 = x_0$. When $\alpha < \delta$, the process eventually reaches equilibrium. Find the limits of $E(X_t)$ and $\text{Var}(X_t)$.

5. In Problem 4 suppose $\nu = 0$. Verify that the process goes extinct with probability $\min\{1, e^{-2\frac{\alpha-\delta}{\alpha+\delta}x_0}\}$ by using equation (11.19) and sending c to 0 and d to ∞.

6. In Problem 4 suppose $\nu > 0$ and $\alpha < \delta$. Show that Wright's formula leads to the equilibrium distribution

$$f(x) = k\left[(\alpha + \delta)x + \nu\right]^{\frac{4\nu\delta}{(\alpha+\delta)^2} - 1} e^{\frac{2(\alpha-\delta)x}{\alpha+\delta}}$$

for some normalizing constant $k > 0$ and $x > 0$.

7. Show that formula (11.23) holds in R^2.

8. Use Stirling's formula to demonstrate that

$$\frac{\Gamma(2N\eta + \frac{1}{2})}{\sqrt{2N(1-f)}\Gamma(2N\eta)} \approx \sqrt{\frac{\eta}{1-f}}$$

when N is large in the Wright-Fisher model for a recessive disease.

9. Consider the Wright-Fisher model with no selection but with mutation from allele A_1 to allele A_2 at rate η_1 and from A_2 to A_1 at rate η_2. With constant population size N, prove that the frequency of the A_1 allele follows the beta distribution

$$f(x) = \frac{\Gamma[4N(\eta_1 + \eta_2)]}{\Gamma(4N\eta_2)\Gamma(4N\eta_1)} x^{4N\eta_2 - 1}(1-x)^{4N\eta_1 - 1}$$

at equilibrium. (Hint: Substitute $p(x) = x$ in formula (11.13) defining the infinitesimal variance $\sigma^2(t, x)$.)

10. Consider the transformed Brownian motion with infinitesimal mean α and infinitesimal variance σ^2 described in Example 11.3.2. If the process starts at $x \in [c, d]$, then prove that it reaches d before c with probability

$$u(x) = \frac{e^{-\beta x} - e^{-\beta c}}{e^{-\beta d} - e^{-\beta c}} \quad \text{for} \quad \beta = \frac{2\alpha}{\sigma^2}.$$

Verify that $u(x)$ reduces to $(x - c)/(d - c)$ when $\alpha = 0$. As noted in the text, this simplification holds for any diffusion process with $\mu(x) = 0$.

11. Suppose the transformed Brownian motion with infinitesimal mean α and infinitesimal variance σ^2 described in Example 11.3.2 has $\alpha \geq 0$. If $c = -\infty$ and $d < \infty$, then demonstrate that equation (11.21) has solution

$$w(x) = e^{\gamma(d-x)} \quad \text{for} \quad \gamma = \frac{\alpha - \sqrt{\alpha^2 + 2\sigma^2\theta}}{\sigma^2}.$$

Simplify $w(x)$ when $\alpha = 0$, and show by differentiation of $w(x)$ with respect to θ that the expected time $E(T)$ to reach the barrier d is infinite. When $\alpha < 0$, show that

$$\Pr(T < \infty) = e^{\frac{2\alpha}{\sigma^2}(d-x)}.$$

(Hints: The variable γ is a root of a quadratic equation. Why do we discard the other root? In general, $\Pr(T < \infty) = \lim_{\theta \downarrow 0} E\left(e^{-\theta T}\right)$.)

12. In Problem 11 find $w(x)$ and $E(T)$ when c is finite. The value $\alpha < 0$ is allowed.

13. Prove that the linear density $f_{ij0} + f_{ij1}x$ is nonnegative throughout the interval $(a_{ij}, a_{i,j+1})$ if and only if its center of mass c_{ij} lies in the middle third of the interval. (Hint: Without loss of generality, take $a_{ij} = 0$.)

12

Poisson Approximation

12.1 Introduction

In the past few years, mathematicians have developed a powerful technique known as the Chen-Stein method for approximating the distribution of a sum of weakly dependent Bernoulli random variables [10, 16, 141]. In contrast to many asymptotic methods, this approximation carries with it explicit error bounds. Let X_α be a Bernoulli random variable with success probability p_α, where α ranges over some finite index set I. As a generalization of the law of rare events discussed in Example 12.3.1, it is natural to speculate that the sum $S = \sum_{\alpha \in I} X_\alpha$ is approximately Poisson with mean $\lambda = \sum_{\alpha \in I} p_\alpha$. The Chen-Stein method estimates the error in this approximation using the total variation distance introduced in equation (7.6) of Chapter 7.

The coupling method is one technique for explicitly bounding the total variation distance between $S = \sum_{\alpha \in I} X_\alpha$ and a Poisson random variable Z with the same mean λ [16, 107]. In many concrete examples, it is possible to construct for each α a random variable V_α on a common probability space with S such that V_α is distributed as $S - 1$ conditional on the event $X_\alpha = 1$. The bound

$$\|\pi_S - \pi_Z\|_{TV} \quad \leq \quad \frac{1 - e^{-\lambda}}{\lambda} \sum_{\alpha \in I} p_\alpha \, \mathrm{E}(|S - V_\alpha|) \tag{12.1}$$

then applies, where π_S and π_Z denote the distributions of S and Z. The size of this bound depends on how tightly S and each V_α are coupled. If

$S \geq V_\alpha$ for all α, then the simplified bound

$$\|\pi_S - \pi_Z\|_{TV} \ \leq \ \frac{1 - e^{-\lambda}}{\lambda}[\lambda - \text{Var}(S)] \tag{12.2}$$

holds. Not only is the upper bound (12.2) easier to evaluate than the upper bound (12.1), but it also makes it clear that the approximate equality $\text{Var}(S) \approx \text{E}(S)$ is nearly a sufficient as well as a necessary condition for S to be approximately Poisson.

The neighborhood method of bounding the total variation distance exploits certain neighborhoods of dependency N_α associated with each $\alpha \in I$ [9]. Here N_α is a subset of I containing α such that X_α is independent of those X_β with $\beta \notin N_\alpha$. In this situation of short-range dependency, the total variation distance between S and its Poisson approximate Z satisfies

$$\|\pi_S - \pi_Z\|_{TV} \ \leq \ \frac{1 - e^{-\lambda}}{\lambda}\Big(\sum_{\alpha \in I}\sum_{\beta \in N_\alpha} p_\alpha p_\beta + \sum_{\alpha \in I}\sum_{\beta \in N_\alpha \setminus \{\alpha\}} p_{\alpha\beta}\Big), \tag{12.3}$$

where again $\lambda = \text{E}(S) = \text{E}(Z)$ and

$$p_{\alpha\beta} \ = \ \text{E}(X_\alpha X_\beta) \ = \ \Pr(X_\alpha = 1, X_\beta = 1).$$

The neighborhood method works best when each N_α is taken as small as possible.

Both Chen-Stein methods are well adapted to solving a myriad of practical problems. The next few sections present a few typical examples. The chapter ends with a mathematical proof of the Chen-Stein bounds. Readers primarily interested in applications can skip this theoretical section. A more comprehensive development of theory and further examples can be found in the references [10, 16, 107].

12.2 Applications of the Coupling Method

Example 12.2.1 *Ménage Problem*

In the classical ménage problem of combinatorics, n married couples are seated around a circular table [17]. If men and women alternate, but husbands and wives are randomly scrambled, then the number of married couples S seated next to each other is approximately Poisson distributed. Given that X_α is the indicator of the event that seats α and $\alpha + 1$ contain a married couple, we can write

$$S \ = \ \sum_{\alpha=1}^{2n} X_\alpha,$$

where $X_{2n+1} = X_1$. Symmetry dictates that $p_\alpha = \frac{1}{n}$ and $\lambda = E(S) = 2$. The total variation distances between S and a Poisson random variable Z with mean λ can be estimated by the coupling method.

To construct the coupled random variable V_α, we exchange the person in seat $\alpha + 1$ with the spouse of the person in seat α and then count the number of adjacent spouse pairs, excluding the pair now occupying seats α and $\alpha+1$. One can show that this construction entails $|S - V_\alpha| \leq 1$. Indeed, suppose the spouse of the person in seat α occupies seat β. If $\beta = \alpha + 1$, then $X_\alpha = 1$ and $V_\alpha = S - 1$. If $\beta \neq \alpha + 1$, then the gain of a matched couple in the pair $\{\alpha, \alpha+1\}$ does not contribute to V_α. The other possible gains and losses of matched couples occur in the three pairs $\{\alpha+1, \alpha+2\}$, $\{\beta - 1, \beta\}$, and $\{\beta, \beta + 1\}$. Although some of these pairs may coincide, it is not hard to see that at most one of the three pairs can suffer a loss and at most one of the three pairs can reap a gain.

We now appeal to the Chen-Stein bound (12.1). To avoid a messy consideration of special cases in calculating $E(|S - V_\alpha|)$, we will bound the probability $Pr(V_\alpha = S)$. The dominant contribution to the event $\{V_\alpha = S\}$ arises when $\beta \notin \{\alpha - 1, \alpha + 1, \alpha + 3\}$ and the person in seat $\alpha + 1$ is not the spouse of any of the people in seats $\alpha + 2$, $\beta - 1$, and $\beta + 1$. Careful consideration of this special case leads to the inequality

$$Pr(V_\alpha = S) \geq \frac{n-3}{n}\left(1 - \frac{3}{n-1}\right)$$

and therefore to the further inequality

$$
\begin{aligned}
E(|S - V_\alpha|) &= Pr(S \neq V_\alpha) \\
&\leq 1 - \frac{n-3}{n}\left(1 - \frac{3}{n-1}\right) \\
&= \frac{6n - 12}{n(n-1)}.
\end{aligned}
$$

The Chen-Stein bound (12.1) now reduces to

$$\|\pi_S - \pi_Z\|_{TV} \leq \frac{2(1 - e^{-\lambda})(6n - 12)}{\lambda n(n-1)},$$

which decreases in n for $n \geq 3$.   ∎

Example 12.2.2 *Birthday Problem*

Consider a multinomial experiment with m categories. The statistic W_d denotes the number of categories with d or more successes after n trials. For example, each category might be a day of the year, and each trial might record the birthday of another random person. If we let q_α be the success rate per trial for category α, then this category accumulates d or more successes with probability

$$p_\alpha = \sum_{k=d}^{n} \binom{n}{k} q_\alpha^k (1 - q_\alpha)^{n-k}.$$

The coupling method provides a bound on the total variation distance between W_d and a Poisson random variable with mean $\lambda = \sum_{\alpha=1}^{m} p_\alpha$.

To validate the coupling bound (12.2) with $S = W_d$, we must construct the coupled random variable V_α. If the number of outcomes Y_α falling in category α satisfies $Y_\alpha \geq d$, then $X_\alpha = 1$, and we set $V_\alpha = \sum_{\beta \neq \alpha} X_\beta$. If $Y_\alpha < d$, then we resample from the conditional distribution of Y_α given the event $Y_\alpha \geq d$. This produces a random variable $Y_\alpha^* > Y_\alpha$, and we redefine the outcomes of the first $Y_\alpha^* - Y_\alpha$ trials falling outside category α so that they now fall in category α. If we let V_α be the number of categories other than α that now exceed their quota d, it is obvious that V_α is distributed as $S - 1$ conditional on the event $X_\alpha = 1$. Because of the redirection of outcomes, it is also clear that $S \geq V_\alpha$. Thus, the conditions for the Chen-Stein bound (12.2) apply. Unfortunately, the variance $\mathrm{Var}(W_d)$ is not entirely trivial to calculate. In the special case $d = 1$, we have

$$E(W_1) = \sum_{\alpha=1}^{m} [1 - (1 - q_\alpha)^n]$$

$$\mathrm{Var}(W_1) = \sum_{\alpha=1}^{m} \mathrm{Var}(1_{\{Y_\alpha \geq 1\}}) + \sum_{\alpha=1}^{m} \sum_{\beta \neq \alpha} \mathrm{Cov}(1_{\{Y_\alpha \geq 1\}}, 1_{\{Y_\beta \geq 1\}})$$

$$= \sum_{\alpha=1}^{m} \mathrm{Var}(1_{\{Y_\alpha = 0\}}) + \sum_{\alpha=1}^{m} \sum_{\beta \neq \alpha} \mathrm{Cov}(1_{\{Y_\alpha = 0\}}, 1_{\{Y_\beta = 0\}})$$

$$= \sum_{\alpha=1}^{m} (1 - q_\alpha)^n [1 - (1 - q_\alpha)^n]$$

$$+ \sum_{\alpha=1}^{m} \sum_{\beta \neq \alpha} [(1 - q_\alpha - q_\beta)^n - (1 - q_\alpha)^n (1 - q_\beta)^n],$$

which are certainly easy to evaluate numerically. Readers can consult reference [16] for various approximations to $E(W_d)$ and $\mathrm{Var}(W_d)$.   ∎

Example 12.2.3 *Biggest Random Gap*

Questions about the spacings of uniformly distributed points crop in many application areas [97, 138]. If we scatter n points randomly on the unit interval [0,1], then it is natural to ask for the distribution of the largest gap between two adjacent points or between either endpoint and its nearest adjacent point. We can attack this problem by the coupling method of Chen-Stein approximation. Corresponding to the order statistics $Y_1, \ldots, Y_n$

of the n points, define indicator random variables $X_1, \ldots, X_{n+1}$ such that $X_\alpha = 1$ when $Y_\alpha - Y_{\alpha-1} \geq d$. At the ends we take $Y_0 = 0$ and $Y_{n+1} = 1$. The sum $S = \sum_{\alpha=1}^{n+1} X_\alpha$ gives the number of gaps of length d or greater.

Because we can circularize the interval, all gaps, including the first and the last, behave symmetrically. Just think of scattering $n + 1$ points on the unit circle and then breaking the circle into an interval at the first random point. It therefore suffices in the coupling method to consider the first Bernoulli variable $X_1 = 1_{\{Y_1 \geq d\}}$. If $Y_1 \geq d$, then define V_1 to be the number of gaps other than Y_1 that exceed d. If, on the other hand, $Y_1 < d$, then resample Y_1 conditional on the event $Y_1 \geq d$ to get Y_1^*. For $\alpha > 1$, replace the gap $Y_\alpha - Y_{\alpha-1}$ by the gap $(Y_\alpha - Y_{\alpha-1})(1 - Y_1^*)/(1 - Y_1)$ so that the points to the right of Y_1 are uniformly chosen from the interval $[Y_1^*, 1]$ rather than from $[Y_1, 1]$. This procedure narrows all remaining gaps but leaves them in the same proportion. If we now define V_1 as the number of remaining gaps that exceed d in length, it is clear that V_1 has the same distribution as $S-1$ conditional on $X_1 = 1$. Because $S \geq V_1$, the Chen-Stein inequality (12.2) applies.

To calculate the mean $\lambda = \mathrm{E}(S)$, we again focus on the first interval. Clearly, $\Pr(X_1 = 1) = \Pr(Y_1 \geq d) = (1 - d)^n$ implies that

$$\lambda = (n+1)(1-d)^n.$$

In similar fashion, we calculate

$$\begin{aligned} \mathrm{Var}(S) &= (n+1)\,\mathrm{Var}(X_1) + (n+1)n\,\mathrm{Cov}(X_1, X_{n+1}) \\ &= (n+1)(1-d)^n - (n+1)(1-d)^{2n} \\ &\quad + (n+1)n\,\mathrm{E}(X_1 X_{n+1}) - (n+1)n(1-d)^{2n}. \end{aligned}$$

To calculate $\mathrm{E}(X_1 X_{n+1}) = \Pr(X_1 = 1, \ X_{n+1} = 1)$ when $2d < 1$, we simply observe that $X_1 = X_{n+1} = 1$ if and only if all n random points are confined to the interval $[d, 1-d]$. It follows that $\mathrm{E}(X_1 X_{n+1}) = (1-2d)^n$ and therefore that

$$\begin{aligned} \mathrm{Var}(S) &= (n+1)(1-d)^n - (n+1)(1-d)^{2n} \\ &\quad + (n+1)n(1-2d)^n - (n+1)n(1-d)^{2n}. \end{aligned}$$

If d is small and n is large, then one can demonstrate that $\mathrm{Var}(S) \approx \mathrm{E}(S)$, and the Poisson approximation is good [16].

It is of some interest to estimate the average number of points required to reduce the largest gap below d. From the Poisson approximation, the median n should satisfy $e^{-(n+1)(1-d)^n} \approx \frac{1}{2}$. This approximate equality can be rewritten as

$$n \approx \frac{-\ln(n+1) + \ln\ln 2}{\ln(1-d)} \tag{12.4}$$

and used iteratively to approximate the median. If one chooses evenly spaced points, it takes only $\frac{1}{d}$ random points to saturate the interval $[0, 1]$.

For the crude guess $n = \frac{1}{d}$, substitution in (12.4) leads to the improved approximation

$$n \;\approx\; \frac{-\ln(\frac{1}{d}+1) + \ln\ln 2}{\ln(1-d)}$$

$$\approx\; \frac{1}{d}\ln\frac{1}{d}.$$

In fact, a detailed analysis shows that the average required number of points is asymptotically similar to $\frac{1}{d}\ln\frac{1}{d}$ for d small [51, 138]. The factor $\ln\frac{1}{d}$ summarizes the penalty exacted for selecting random points rather than evenly spaced points. ∎

12.3   Applications of the Neighborhood Method

Example 12.3.1 *The Law of Rare Events*

Suppose that $X_1, \ldots, X_n$ are independent Bernoulli random variables with success probabilities $p_1, \ldots, p_n$. If the p_α are small and $\lambda = \sum_{\alpha=1}^{n} p_\alpha$ is moderate in size, then the law of rare events declares that the sum $S = \sum_{\alpha=1}^{n} X_\alpha$ is approximately Poisson distributed. The neighborhood method provides an easy verification of this result. If we let N_α be the singleton set $\{\alpha\}$ and Z be a Poisson random variable with mean λ, then inequality (12.3) reduces to

$$\|\pi_S - \pi_Z\|_{TV} \;\leq\; \frac{1 - e^{-\lambda}}{\lambda} \sum_{\alpha=1}^{n} p_\alpha^2$$

because the sum $\sum_{\beta \in N_\alpha \setminus \{\alpha\}} p_{\alpha\beta}$ is empty. ∎

Example 12.3.2 *Construction of Somatic Cell Hybrid Panels*

Somatic cell hybrids are routinely used to assign particular human genes to particular human chromosomes [39, 148]. In brief outline, somatic cell hybrids are constructed by fusing normal human cells with permanently transformed rodent cells. The resulting hybrid cells retain all of the rodent chromosomes while losing random subsets of the human chromosomes. A few generations after cell fusion, clones of cells can be identified with stable subsets of the human chromosomes. All chromosomes, human and rodent, normally remain functional. With a broad enough collection of different hybrid clones, it is possible to establish a correspondence between the presence or absence of a given human gene and the presence or absence of each of the 24 distinct human chromosomes in each clone. From this pattern one can assign the gene to a particular chromosome.

For this program of gene assignment to be successful, certain critical assumptions must be satisfied. First, the human gene should be present on a

FIGURE 12.1. A Somatic Cell Hybrid Panel

```
0 1 0 1 0 0 0 1 0 0 0 0 0 0 1 0 1 1 0 1 1 1 1
1 0 1 0 1 1 0 0 1 0 0 0 0 1 0 0 1 0 1 0 1 1 1
0 1 1 1 1 0 1 0 0 0 0 0 1 0 0 1 1 0 1 1 0 1 1
1 1 1 0 0 1 1 0 0 1 0 1 0 0 0 1 1 1 0 0 1 0 1
0 0 0 1 1 1 1 0 0 0 1 1 1 1 1 0 1 0 0 0 1 1 0
0 1 1 1 1 1 1 1 1 1 0 0 0 0 0 0 1 0 0 0 0 0 0
0 0 1 0 1 0 1 1 0 1 1 1 0 0 0 0 1 1 1 1 1 0 0
0 0 0 1 0 1 1 1 0 0 0 1 0 1 1 1 1 0 1 0 1 0 1
1 0 0 0 1 1 0 0 0 1 0 1 1 0 1 0 1 0 1 1 0 0 1
```

single human chromosome or on a single pair of homologous human chromosomes. Second, the human gene should be detectable when present in a clone and should be distinguishable from any rodent analog of the human gene in the clone. Genes are usually detected by electrophoresis of their protein products or by annealing of an appropriate DNA probe directly to part of the gene. Third, each of the 24 distinct human chromosomes should be either absent from a clone or cytologically or biochemically detectable in the clone. Chromosomes can be differentiated cytologically by size, by the position of their centromeres, and by their distinctive banding patterns under appropriate stains. It is also possible to distinguish chromosomes by in situ hybridization of large, fluorescent DNA probes or by isozyme assays that detect unique proteins produced by genes on the chromosomes.

In this application of the Chen-Stein method, we consider the information content of a panel of somatic cell hybrids [61]. Let n denote the number of hybrid clones in a panel. Since the Y chromosome bears few genes of interest, hybrids are usually created from human female cells. This gives a total of 23 different chromosome types—22 autosomes and the X chromosome. Figure 12.1 depicts a hybrid panel with $n = 9$ clones. Each row of this panel corresponds to a particular clone. Each of the 23 columns corresponds to a particular chromosome. A 1 in row i and column j of the panel indicates the presence of chromosome j in clone i. A 0 indicates the absence of a chromosome in a clone. An additional test column of 0's and 1's is constructed when each clone is assayed for the presence of a given human gene. Barring assay errors or failure of one of the critical assumptions, the test column will uniquely match one of the columns of the panel. In this case the gene is assigned to the corresponding chromosome.

If two columns of a panel are identical, then gene assignment becomes ambiguous for any gene residing on one of the two corresponding chromosomes. Fortunately, the columns of the panel in Figure 12.1 are unique. This panel has the unusual property that every pair of columns differs in at least three entries. This level of redundancy is useful. If a single assay error is made in creating a test column for a human gene, then the gene can still be successfully assigned to a particular human chromosome because it will differ from one column of the panel in one entry and from all other columns of the panel in at least two entries. This consideration suggests that built-in redundancy of a panel is desirable. In practice, the chromosome constitution of a clone cannot be predicted in advance, and the level of redundancy is random. Minimum Hamming distance is a natural measure of the redundancy of a panel. The Hamming distance $\rho(c_s, c_t)$ between two columns c_s and c_t is just the number of entries in which they differ. The minimum Hamming distance of a panel is defined as $\min_{\{s,t\}} \rho(c_s, c_t)$, where $\{s,t\}$ ranges over all pairs of columns from the panel.

When somatic cell hybrid panels are randomly created, it is reasonable to make three assumptions. First, each human chromosome is lost or retained independently during the formation of a stable clone. Second, there is a common retention probability p applying to all chromosome pairs. This means that at least one member of each pair of homologous chromosomes is retained with probability p. Rushton [134] estimates a range of p from .07 to .75. The value $p = \frac{1}{2}$ simplifies our theory considerably. Third, different clones behave independently in their retention patterns.

Now denote column s of a random panel of n clones by C_s^n. For any two distinct columns C_s^n and C_t^n, define $X_{\{s,t\}}^n$ to be the indicator of the event $\rho(C_s^n, C_t^n) < d$, where d is some fixed Hamming distance. The random variable $Y_d^n = \sum_{\{s,t\}} X_{\{s,t\}}^n$ is 0 precisely when the minimum Hamming distance equals or exceeds d. There are $\binom{23}{2}$ pairs $\alpha = \{s,t\}$ in the index set I, and each of the associated X_α^n has the same mean

$$p_\alpha = \sum_{i=0}^{d-1} \binom{n}{i} q^i (1-q)^{n-i},$$

where $q = 2p(1-p)$ is the probability that C_s^n and C_t^n differ in any entry. This gives the mean of Y_d^n as $\lambda = \binom{23}{2} p_\alpha$.

The Chen-Stein heuristic suggests estimating $\Pr(Y_d^n > 0)$ by the Poisson tail probability $1 - e^{-\lambda}$. The error bound (12.3) on this approximation can be computed by defining the neighborhoods $N_\alpha = \{\beta : |\beta| = 2, \beta \cap \alpha \neq \emptyset\}$, where vertical bars enclosing a set indicate the number of elements in the set. It is clear that X_α^n is independent of those X_β^n with β outside N_α. Straightforward counting arguments give

$$\sum_{\alpha \in I} \sum_{\beta \in N_\alpha} p_\alpha p_\beta = \binom{23}{2} |N_\alpha| p_\alpha^2$$

and

$$|N_\alpha| = \binom{23}{2} - \binom{21}{2} = 43.$$

TABLE 12.1. Chen-Stein Estimate of $\Pr(Y_d^n > 0)$

d	n	Estimate	Lower Bound	Upper Bound
1	10	0.2189	0.1999	0.2379
1	15	0.0077	0.0077	0.0077
1	20	0.0002	0.0002	0.0002
1	25	0.0000	0.0000	0.0000
2	10	0.9340	0.0410	1.0000
2	15	0.1162	0.1112	0.1213
2	20	0.0051	0.0050	0.0051
2	25	0.0002	0.0002	0.0002
3	10	1.0000	0.0410	1.0000
3	15	0.6071	0.4076	0.8066
3	20	0.0496	0.0487	0.0505
3	25	0.0025	0.0025	0.0025

Since the joint probability $p_{\alpha\beta}$ does not depend on the particular column pair $\beta \in N_\alpha \backslash \{\alpha\}$ chosen, we also deduce that

$$\sum_{\alpha \in I} \sum_{\beta \in N_\alpha \backslash \{\alpha\}} p_{\alpha\beta} \;=\; \binom{23}{2}(|N_\alpha| - 1)p_{\alpha\beta}.$$

Fortunately, $p_{\alpha\beta} = p_\alpha^2$ when $p = 1/2$. Indeed, upon conditioning on the value of the common column shared by α and β, it is obvious in this special case that the events $X_\alpha^n = 1$ and $X_\beta^n = 1$ are independent and occur with constant probability p_α. The case $p \neq 1/2$ is more subtle, and we defer the details of computing $p_{\alpha\beta}$ to Problem 8. Table 12.1 provides some representative estimates of the probabilities $\Pr(Y_d^n > 0)$ for $p = 1/2$. Because the Chen-Stein method also provides upper and lower bounds on the estimates, we can be confident that the estimates are accurate for large n. In two cases in Table 12.1, the Chen-Stein upper bound is truncated to the more realistic value 1. ∎

12.4 Proof of the Chen-Stein Estimates

Verification of the Chen-Stein estimates depends on forging a subtle connection between Chen's lemma in Example 2.7.3 and the definition of the total variation norm in equation (7.6). If the sum $S = \sum_\alpha X_\alpha$ were actually Poisson, then Chen's lemma of Example 2.7.3 would entail the identity

$$\lambda\,\mathrm{E}[g(S+1)] - \mathrm{E}[Sg(S)] \;=\; 0$$

for every bounded function $g(s)$. To the extent that S is approximately Poisson, the left-hand side of this equality should be approximately 0. The total variation distance between S and a Poisson random variable Z with the same mean λ is given by

$$\|\pi_S - \pi_Z\|_{TV} \quad = \quad \sup_{A \subset \mathcal{Z}} \left| \Pr(S \in A) - \sum_{j \in A} e^{-\lambda} \frac{\lambda^j}{j!} \right|.$$

The key to proving the Chen-Stein estimates is to concoct a particular bounded function $g(s)$ satisfying

$$\lambda E[g(S+1)] - E[Sg(S)] \quad = \quad \Pr(S \in A) - \sum_{j \in A} e^{-\lambda} \frac{\lambda^j}{j!} \qquad (12.5)$$

and then to bound the difference in expectations on the left-hand side of this equality.

The easiest way of securing equality (12.5) is to force $g(s)$ to satisfy the identity

$$\lambda g(s+1) - sg(s) \quad = \quad 1_A(s) - \sum_{j \in A} e^{-\lambda} \frac{\lambda^j}{j!} \qquad (12.6)$$

for all nonnegative integers s. Indeed, if equation (12.6) holds, then we simply substitute the random variable S for the integer s and take expectations. Fortunately, equation (12.6) can be viewed as a recurrence relation for calculating $g(s+1)$ from $g(s)$. The value $g(0)$ is irrelevant in determining $g(1)$, so we adopt the usual convention $g(0) = 0$. One can explicitly solve the difference equation (12.6) by multiplying it by $e^{-\lambda}\lambda^{s-1}/s!$ and defining the new function $f(s) = e^{-\lambda}\lambda^s g(s+1)/s!$. These maneuvers yield the difference equation

$$f(s) - f(s-1) \quad = \quad e^{-\lambda}\frac{\lambda^{s-1}}{s!}1_A(s) - e^{-\lambda}\frac{\lambda^{s-1}}{s!}\sum_{j \in A} e^{-\lambda}\frac{\lambda^j}{j!}$$

with the initial condition $f(-1) = 0$. One can now find $f(s)$ via the telescoping sum

$$f(s) \quad = \quad \sum_{k=0}^{s} [f(k) - f(k-1)]$$

$$= \quad \lambda^{-1} \left[\sum_{k=0}^{s} e^{-\lambda}\frac{\lambda^k}{k!}1_A(k) - \sum_{k=0}^{s} e^{-\lambda}\frac{\lambda^k}{k!}\sum_{j \in A} e^{-\lambda}\frac{\lambda^j}{j!} \right].$$

This translates into the solution

$$g(s+1) \quad = \quad \frac{e^{\lambda}s!}{\lambda^{s+1}} \left[\sum_{k=0}^{s} e^{-\lambda}\frac{\lambda^k}{k!}1_A(k) - \sum_{k=0}^{s} e^{-\lambda}\frac{\lambda^k}{k!}\sum_{j \in A} e^{-\lambda}\frac{\lambda^j}{j!} \right] \qquad (12.7)$$

for $g(s+1)$. Although it is not immediately evident from formula (12.7), we will demonstrate that $g(s)$ is bounded and satisfies the Lipschitz inequality $|g(s+1) - g(s)| \leq (1 - e^{-\lambda})/\lambda$ for all s.

Before attending to these important details, let us return to the main line of argument. We first note that for any random variable R

$$
\begin{aligned}
\mathrm{E}(RS) &= \sum_\alpha \mathrm{E}(RX_\alpha) \\
&= \sum_\alpha p_\alpha \mathrm{E}(RX_\alpha \mid X_\alpha = 1) \qquad (12.8) \\
&= \sum_\alpha p_\alpha \mathrm{E}(R \mid X_\alpha = 1).
\end{aligned}
$$

We now apply this identity to $R = g(S)$ and invoke the coupling-method premise that $V_\alpha + 1$ has the same distribution as S conditional on the event $X_\alpha = 1$. These considerations imply that

$$
\begin{aligned}
\mathrm{E}[Sg(S)] &= \sum_\alpha p_\alpha \mathrm{E}[g(S) \mid X_\alpha = 1] \\
&= \sum_\alpha p_\alpha \mathrm{E}[g(V_\alpha + 1)]. \qquad (12.9)
\end{aligned}
$$

We also observe that the Lipschitz condition on $g(s)$ can be extended to

$$
\begin{aligned}
|g(t) - g(s)| &\leq \sum_{j=s}^{t-1} |g(j+1) - g(j)| \\
&\leq \frac{1 - e^{-\lambda}}{\lambda} |t - s|
\end{aligned}
$$

for any $t > s$. Mindful of these facts, we infer from equality (12.5) that

$$
\begin{aligned}
\left| \Pr(S \in A) - \sum_{j \in A} e^{-\lambda} \frac{\lambda^j}{j!} \right| &= |\lambda \mathrm{E}[g(S+1)] - \mathrm{E}[Sg(S)]| \\
&= \left| \sum_\alpha p_\alpha \{ \mathrm{E}[g(S+1)] - \mathrm{E}[g(V_\alpha + 1)] \} \right| \\
&\leq \sum_\alpha p_\alpha \mathrm{E}[|g(S+1) - g(V_\alpha + 1)|] \\
&\leq \frac{1 - e^{-\lambda}}{\lambda} \sum_\alpha p_\alpha \mathrm{E}(|S - V_\alpha|).
\end{aligned}
$$

Taking the supremum over A now yields the Chen-Stein bound (12.1).

If $S \geq V_\alpha$ for all α, then equation (12.8) with $R = S$ implies

$$
\sum_\alpha p_\alpha \mathrm{E}(|S - V_\alpha|) = \sum_\alpha p_\alpha \mathrm{E}(S) + \lambda - \sum_\alpha p_\alpha \mathrm{E}(V_\alpha + 1)
$$

$$= \lambda^2 + \lambda - \sum_\alpha p_\alpha \, \mathrm{E}(S \mid X_\alpha = 1)$$

$$= \lambda^2 + \lambda - \mathrm{E}(S^2)$$

$$= \lambda - \mathrm{Var}(S).$$

This establishes the Chen-Stein bound (12.2).

In the neighborhood method, it is convenient to define the random variable $U_\alpha = \sum_{\beta \notin N_\alpha} X_\beta$, which is independent of X_α. Because

$$g(S - X_\alpha + 1) = \begin{cases} g(S+1), & X_\alpha = 0 \\ g(S), & X_\alpha = 1, \end{cases}$$

we have

$$\lambda \, \mathrm{E}[g(S+1)] - \mathrm{E}[Sg(S)]$$

$$= \sum_\alpha \mathrm{E}[p_\alpha g(S+1) - X_\alpha g(S)]$$

$$= \sum_\alpha \mathrm{E}[p_\alpha g(S+1) - p_\alpha g(S - X_\alpha + 1)]$$

$$+ \sum_\alpha \mathrm{E}[p_\alpha g(S - X_\alpha + 1) - X_\alpha g(S)] \qquad (12.10)$$

$$= \sum_\alpha \mathrm{E}\{p_\alpha X_\alpha [g(S+1) - g(S)]\}$$

$$+ \sum_\alpha \mathrm{E}\{(p_\alpha - X_\alpha)[g(S - X_\alpha + 1) - g(U_\alpha + 1)]\}$$

$$+ \sum_\alpha \mathrm{E}[(p_\alpha - X_\alpha)g(U_\alpha + 1)].$$

We will bound each of the sums defining the final quantity in the string of equalities (12.10). The third sum equals 0 since $\mathrm{E}[(p_\alpha - X_\alpha)g(U_\alpha + 1)] = 0$ by independence. The first sum is bounded in absolute value by

$$\sum_\alpha \mathrm{E}[p_\alpha X_\alpha | g(S+1) - g(S)|] \leq \frac{1 - e^{-\lambda}}{\lambda} \sum_\alpha p_\alpha \, \mathrm{E}(X_\alpha)$$

$$= \frac{1 - e^{-\lambda}}{\lambda} \sum_\alpha p_\alpha^2 \qquad (12.11)$$

owing to the Lipschitz property of $g(s)$. The middle sum is bounded in absolute value by

$$\sum_\alpha \mathrm{E}[(p_\alpha + X_\alpha)|g(S - X_\alpha + 1) - g(U_\alpha + 1)|]$$

$$\leq \frac{1 - e^{-\lambda}}{\lambda} \sum_\alpha \sum_{\beta \in N_\alpha \setminus \{\alpha\}} \mathrm{E}[(p_\alpha + X_\alpha)X_\beta] \qquad (12.12)$$

$$= \frac{1 - e^{-\lambda}}{\lambda} \sum_{\alpha} \sum_{\beta \in N_\alpha \setminus \{\alpha\}} (p_\alpha p_\beta + p_{\alpha\beta})$$

based on the extended Lipschitz property. Combining inequalities (12.11) and (12.12) with equalities (12.5) and (12.10) now produces the Chen-Stein bound (12.3).

Returning now to the question of whether $g(s)$ is a bounded function, we rearrange equation (12.7) by subtracting and adding the same quantity $e^{\lambda} s! \lambda^{-s-1} \sum_{k=0}^{s} e^{-\lambda} \frac{\lambda^k}{k!} 1_A(k) \sum_{j=0}^{s} e^{-\lambda} \frac{\lambda^j}{j!}$. This tactic gives

$$g(s+1) = e^{\lambda} s! \lambda^{-s-1} \sum_{k=0}^{s} e^{-\lambda} \frac{\lambda^k}{k!} 1_A(k) \sum_{j=s+1}^{\infty} e^{-\lambda} \frac{\lambda^j}{j!}$$

$$- e^{\lambda} s! \lambda^{-s-1} \sum_{k=s+1}^{\infty} e^{-\lambda} \frac{\lambda^k}{k!} 1_A(k) \sum_{j=0}^{s} e^{-\lambda} \frac{\lambda^j}{j!}.$$

From this representation and Proposition 4.3, we deduce the integral bound

$$|g(s+1)| \leq e^{\lambda} s! \lambda^{-s-1} \sum_{k=s+1}^{\infty} e^{-\lambda} \frac{\lambda^k}{k!}$$

$$= e^{\lambda} s! \lambda^{-s-1} \int_0^\lambda e^{-r} \frac{r^s}{s!} dr$$

$$= \lambda^{-1} \int_0^\lambda e^{\lambda - r} \left(\frac{r}{\lambda}\right)^s dr$$

$$\leq \lambda^{-1} \int_0^\lambda e^{\lambda - r} dr.$$

Finally, let us tackle the delicate issue of proving that $g(s)$ is Lipschitz. We first note that $g(s)$ can be decomposed as the sum $g(s) = \sum_{j \in A} g_j(s)$ of solutions $g_j(s)$ to the difference equation (12.6) with the singletons $\{j\}$ substituting for the set A. This fact is immediately obvious from formula (12.7). Furthermore, the function $g_j(s)$ can be expressed as

$$g_j(s+1) = \begin{cases} 0, & s = -1 \\ -\frac{s!}{\lambda^{s+1-j} j!} \sum_{k=0}^{s} e^{-\lambda} \frac{\lambda^k}{k!}, & 0 \leq s < j \\ \frac{s!}{\lambda^{s+1-j} j!} \sum_{k=s+1}^{\infty} e^{-\lambda} \frac{\lambda^k}{k!}, & s \geq j. \end{cases}$$

For $s < j$ the difference $g_j(s+1) - g_j(s) \leq 0$ because

$$\frac{s}{\lambda} \sum_{k=0}^{s} e^{-\lambda} \frac{\lambda^k}{k!} \geq \sum_{k=0}^{s-1} e^{-\lambda} \frac{\lambda^k}{k!}$$

owing to the inequality $\lambda^k/k! \geq \lambda^k/[(k-1)!s]$ for $k = 1, \dots, s$. For $s > j$ again the difference $g_j(s+1) - g_j(s) \leq 0$ because

$$\frac{s}{\lambda} \sum_{k=s+1}^{\infty} e^{-\lambda} \frac{\lambda^k}{k!} \leq \sum_{k=s}^{\infty} e^{-\lambda} \frac{\lambda^k}{k!}$$

owing to the opposite inequality $\lambda^k/k! \leq \lambda^k/[(k-1)!s]$ for $k \geq s+1$. Only the difference $g_j(j+1) - g_j(j) \geq 0$, and this difference is bounded above by

$$
\begin{aligned}
g_j(j+1) - g_j(j) &= \frac{1}{\lambda} \sum_{k=j+1}^{\infty} e^{-\lambda} \frac{\lambda^k}{k!} + \frac{1}{j} \sum_{k=0}^{j-1} e^{-\lambda} \frac{\lambda^k}{k!} \\
&= \frac{e^{-\lambda}}{\lambda} \left[\sum_{k=j+1}^{\infty} \frac{\lambda^k}{k!} + \sum_{k=1}^{j} \frac{\lambda^k}{k!} \frac{k}{j} \right] \\
&\leq \frac{e^{-\lambda}}{\lambda} (e^\lambda - 1) \\
&= \frac{1 - e^{-\lambda}}{\lambda}.
\end{aligned}
$$

This upper-bound inequality carries over to

$$g(s+1) - g(s) = \sum_{j \in A} [g_j(s+1) - g_j(s)]$$

since only one difference on its right-hand sum is nonnegative for any given s. Finally, inspection of the solution (12.7) makes it evident that the function $h(s) = -g(s)$ solves the difference equation (12.6) with the complement A^c replacing A. It follows that

$$g(s) - g(s+1) = h(s+1) - h(s) \leq \frac{1 - e^{-\lambda}}{\lambda},$$

and this completes the proof that $g(s)$ satisfies the Lipschitz condition.

12.5 Problems

1. For a random permutation $\sigma_1, \dots, \sigma_n$ of $\{1, \dots, n\}$, let $X_\alpha = 1_{\{\sigma_\alpha = \alpha\}}$ be the indicator of a match at position α. Show that the total number of matches $S = \sum_{\alpha=1}^{n} X_\alpha$ satisfies the coupling bound

$$\|\pi_S - \pi_Z\|_{TV} \leq \frac{2(1 - e^{-1})}{n},$$

where Z follows a Poisson distribution with mean 1.

2. In the ménage problem, prove that $\mathrm{Var}(S) = 2 - 2/(n-1)$.

3. In certain situations the hypergeometric distribution can be approximated by a Poisson distribution. Suppose that w white balls and b black balls occupy a box. If you extract $n < w + b$ balls at random, then the number of white balls S extracted follows a hypergeometric distribution. Note that if we label the white balls $1, \ldots, w$, and let X_α be the random variable indicating whether white ball α is chosen, then $S = \sum_{\alpha=1}^{w} X_\alpha$. One can construct a coupling between S and V_α by the following device. If white ball α does not show up, then randomly take one of the balls extracted and exchange it for white ball α. Calculate an explicit Chen-Stein bound, and give conditions under which the Poisson approximation to S will be good.

4. In the context of Example 12.3.1 on the law of rare events, prove the less stringent bound

$$\|\pi_S - \pi_Z\|_{TV} \;\leq\; \sum_{\alpha=1}^{n} p_\alpha^2$$

by invoking Problems 14 and 15 of Chapter 7.

5. Consider the n-dimensional unit cube $[0,1]^n$. Suppose that each of its $n2^{n-1}$ edges is independently assigned one of two equally likely orientations. Let S be the number of vertices at which all neighboring edges point toward the vertex. The Chen-Stein method implies that S has an approximate Poisson distribution Z with mean 1. Use the neighborhood method to verify the estimate

$$\|\pi_S - \pi_Z\|_{TV} \leq (n+1)2^{-n}(1 - e^{-1}).$$

(Hints: Let I be the set of all 2^n vertices, X_α the indicator that vertex α has all of its edges directed toward α, and $N_\alpha = \{\beta : \|\beta - \alpha\| \leq 1\}$. Note that X_α is independent of those X_β with $\|\beta - \alpha\| > 1$. Also, $p_{\alpha\beta} = 0$ for $\|\beta - \alpha\| = 1$.)

6. A graph with n nodes is created by randomly connecting some pairs of nodes by edges. If the connection probability per pair is p, then all pairs from a triple of nodes are connected with probability p^3. For p small and $\lambda = \binom{n}{3}p^3$ moderate in size, the number of such triangles in the random graph is approximately Poisson with mean λ. Use the neighborhood method to estimate the total variation error in this approximation.

7. Suppose n balls (people) are uniformly and independently distributed into m boxes (days of the year). The birthday problem involves finding the approximate distribution of the number of boxes that receive

d or more balls for some fixed positive integer d. This is a special case of the Poisson approximation treated in Example 12.2.2 by the coupling method. In this exercise we attack the birthday problem by the neighborhood method. To get started, let the index set I be the collection of all sets of trials $\alpha \subset \{1, \ldots, n\}$ having $|\alpha| = d$ elements. Let X_α be the indicator of the event that the balls indexed by α all fall into the same box. Argue $\Pr(S = 0) \approx e^{-\lambda}$ with $S = \sum_\alpha X_\alpha$ and

$$\lambda = \binom{n}{d} \frac{1}{m^{d-1}}$$

is plausible. Now define the neighborhoods N_α so that X_α is independent of those X_β with β outside N_α. Demonstrate that

$$\sum_{\alpha \in I} \sum_{\beta \in N_\alpha} p_\alpha p_\beta = \binom{n}{d} \left[\binom{n}{d} - \binom{n-d}{d} \right] \left(\frac{1}{m} \right)^{2d-2}$$

$$\sum_{\alpha \in I} \sum_{\beta \in N_\alpha \setminus \{\alpha\}} p_{\alpha\beta} = \binom{n}{d} \sum_{i=1}^{d-1} \binom{d}{i} \binom{n-d}{d-i} \left(\frac{1}{m} \right)^{2d-i-1}.$$

When $d = 2$, calculate the total variation bound

$$\| \pi_S - \pi_Z \|_{TV} \leq \frac{1 - e^{-\lambda}}{\lambda} \frac{\binom{n}{2}(4n - 7)}{m^2}.$$

8. In the somatic cell hybrid model, suppose that the retention probability $p \neq \frac{1}{2}$. Define $w_{n,d_{12},d_{13}} = \Pr[\rho(C_1^n, C_2^n) = d_{12}, \ \rho(C_1^n, C_3^n) = d_{13}]$ for a random panel with n clones. Show that

$$p_{\alpha\beta} = \sum_{d_{12}=0}^{d-1} \sum_{d_{13}=0}^{d-1} w_{n,d_{12},d_{13}},$$

regardless of which $\beta \in N_\alpha \setminus \{\alpha\}$ is chosen [60]. Setting $r = p(1 - p)$, verify the recurrence relation

$$w_{n+1,d_{12},d_{13}} = r(w_{n,d_{12}-1,d_{13}} + w_{n,d_{12},d_{13}-1} + w_{n,d_{12}-1,d_{13}-1})$$
$$+ (1 - 3r)w_{n,d_{12},d_{13}}.$$

Under the natural initial conditions, $w_{0,d_{12},d_{13}}$ is 1 when $d_{12} = d_{13} = 0$ and 0 otherwise.

9. In the somatic cell hybrid model, suppose that one knows a priori that the number of assay errors does not exceed some positive integer d. Prove that assay error can be detected if the minimum Hamming distance of the panel is strictly greater than d. Prove that the locus can still be correctly assigned to a single chromosome if the minimum Hamming distance is strictly greater than $2d$.

10. Consider an infinite sequence $W_1, W_2, \ldots$ of independent, Bernoulli random variables with common success probability p. Let X_α be the indicator of the event that a success run of length t or longer begins at position α. Note that $X_1 = \prod_{k=1}^{t} W_k$ and

$$X_j = (1 - W_{j-1}) \prod_{k=j}^{j+t-1} W_k$$

for $j > 1$. The number of such success runs starting in the first n positions is given by $S = \sum_{\alpha \in I} X_\alpha$, where the index set $I = \{1, \ldots, n\}$. The Poisson heuristic suggests the S is approximately Poisson with mean $\lambda = p^t[(n-1)(1-p)+1]$. Let $N_\alpha = \{\beta \in I : |\beta - \alpha| \leq t\}$. Show that X_α is independent of those X_β with β outside N_α. In the Chen-Stein bound (12.3), prove that $\sum_{\alpha \in I} \sum_{\beta \in N_\alpha \setminus \{\alpha\}} p_{\alpha\beta} = 0$. Finally, show that $b_1 = \sum_{\alpha \in I} \sum_{\beta \in N_\alpha} p_\alpha p_\beta \leq \lambda^2(2t+1)/n + 2\lambda p^t$. (Hint:

$$
\begin{aligned}
b_1 &= p^{2t} + 2tp^{2t}(1-p) \\
&\quad + [2nt - t^2 + n - 3t - 1]p^{2t}(1-p)^2
\end{aligned}
$$

exactly. Note that the pairs α and β entering into the double sum for b_1 are drawn from the integer lattice points $\{(i, j) : 1 \leq i, j \leq n\}$. An upper left triangle and a lower right triangle of lattice points from this square do not qualify for the double sum defining b_1. The term p^{2t} in b_1 corresponds to the lattice point $(1, 1)$.)

11. In the coupling method demonstrate the bound

$$\Pr(S > 0) \geq \sum_{\alpha \in I} \frac{p_\alpha}{1 + E(V_\alpha)}.$$

See reference [130] for some numerical examples. (Hints: Choose R appropriately in equality (12.8) and apply Jensen's inequality.)

13
Number Theory

13.1 Introduction

Number theory is one of the richest and oldest branches of mathematics. It is notable for its many unsolved but easily stated conjectures. The current chapter touches on issues surrounding prime numbers and their density. In particular, the chapter and book culminate with a proof of the prime number theorem. This highlight of 19th century mathematics was surmised by Legendre and Gauss, attacked by Riemann and Chebyshev, and finally proved by Hadamard and de la Vallée Poussin. These mathematicians created a large part of analytic function theory in the process. In the mid-20th century, Erdös and Selberg succeeded in crafting a proof that avoids analytic functions. Even so, their elementary proof is longer and harder to comprehend than the classical proofs. Our treatment follows the recent trail blazed by Newman [115] and Zagier [155] that uses a minimum of analytic function theory. We particularly stress the connections and insight provided by probability.

In our exposition, we will take several mathematical facts for granted. For example provided no $a_n = -1$, the absolute convergence of the infinite product $\prod_{n=1}^{\infty}(1 + a_n)$ to a nonzero number is equivalent to the absolute convergence of the infinite series $\sum_{n=1}^{\infty} a_n$. Absolute convergence of either an infinite series or an infinite product implies convergence of the corresponding series or product [6, 77].

The necessary background in number theory is even more slender [7, 67, 80, 116]. Multiplicative number theory deals with the set of positive integers (or natural numbers). The integer a divides the integer b, written $a \mid b$, if $b = ac$ for some integer c. A natural number $p > 1$ is said to be prime if its only positive divisors are 1 and itself. Thus, 2, 3, 5, 7, 11, 13, 17, 19, and so forth are primes; 1 is not prime. The number of primes is infinite. The classical proof of Euclid proceeds by contradiction. Suppose $p_1, \ldots, p_n$ is a complete list of the primes. Then the number $1 + \prod_{i=1}^{n} p_i$ is not divisible by any of the primes in the list and consequently must itself be prime. Equally important is the fundamental theorem of arithmetic. This theorem says that every natural number can be factored into a product of primes. Such a representation is unique except for the order of the factors. For example, the composite number $60 = 2^2 3^1 5^1$. Two integers are said to be relatively prime if they possess no common factors.

13.2   Zipf's Distribution and Euler's Theorem

Riemann's zeta function $\zeta(s) = \sum_{n=1}^{\infty} n^{-s}$ converges for $s > 1$. Thus, Zipf's probability measure

$$\omega_s(A) = \frac{1}{\zeta(s)} \sum_{n \in A} n^{-s}$$

on the natural numbers makes sense. It obviously puts more weight on small numbers than on large numbers. If p and q are prime numbers, then one can also show by direct calculation that the sets $D_p = \{kp : k \geq 1\}$ and $D_q = \{kq : k \geq 1\}$ of integers divisible by p and q are independent under ω_s.

It is more illuminating to reverse the process, start from independence, and construct ω_s indirectly. Toward this end, consider a sequence of independent, geometrically distributed random variables X_p indexed by the prime numbers p. Here X_p counts the number of failures until success in a sequence of Bernoulli trials with failure probability p^{-s}. Straightforward calculations demonstrate that X_p has mean

$$\frac{p^{-s}}{1 - p^{-s}} = \frac{1}{p^s - 1}$$

and that the event $A_p = \{X_p > 0\}$ has probability p^{-s}. Because

$$\sum_p \Pr(A_p) = \sum_p p^{-s} < \sum_{n=1}^{\infty} n^{-s} < \infty,$$

the Borel-Cantelli lemma implies that only finitely many of the A_p occur. This means that all but a finite number of the factors of the infinite product $N = \prod_p p^{X_p}$ reduce to 1.

We now claim that N has Zipf's distribution. Indeed, if the integer n has unique prime factorization $n = \prod_p p^{x_p}$, then the continuity of probability on a decreasing sequence of events implies

$$
\begin{aligned}
\Pr(N = n) &= \prod_p \Pr(X_p = x_p) \\
&= \prod_p \left(1 - p^{-s}\right) p^{-x_p s} \\
&= n^{-s} \prod_p \left(1 - p^{-s}\right).
\end{aligned}
$$

The equation

$$
\prod_p \left(1 - p^{-s}\right) = \frac{1}{\zeta(s)} \tag{13.1}
$$

identifying the normalizing constant figures in the proof of Proposition 13.2.1. More to the point, independence of the events D_p and D_q is now trivial because in this setting D_p reduces to A_p and D_q to A_q.

Number theory, like many branches if mathematics, has its own jargon. A real or complex-valued function defined on the natural numbers is called an arithmetic function. From our perspective, an arithmetic function is a random variable $Y = f(N)$ with value $f(n)$ at the sample point n. To avoid confusion, we will subscript our expectation signs by the parameter s so that

$$
E_s(Y) = \int f(n)\, d\omega_s(n) = \zeta(s)^{-1} \sum_{n=1}^{\infty} f(n) n^{-s}.
$$

The best behaved arithmetic functions Y are completely multiplicative in the sense that $f(mn) = f(m)f(n)$ for all m and n. The arithmetic function $f(n) = n^r$ furnishes an example. A multiplicative function $f(n)$ is only required to satisfy $f(mn) = f(m)f(n)$ when m and n are relatively prime. Excluding the trivial case $f(n) \equiv 0$, both definitions require $f(1) = 1$. Indeed, if $f(n) \neq 0$ for some n, then the equation $f(n) = f(n)f(1)$ is only possible if $f(1) = 1$. The sets of completely multiplicative and multiplicative functions are closed under the formation of pointwise products and, provided division by 0 is not involved, pointwise quotients. A random variable Y defined by a multiplicative function $f(n)$ splits into a product $Y = \prod_p f(p^{X_p})$ of independent random variables depending on the prime powers X_p. With probability 1 only a finite number of factors of the infinite product $\prod_p f(p^{X_p})$ differ from 1. The importance of multiplicative functions stems from the following result.

Proposition 13.2.1 (Euler) *Suppose the multiplicative arithmetic function* $Y = f(N)$ *has finite expectation. Then*

$$\mathrm{E}_s(Y) \;=\; \prod_p \mathrm{E}_s\left[f\left(p^{X_p}\right)\right], \tag{13.2}$$

where p extends over all primes. If $f(n)$ is completely multiplicative, then

$$\mathrm{E}_s\left[f\left(p^{X_p}\right)\right] \;=\; \mathrm{E}_s\left[f(p)^{X_p}\right] \;=\; \frac{1 - p^{-s}}{1 - f(p)p^{-s}}. \tag{13.3}$$

Proof: In view of equation (13.1), it suffices to prove

$$\sum_{n=1}^{\infty} \frac{f(n)}{n^s} \;=\; \prod_p \left[1 + \sum_{m=1}^{\infty} \frac{f(p^m)}{p^{ms}}\right].$$

If we let $g(n) = f(n)n^{-s}$, then $g(n)$ is multiplicative whenever $f(n)$ is multiplicative. In other words, it is enough to prove that

$$\sum_{n=1}^{\infty} g(n) \;=\; \prod_p \left[1 + \sum_{m=1}^{\infty} g(p^m)\right] \tag{13.4}$$

for $g(n)$ multiplicative. The infinite sum on the left of equation (13.4) converges absolutely by assumption. The infinite product on the right is well defined and converges absolutely because

$$\sum_{p \leq q} \left|\sum_{m=1}^{\infty} g(p^m)\right| \;\leq\; \sum_{p \leq q} \sum_{m=1}^{\infty} |g(p^m)|$$

$$\leq\; \sum_{n=1}^{\infty} |g(n)|$$

for all primes q. Therefore, both sides of the proposed equality (13.4) make sense.

We can rearrange the terms in the finite product

$$\pi_q \;=\; \prod_{p \leq q} \left[1 + \sum_{m=1}^{\infty} g(p^m)\right]$$

of absolutely convergent series without altering the value of the product. Exploiting the multiplicative nature of $g(n)$, we find that

$$\pi_q \;=\; \sum_{n \in B_q} g(n),$$

where B_q consists of all natural numbers having no prime factor strictly greater than q. It follows that

$$\left| \sum_{n=1}^{\infty} g(n) - \pi_q \right| \leq \sum_{n \notin B_q} |g(n)|$$

$$\leq \sum_{n>q} |g(n)|.$$

This last sum can be made arbitrarily small by taking q large enough. This proves identity (13.2).

To verify identity (13.3), we calculate

$$
\begin{aligned}
\mathrm{E}_s \left[f \left(p^{X_p} \right) \right] &= \mathrm{E}_s [f(p)^{X_p}] \\
&= \frac{\sum_{m=0}^{\infty} f(p)^m p^{-ms}}{\sum_{m=0}^{\infty} p^{-ms}} \\
&= \frac{1 - p^{-s}}{1 - f(p)p^{-s}}
\end{aligned}
$$

by summing the indicated geometric series.   ∎

Example 13.2.1 *A Scaled Version of Euler's Totient*

Euler's totient function $\varphi(n)$, mentioned in Example 4.2.2, counts the numbers between 1 and n that are relatively prime to n. Because $\varphi(n)$ satisfies

$$\frac{\varphi(n)}{n} = \prod_p \left(1 - \frac{1}{p} \right)^{1\{X_p > 0\}},$$

we have

$$
\begin{aligned}
\mathrm{E}_s \left[\frac{\varphi(N)}{N} \right] &= \prod_p \mathrm{E}_s \left[\left(1 - p^{-1} \right)^{1\{X_p > 0\}} \right] \\
&= \prod_p \left[\left(1 - p^{-s} \right) \left(1 - p^{-1} \right)^0 + p^{-s} \left(1 - p^{-1} \right)^1 \right] \\
&= \prod_p \left(1 - p^{-s-1} \right) \\
&= \frac{1}{\zeta(s+1)}.
\end{aligned}
$$

It is noteworthy that this expectation tends to the limit $\zeta(2)^{-1} = 6/\pi^2$ as s tends to 1. Section 13.5 and Problem 13 explore this phenomenon in more detail.   ∎

Example 13.2.2 *Expected Number of Divisors*

A moment's reflection shows that the number of divisors of a random natural number N can be expressed as $\tau(N) = \prod_p (1+X_p)$. Euler's formula (13.2) gives

$$E_s[\tau(N)] \;=\; \prod_p E_s(1+X_p) \;=\; \prod_p \left(1 + \frac{1}{p^s - 1}\right).$$

This infinite product converges because

$$\sum_p \frac{1}{p^s - 1} \;\leq\; \sum_{n=2}^{\infty} \frac{1}{n^s - 1} \;<\; \infty.$$

Of course, exact evaluation of $E_s[\tau(N)]$ is much harder. ∎

Example 13.2.3 *Evaluation of* $E_s(\ln N)$

To avoid the impression that Euler's formula is the only method of calculating expectations, consider

$$-\frac{d}{ds} \ln \zeta(s) \;=\; -\zeta(s)^{-1}\zeta'(s)$$

$$= \;\zeta(s)^{-1} \sum_{n=1}^{\infty} \frac{\ln n}{n^s}$$

$$= \;E_s(\ln N).$$

Symbolic algebra programs such as Maple are capable of numerically evaluating and differentiating $\zeta(s)$. ∎

13.3   Dirichlet Products and Möbius Inversion

Many of the arithmetic functions pursued in analytic number theory are multiplicative. A few examples are

$$\delta(n) \;=\; \begin{cases} 1 & n = 1 \\ 0 & n > 1 \end{cases}$$

$$1(n) \;=\; 1$$

$$\mathrm{id}(n) \;=\; n$$

$$\varphi(n) \;=\; n\prod_{p|n} \left(1 - \frac{1}{p}\right)$$

$$\sigma_k(n) \;=\; \sum_{d|n} d^k.$$

In this list, n is any natural number, p any prime, and k any nonnegative integer. We have already met the two arithmetic functions $\tau(n) = \sigma_0(n)$ and $\varphi(n)$. The sum of the divisors of n has the alias $\sigma(n) = \sigma_1(n)$. One of the goals of this section is to prove that all of the arithmetic functions $\sigma_k(n)$ are multiplicative. The von Mangoldt function

$$\Lambda(n) \;=\; \begin{cases} \ln p & \text{if } n = p^m \text{ for some prime } p \\ 0 & \text{otherwise} \end{cases}$$

is a prominent arithmetic function that fails to be multiplicative. Its relevance arises from the representation

$$\ln n \;=\; \sum_{d|n} \Lambda(d). \tag{13.5}$$

To recognize multiplicative functions, it helps to define a convolution operation sending a pair of arithmetic functions $f(n)$ and $g(n)$ into the new arithmetic function

$$f * g(n) \;=\; \sum_{d|n} f(d)g(n/d)$$

termed their Dirichlet product. Among the virtues of this definition are the formulas

$$f * \delta(n) \;=\; f(n)$$
$$f * g(n) \;=\; \sum_{ab=n} f(a)g(b)$$
$$\;=\; g * f(n)$$
$$f * (g * h)(n) \;=\; \sum_{abc=n} f(a)g(b)h(c)$$
$$\;=\; (f * g) * h(n).$$

Except for the existence of an inverse, these are precisely the axioms characterizing a commutative group [7]. Because δ serves as an identity element, the inverse $f^{[-1]}$ of an arithmetic function f must satisfy the equation $f * f^{[-1]} = \delta$. In particular, $1 = f * f^{[-1]}(1) = f(1)f^{[-1]}(1)$. Thus, we must restrict our attention to arithmetic functions with $f(1) \neq 0$ to achieve a group structure. With this proviso, it remains to specify the inverse $f^{[-1]}$ of f. Fortunately, this is accomplished inductively through the formula

$$f^{[-1]}(n) \;=\; -f(1)^{-1} \sum_{d|n;\, d>1} f(d)f^{[-1]}(n/d),$$

beginning with $f^{[-1]}(1) = f(1)^{-1}$.

The multiplicative functions form a subgroup of this group. To prove part of this assertion, consider two such functions f and g and two relatively prime natural numbers m and n. In the expression

$$f * g(mn) \;=\; \sum_{d|mn} f(d)g\left(\frac{mn}{d}\right),$$

simply observe that every divisor d of mn can be expressed as a product $d = ab$ of two relatively prime numbers a and b such that $a \mid m$, $b \mid n$, and m/a and n/b are relatively prime. It follows that

$$
\begin{aligned}
\sum_{d|mn} f(d)g\left(\frac{mn}{d}\right) &= \sum_{a|m;\, b|n} f(a)f(b)g(m/a)g(n/b)\\
&= \sum_{a|m} f(a)g(m/a) \sum_{b|n} f(b)g(n/b)\\
&= f * g(m) f * g(n),
\end{aligned}
$$

completing the proof that $f * g$ is multiplicative. Problem 6 asks the reader to check that $f^{[-1]}$ is multiplicative whenever f is multiplicative.

The forgoing is summarized by the next proposition.

Proposition 13.3.1 *The set of arithmetic functions $f(n)$ with $f(1) \neq 0$ forms a commutative group. The subset of multiplicative functions constitutes a subgroup of this group.*

Proof: See the above arguments. ∎

Example 13.3.1 *The Möbius Function $\mu(n)$*

The Möbius function $\mu(n)$ is the inverse of $\mathbf{1}(n)$. We claim that

$$
\mu(n) \;=\;
\begin{cases}
1, & n = 1\\
(-1)^k, & n = p_1 \cdots p_k\\
0, & \text{otherwise,}
\end{cases}
$$

where $p_1, \ldots, p_k$ are distinct primes. It suffices to verify that $\mu * \mathbf{1}(n) = \delta(n)$. This is clear for $n = 1$, and for $n = p_1^{e_1} \cdots p_k^{e_k} > 1$ we have

$$
\begin{aligned}
\sum_{d|n} \mu(d) &= \mu(1) + \sum_i \mu(p_i) + \sum_{i<j} \mu(p_i p_j) + \cdots + \mu(p_1 \cdots p_k)\\
&= \sum_{i=0}^{k} \binom{k}{i}(-1)^i\\
&= (1-1)^k\\
&= 0.
\end{aligned}
$$

The logical equivalence of the Möbius relations

$$g(n) = \sum_{d|n} f(d)$$

$$f(n) = \sum_{d|n} \mu(d) g(n/d)$$

just restates the equivalence of the relations $g = \mathbf{1} * f$ and $f = \mu * g$.   ∎

Example 13.3.2 *Examples of Möbius Inversion*

Applying the identity $\mu * \mathbf{1}(n) \ln n = 0$, we find that Equation (13.5) has inverse

$$\Lambda(n) = \sum_{d|n} \mu(d) \ln(n/d)$$

$$= -\sum_{d|n} \mu(d) \ln d.$$

Euler's totient satisfies the pair of relations

$$n = \sum_{d|n} \varphi(d)$$

$$\varphi(n) = \sum_{d|n} \mu(d) \frac{n}{d} .$$

To prove the first of these, it helps to note that both sides are multiplicative functions. If $n = p^k$ is a power of a prime, then

$$\sum_{d|n} \varphi(d) = \sum_{l=0}^{k} \varphi(p^l)$$

$$= 1 + \sum_{l=1}^{k} p^l(1 - p^{-1})$$

$$= 1 + \sum_{l=1}^{k} (p^l - p^{l-1})$$

$$= p^k.$$

Since n and $\sum_{d|n} \varphi(d)$ agree for every power of a prime, they agree for every natural number n.   ∎

If $f(n)$ and $g(n)$ are two arithmetic functions for which $E_s[f(N)]$ and $E_s[g(N)]$ exist, then

$$\left[\sum_{l=1}^{\infty} \frac{f(l)}{l^s}\right]\left[\sum_{m=1}^{\infty} \frac{g(m)}{m^s}\right] = \sum_{n=1}^{\infty} \frac{1}{n^s} \sum_{lm=n} f(l)g(m) = \sum_{n=1}^{\infty} \frac{f * g(n)}{n^s}.$$

Here all series converge absolutely. This result can be restated as

$$E_s[f * g(N)] \ = \ \zeta(s) \, E_s[f(N)] \, E_s[g(N)].$$

For instance, Example 13.2.3 and equation (13.5) together imply

$$-\frac{d}{ds} \ln \zeta(s) \ = \ E_s(\ln N)$$
$$= \ \zeta(s) \, E_s[\Lambda(N)] \, E_s(1) \qquad (13.6)$$
$$= \ \sum_{n=1}^{\infty} \frac{\Lambda(n)}{n^s}.$$

13.4   Averages of Arithmetic Functions

Most questions in analytic number theory involve number density rather than Zipf probability. This is certainly the case for the prime number theorem [81, 144]. The Zipf distribution is easier to work with than number density because it is countably additive. Fortunately, the two notions are intimately connected. The connections can best be exposed by defining the long-run average

$$A(f) \ = \ \lim_{n \to \infty} \frac{1}{n} \sum_{m=1}^{n} f(m)$$

of an arithmetic function $f(n)$ and contrasting it to $\lim_{s \to 1} E_s[f(N)]$. It turns out that in many cases these two limits coincide. When $f(n)$ is the indicator function of a set of natural numbers, this surprising insight helps in computing the number density of the set.

Before tackling this issue and how it bears on the prime number theorem, some preliminary spadework is needed. Our first result is Abel's summation formula.

Proposition 13.4.1 (Abel) *Suppose $f(n)$ is an arithmetic function and $g(t)$ is a continuously differentiable function. If $F(t) = \sum_{n=1}^{[t]} f(n)$, then*

$$\sum_{m=1}^{n} f(m)g(m) \ = \ F(n)g(n) - \int_{1}^{n} F(t)g'(t) \, dt.$$

Proof: With the convention $F(0) = 0$, the fundamental theorem of calculus implies

$$\sum_{m=1}^{n} f(m)g(m) \ = \ \sum_{m=1}^{n} [F(m) - F(m-1)]g(m)$$

$$
\begin{aligned}
&= \sum_{m=1}^{n} F(m)g(m) - \sum_{m=1}^{n-1} F(m)g(m+1) \\
&= F(n)g(n) - \sum_{m=1}^{n-1} F(m)[g(m+1) - g(m)] \\
&= F(n)g(n) - \sum_{m=1}^{n-1} F(m) \int_{m}^{m+1} g'(t)\, dt \\
&= F(n)g(n) - \sum_{m=1}^{n-1} \int_{m}^{m+1} F(t)g'(t)\, dt \\
&= F(n)g(n) - \int_{1}^{n} F(t)g'(t)\, dt.
\end{aligned}
$$

This completes the proof.   ∎

Our next result shows that $\zeta(s)$ has a removable singularity at $s = 1$.

Proposition 13.4.2 *Riemann's zeta function can be written as*

$$
\zeta(s) \;=\; s \int_{1}^{\infty} \frac{[t] - t + 1/2}{t^{s+1}}\, dt + \frac{1}{2} + \frac{1}{(s-1)} \tag{13.7}
$$

for $s > 1$. Hence, $\lim_{s \to 1}(s-1)\zeta(s) = 1$.

Proof: Straightforward integration gives

$$
s \int_{n}^{n+1} \frac{n - t + 1/2}{t^{s+1}}\, dt \;=\; \left(n + \frac{1}{2}\right)\left[\frac{1}{n^{s}} - \frac{1}{(n+1)^{s}}\right]
$$
$$
- \frac{s}{s-1}\left[\frac{1}{n^{s-1}} - \frac{1}{(n+1)^{s-1}}\right].
$$

Adding this result over n produces

$$
\begin{aligned}
s \int_{1}^{\infty} \frac{[t] - t + 1/2}{t^{s+1}}\, dt
&= \sum_{n=1}^{\infty}\left(n + \frac{1}{2}\right)\frac{1}{n^{s}} - \sum_{n=2}^{\infty}\left(n - \frac{1}{2}\right)\frac{1}{n^{s}} - \frac{s}{s-1} \\
&= \sum_{n=2}^{\infty}\frac{1}{n^{s}} + \frac{3}{2} - \frac{s}{s-1} \\
&= \zeta(s) - 1 + \frac{3}{2} - \frac{s-1}{s-1} - \frac{1}{s-1} \\
&= \zeta(s) - \frac{1}{2} - \frac{1}{s-1}.
\end{aligned}
$$

Formula (13.7) now follows by rearrangement.   ∎

We are now in position to establish one connection between $A(f)$ and the $E_s[f(N)]$.

Proposition 13.4.3 *Let $f(n)$ is an arithmetic function such that $A(f)$ and all $E_s[f(N)]$ exist. Then $\lim_{s\to 1} E_s[f(N)]$ exists and equals $A(f)$.*

Proof: Let $F(t) = \sum_{n=1}^{[t]} f(n)$ and $G(t) = t^{-1}F(t)$. Abel's summation formula gives

$$\sum_{m=1}^{n} \frac{f(m)}{m^s} = \frac{1}{n^{s-1}} \frac{F(n)}{n} + \int_1^n \frac{F(t)}{t} \frac{s}{t^s} dt.$$

$$= \frac{G(n)}{n^{s-1}} + s \int_0^n \frac{G(t)}{t^s} dt.$$

Taking limits on n and invoking the existence of $A(f)$ therefore imply

$$\sum_{m=1}^{\infty} \frac{f(m)}{m^s} = s \int_0^{\infty} \frac{G(t)}{t^s} dt.$$

for each $s > 1$. In view of Proposition 13.4.2, it now suffices to prove that

$$\lim_{s\to 1}(s-1) \sum_{m=1}^{\infty} \frac{f(m)}{m^s} = \lim_{s\to 1}(s-1)s \int_0^{\infty} \frac{G(t)}{t^s} dt = A(f).$$

To achieve this end, we exploit the integral $(s-1)\int_1^{\infty} t^{-s} dt = 1$ and the inequality

$$\left| (s-1)s \int_0^{\infty} \frac{G(t)}{t^s} dt - sA(f) \right| = \left| (s-1)s \int_0^{\infty} \frac{G(t) - A(t)}{t^s} dt \right|$$

$$\leq (s-1)s \int_1^{\infty} \left| \frac{G(t) - A(t)}{t^s} \right| dt.$$

For $\epsilon > 0$ small, choose $\delta \geq 1$ so that $|G(t) - A(f)| < \epsilon$ for $t \geq \delta$. This permits us to form the bound

$$(s-1)s \int_1^{\infty} \left| \frac{G(t) - A(t)}{t^s} \right| dt \leq (s-1)s \int_1^{\delta} \left| \frac{G(t) - A(t)}{t^s} \right| dt$$

$$+ \epsilon(s-1)s \int_{\delta}^{\infty} \frac{1}{t^s} dt$$

$$= (s-1)s \int_1^{\delta} \left| \frac{G(t) - A(t)}{t^s} \right| dt + \epsilon s \delta^{-s+1}.$$

For s sufficiently close to 1, this bound can be made less than 2ϵ. To finish the proof, we merely note that $\lim_{s\to 1}(s-1)A(f) = 0$. ∎

Example 13.4.1 *Periodic Arithmetic Functions*

A periodic arithmetic function $f(n)$ satisfies $f(n + r) = f(n)$ for all n and some fixed r. A brief calculation shows that

$$A(f) \;=\; \frac{1}{r} \sum_{n=1}^{r} f(n).$$

Proposition 13.4.3 guarantees that $\lim_{s \to 1} E_s[f(N)] = A(f)$. We can also deduce this result by bringing in Hurwitz's zeta function

$$\zeta(s, a) \;=\; \sum_{n=0}^{\infty} \frac{1}{(n + a)^s}$$

for $a > 0$. The techniques of Proposition 13.4.2 yield the expansion

$$\zeta(s, a) \;=\; s \int_{a}^{\infty} \frac{[t - a] + a - t + 1/2}{t^{s+1}} \, dt + \frac{1}{2a^s} + \frac{1}{(s - 1)a^{s-1}}, \quad (13.8)$$

with the obvious implication $\lim_{s \to 1}(s - 1)\zeta(s, a) = 1$. It follows that

$$\frac{1}{\zeta(s)} \sum_{n=1}^{\infty} \frac{f(n)}{n^s} \;=\; \frac{1}{\zeta(s)} \sum_{n=1}^{r} f(n) \sum_{m=0}^{\infty} \frac{1}{(mr + n)^s}$$

$$\;=\; \frac{1}{(s - 1)\zeta(s)} \sum_{n=1}^{r} f(n) \frac{s - 1}{r^s} \zeta(s, n/r).$$

Setting $a = n/r$ and sending s to 1 now produce the desired limit $A(f)$. ∎

13.5   The Prime Number Theorem

Let $\pi(n)$ be the number of primes less than or equal to n. The prime number theorem says

$$\lim_{n \to \infty} \frac{\pi(n) \ln n}{n} \;=\; 1.$$

We would like to rephrase this celebrated result as an long-run arithmetic average involving the von Mangoldt function $\Lambda(n)$ and its summatory function $\psi(t) = \sum_{m=1}^{[t]} \Lambda(m)$. For a given prime p, there are $[\ln n / \ln p]$ integers of the form p^k satisfying $p^k \le n$. Therefore,

$$\psi(n) \;=\; \sum_{p \le n} \left[\frac{\ln n}{\ln p} \right] \ln p$$

$$\le\; \sum_{p \le n} \frac{\ln n}{\ln p} \ln p$$

$$\le\; \pi(n) \ln n.$$

To derive a bound from the other side, choose δ so that $0 < \delta < 1$ and set $m = [n^{1-\delta}] + 1$. Then

$$
\begin{aligned}
\pi(n) &= \pi(m) + \sum_{m < p \le n} 1 \\
&\le m + \frac{1}{\ln m} \psi(n) \\
&\le n^{1-\delta} + 1 + \frac{\psi(n)}{(1-\delta)\ln n}.
\end{aligned}
$$

These two bounds make it evident that the prime number theorem is equivalent to the equality $A(\Lambda) = \lim_{n \to \infty} n^{-1} \psi(n) = 1$.

The prime number theorem can be deduced from a more general proposition giving a sufficient condition for the existence of an arithmetic average $A(f)$. In deriving this result, we will rely on an expanded definition of the integral on the real line known as the gauge integral or generalized Riemann integral [112, 154]. The gauge integral subsumes both the Lebesgue integral and the improper integrals met in advanced calculus. The integrands of the gauge integral are not necessarily absolutely integrable. An integral of a function $g(t)$ over a infinite interval such as $[0, \infty)$ exists if and only $\int_0^u g(t)\, dt$ exists for all finite $u \ge 0$ and

$$
\lim_{u \to \infty} \int_0^u g(t)\, dt = \int_0^\infty g(t)\, dt.
$$

We will employ the following Tauberian result on Laplace transforms.

Proposition 13.5.1 *Let the bounded function $g(t)$ be integrable over every finite interval $[0, u]$. If its Laplace transform*

$$
\hat{g}(s) = \int_0^\infty g(t) e^{-st}\, dt
$$

exists for all $s > 0$ and has limit $\hat{g}(0) = \lim_{s \to 0} \hat{g}(s)$, then the integral $\int_0^\infty g(t)\, dt$ also exists and equals $\hat{g}(0)$.

Proof: The references [115, 155] demonstrate this result using Cauchy's integral formula from analytic function theory. ∎

With these preliminaries under our belt, we state our general proposition.

Proposition 13.5.2 *Suppose $f(n)$ is a nonnegative arithmetic function such that $\mathrm{E}_s[f(N)]$ exists for all $s > 1$,*

$$
\lim_{n \to \infty} \frac{1}{n^s} \sum_{m=1}^n f(m) = 0 \tag{13.9}
$$

holds for all $s > 1$, and

$$
\lim_{s \to 1} \left[\frac{1}{s} \sum_{n=1}^\infty \frac{f(n)}{n^s} - \frac{a}{s-1} \right] = b \tag{13.10}
$$

holds for constants a and b. Then both $\lim_{s \to 1} E_s[f(N)]$ *and* $A(f)$ *exist and equal a.*

Proof: Hypothesis (13.10), the identity

$$\frac{1}{s} \sum_{n=1}^{\infty} \frac{f(n)}{n^s} - \frac{a}{s-1} = \frac{1}{s(s-1)} \left[(s-1) \sum_{n=1}^{\infty} \frac{f(n)}{n^s} - a \right] - \frac{a}{s},$$

and Proposition 13.4.2 show that $\lim_{s \to 1} E_s[f(N)]$ exists and equals a.

Now let $F(t) = \sum_{n=1}^{[t]} f(n)$. Abel's summation formula implies

$$\sum_{m=1}^{n} \frac{f(m)}{m^s} = \frac{F(n)}{n^s} + s \int_1^n \frac{F(t)}{t^{s+1}} dt.$$

Because of hypothesis (13.9), this gives in the limit

$$\sum_{m=1}^{\infty} \frac{f(m)}{m^s} = s \int_1^{\infty} \frac{F(t)}{t^{s+1}} dt.$$

Dividing by s and subtracting $a \int_1^{\infty} t^{-s} dt = a(s-1)^{-1}$ therefore yield

$$\frac{1}{s} \sum_{n=1}^{\infty} \frac{f(n)}{n^s} - \frac{a}{s-1} = \int_1^{\infty} \left[\frac{1}{t} F(t) - a \right] \frac{1}{t^s} dt$$

$$= \int_0^{\infty} \left[e^{-u} F(e^u) - a \right] e^{-(s-1)u} du$$

Finally, invoking hypothesis (13.10) and Proposition 13.5.1 guarantees the existence of the integral $\int_1^{\infty} \left[t^{-1} F(t) - a \right] t^{-1} dt$.

To use this conclusion, suppose that $u^{-1} F(u) - a \geq \delta > 0$ for arbitrarily large u. Then with $\lambda > 1$ chosen so that $a\lambda \leq a + \delta$, we have for such u

$$\int_u^{\lambda u} \left[\frac{1}{t} F(t) - a \right] \frac{1}{t} dt = \int_u^{\lambda u} \frac{1}{t^2} [F(t) - at] dt$$

$$\geq \int_u^{\lambda u} \frac{1}{t^2} [F(u) - at] dt$$

$$\geq \int_u^{\lambda u} \frac{1}{t^2} [au + \delta u - at] dt$$

$$= \int_1^{\lambda} \frac{1}{(ur)^2} [au + \delta u - aur] u \, dr$$

$$= \int_1^{\lambda} \frac{1}{r^2} [a + \delta - ar] dr$$

$$> 0.$$

The fact that $\int_u^{\lambda u} \left[t^{-1}F(t) - a\right]t^{-1}dt$ does not converge to 0 as u tends to ∞ contradicts the existence of $\int_1^\infty \left[t^{-1}F(t) - a\right]t^{-1}dt$. Hence, the inequality $u^{-1}F(u) - a < \delta$ must hold for all sufficiently large u.

One can likewise demonstrate that $u^{-1}F(u) - a > -\delta$ for all large u by assuming the contrary. If u is a point where $u^{-1}F(u) - a \leq -\delta$, then choose $0 < \theta < 1$ so that $a - \delta < a\theta$. The inequality

$$\int_{\theta u}^u \left[\frac{1}{t}F(t) - a\right]\frac{1}{t}\,dt \;<\; \int_\theta^1 \frac{1}{r^2}\left[a - \delta - ar\right]dr$$

for arbitrarily large u again contradicts the existence of the integral

$$\int_1^\infty \left[\frac{1}{t}F(t) - a\right]t^{-1}dt$$

and completes the proof of the proposition.   ∎

Example 13.5.1 *Application to the Prime Number Theorem*

Let us verify that the assumptions of Proposition 13.5.2 apply. Certainly, the function $f(n) = \Lambda(n)$ is nonnegative. It satisfies hypothesis (13.9) because $\sum_{m=1}^n \Lambda(m) \leq n \ln n$. Hypothesis (13.10) is more delicate. In view of equation (13.6), we have

$$\frac{1}{s}\sum_{n=1}^\infty \frac{\Lambda(n)}{n^s} - \frac{1}{s-1} = -\frac{\zeta'(s)}{s\zeta(s)} - \frac{1}{s-1}$$

$$= -\frac{(s-1)\zeta'(s) + s\zeta(s)}{s(s-1)\zeta(s)}.$$

This can be rewritten as

$$\frac{1}{s}\sum_{n=1}^\infty \frac{\Lambda(n)}{n^s} - \frac{1}{s-1} = -\frac{(s-1)\eta'(s) + s\eta(s) + 1}{s(s-1)\zeta(s)}$$

using the function $\eta(s) = \zeta(s) - (s-1)^{-1}$, which is continuously differentiable in a neighborhood of $s = 1$ by Proposition 13.4.2. Because $(s-1)\zeta(s)$ is also well behaved around $s = 1$, the limit

$$\lim_{s\to 1}\left[\frac{1}{s}\sum_{n=1}^\infty \frac{\Lambda(n)}{n^s} - \frac{1}{s-1}\right]$$

exists, and this completes the proof of the prime number theorem   ∎

13.6   Problems

1. Let N follow Zipf's distribution. Demonstrate that

$$E_s(N^k) = \frac{\zeta(s-k)}{\zeta(s)}$$

$$E_s(N^k \mid q \text{ divides } N) \;=\; \frac{q^k \zeta(s-k)}{\zeta(s)}$$

$$E_s(N^k \mid N \text{ prime}) \;=\; \frac{\sum_p p^{k-s}}{\sum_p p^{-s}}$$

for k a natural number with $s > k + 1$.

2. Choose two independent random numbers M and N according to Zipf's distribution. Prove that M and N are relatively prime with probability $\zeta(2s)^{-1}$. (Hints: Let X_p and Y_p be the powers of the prime p occurring in M and N, respectively. Calculate $E_s(\prod_p 1_{B_p})$, where B_p is the event $\{X_p Y_p = 0\}$.)

3. Suppose the arithmetic function $Y = f(N)$ satisfies $E_s(Y) = 0$ for all $s \geq r > 1$. Show that $Y \equiv 0$. (Hint: Prove that $f(n) = 0$ for all n by induction and sending s to ∞ in the equation $n^s \, E_s(Y) = 0$.)

4. Let $N_1, \ldots, N_m$ be an i.i.d. sample from the Zipf distribution with values $n_1, \ldots, n_m$. If at least one $n_i > 1$, then prove that the maximum likelihood estimate of s is uniquely determined by the equation

$$-\frac{d}{ds} \ln \zeta(s) \;=\; E_s(\ln N) \;=\; \frac{1}{m} \sum_{i=1}^{m} \ln n_i.$$

What happens in the exceptional case when all $n_i = 1$?

5. Suppose the independent realizations M and N of Zipf's distribution generate arithmetic functions $Y = f(M)$ and $Z = g(N)$ with finite expectations. Show that the random variables $L = MN$ and $W = YZ$ satisfy

$$\Pr(L = l) \;=\; \frac{\tau(l)}{l^s \zeta(s)^2}$$

$$E_s(W \mid L = l) \;=\; \tau(l)^{-1} f * g(l).$$

Recall that $\tau(l)$ is the number of divisors of l. Use these results to demonstrate that $E_s(W) = E_s(Y) E_s(Z)$ entails

$$\sum_{l=1}^{\infty} \frac{f * g(l)}{l^s} \;=\; \left[\sum_{m=1}^{\infty} \frac{f(m)}{m^s} \right] \left[\sum_{n=1}^{\infty} \frac{g(n)}{n^s} \right].$$

6. Check that the Dirichlet inverse $f^{[-1]}$ of a multiplicative arithmetic function f is multiplicative. (Hints: Assume otherwise, and consider the least product mn of two relatively prime natural numbers m and n with $f^{[-1]}(mn) \neq f^{[-1]}(m) f^{[-1]}(n)$. Now apply the inductive definition of $f^{[-1]}(mn)$.)

7. Verify that the identities

$$1 \;=\; \sum_{d|n} \mu(d)\tau(n/d)$$

$$n \;=\; \sum_{d|n} \mu(d)\sigma(n/d)$$

hold for all natural numbers n.

8. Demonstrate that neither $\varphi(n)$ nor $\mu(n)$ is completely multiplicative. Show that the Dirichlet convolution of two completely multiplicative functions need not be completely multiplicative. (Hint: $\sigma = id * \mathbf{1}$.)

9. Liouville's arithmetic function is defined by

$$\lambda(n) \;=\; \begin{cases} 1, & n = 1 \\ (-1)^{e_1 + \cdots + e_k}, & n = p_1^{e_1} \cdots p_k^{e_k} . \end{cases}$$

Prove that

$$\sum_{d|n} \lambda(d) \;=\; \begin{cases} 1 & \text{if } n \text{ is a square} \\ 0 & \text{otherwise.} \end{cases}$$

10. If $\lim_{n\to\infty} f(n) = c$, then demonstrate that

$$A(f) \;=\; \lim_{s\to 1} \mathrm{E}_s[f(N)] \;=\; c.$$

11. Derive formula (13.8).

12. In Proposition 13.5.2, show that the assumption that $f(n)$ is nonnegative can be relaxed to $\inf_n f(n) > -\infty$ or $\sup_n f(n) < \infty$.

13. Apply Proposition 13.5.2 and prove that $A[\varphi(N)/N] = \zeta(2)^{-1}$.

References

[1] Aigner M, Ziegler GM (1999) *Proofs from the Book*. Springer-Verlag, New York

[2] Aldous D (1989) *Probability Approximations via the Poisson Clumping Heuristic*. Springer-Verlag, New York

[3] Aldous D, Diaconis P (1986) Shuffling cards and stopping times. *Amer Math Monthly* 93:333–348

[4] Alon N, Spencer JH, Erdös P (1992) *The Probabilistic Method*. Wiley, New York

[5] Anděl J (2001) *Mathematics of Chance*. Wiley, New York

[6] Apostol T (1974) *Mathematical Analysis*, 2nd ed. Addison-Wesley, Reading, MA

[7] Apostol T (1976) *Introduction to Analytic Number Theory*. Springer-Verlag, New York

[8] Arnold BC, Balakrishnan N, Nagaraja HN (1992) *A First Course in Order Statistics*. Wiley, New York

[9] Arratia R, Goldstein L, Gordon L (1989) Two moments suffice for Poisson approximations: the Chen-Stein method. *Ann Prob* 17:9–25

[10] Arratia R, Goldstein L, Gordon L (1990) Poisson approximation and the Chen-Stein method. *Statist Sci* 5:403–434

[11] Asmussen S, Hering H (1983) *Branching Processes.* Birkhäuser, Boston

[12] Athreya KB, Ney PE (1972) *Branching Processes.* Springer-Verlag, New York

[13] Baclawski K, Rota G-C, Billey S (1989) *An Introduction to the Theory of Probability.* Massachusetts Institute of Technology, Cambridge, MA

[14] Baker JA (1997) Integration over spheres and the divergence theorem for balls. *Amer Math Monthly* 64:36–47

[15] Balasubramanian K, Balakrishnan N (1993) Duality principle in order statistics. *J Roy Statist Soc B* 55:687–691

[16] Barbour AD, Holst L, Janson S (1992) *Poisson Approximation.* Oxford University Press, Oxford

[17] Berge C (1971) *Principles of Combinatorics.* Academic Press, New York

[18] Bhattacharya RN, Waymire EC (1990) *Stochastic Processes with Applications.* Wiley, New York

[19] Billingsley P (1986) *Probability and Measure,* 2nd ed. Wiley, New York

[20] Blom G, Holst L (1991) Embedding procedures for discrete problems in probability. *Math Scientist* 16:29–40

[21] Blom G, Holst L, Sandell D (1994) *Problems and Snapshots from the World of Probability.* Springer-Verlag, New York

[22] Blyth CR, Pathak PK (1986) A note on easy proofs of Stirling's formula. *Amer Math Monthly* 93:376–379

[23] Bradley RA, Terry ME (1952), Rank analysis of incomplete block designs, *Biometrika* 39:324–345

[24] Brezeźniak Z, Zastawniak T (1999) *Basic Stochastic Processes.* Springer-Verlag, New York

[25] Brualdi RA (1977) *Introductory Combinatorics.* North-Holland, New York

[26] Casella G, Berger RL (1990) *Statistical Inference.* Wadsworth and Brooks/Cole, Pacific Grove, CA

[27] Casella G, George EI (1992) Explaining the Gibbs sampler. *Amer Statistician* 46:167–174

[28] Chen LHY (1975) Poisson approximation for dependent trials. *Ann Prob* 3:534–545

[29] Chib S, Greenberg E (1995) Understanding the Metropolis-Hastings algorithm. *Amer Statistician* 49:327–335

[30] Chung KL, Williams RJ (1990) *Introduction to Stochastic Integration,* 2nd ed. Birkhäuser, Boston

[31] Ciarlet PG (1989) *Introduction to Numerical Linear Algebra and Optimization.* Cambridge University Press, Cambridge

[32] Cochran WG (1977) *Sampling Techniques*, 3rd ed. Wiley, New York

[33] Comet L (1974) *Advanced Combinatorics*, Reidel, Dordrecht

[34] Cressie N, Davis AS, Folks JL, Polocelli GE Jr (1981) The moment-generating function and negative integer moments. *Amer Statistician* 35:148–150

[35] Crow, JF, Kimura M (1970) *An Introduction to Population Genetics Theory.* Harper & Row, New York

[36] David HA (1993) A note on order statistics for dependent variates. *Amer Statistician* 47:198–199

[37] De Pierro AR (1993) On the relation between the ISRA and EM algorithm for positron emission tomography. *IEEE Trans Med Imaging* 12:328–333

[38] Diaconis, P (1988) *Group Representations in Probability and Statistics.* Institute of Mathematical Statistics, Hayward, CA

[39] D'Eustachio P, Ruddle FH (1983) Somatic cell genetics and gene families. *Science* 220:919–924

[40] Dieudonné J (1971) *Infinitesimal Calculus.* Houghton-Mifflin, Boston

[41] Doyle PG, Snell JL (1984) *Random Walks and Electrical Networks.* The Mathematical Association of America, Washington, DC

[42] Durrett R (1991) *Probability: Theory and Examples.* Wadsworth & Brooks/Cole, Pacific Grove, CA

[43] Dym H, McKean HP (1972) *Fourier Series and Integrals.* Academic Press, New York

[44] Erdös P, Füredi Z (1981) The greatest angle among n points in the d-dimensional Euclidean space. *Ann Discrete Math* 17:275–283

[45] Ewens, W.J. (1979). *Mathematical Population Genetics.* Springer-Verlag, New York

[46] Fan R, Lange K (1999) Diffusion process calculations for mutant genes in nonstationary populations. In *Statistics in Molecular Biology and Genetics.* Edited by Seillier-Moiseiwitsch F, The Institute of Mathematical Statistics and the American Mathematical Society, Providence, RI, pp 38-55

[47] Fan R, Lange K, Peña EA (1999) Applications of a formula for the variance function of a stochastic process. *Statist Prob Letters* 43:123–130

[48] Feller W (1968) *An Introduction to Probability Theory and its Applications, Vol 1,* 3rd ed. Wiley, New York

[49] Feller W (1971) *An Introduction to Probability Theory and its Applications, Vol 2,* 2nd ed. Wiley, New York

[50] Fix E, Neyman J (1951) A simple stochastic model of recovery, relapse, death and loss of patients. *Hum Biol* 23:205–241

[51] Flatto L, Konheim AG (1962) The random division of an interval and the random covering of a circle. *SIAM Rev* 4:211–222

[52] Friedlen DM (1990) More socks in the laundry. Problem E 3265. *Amer Math Monthly* 97:242–244

[53] Galambos J, Simonelli I (1996) *Bonferroni-type Inequalities with Applications.* Springer-Verlag, New York

[54] Gani J, Saunders IW (1977) Fitting a model to the growth of yeast colonies. *Biometrics* 33:113–120

[55] Gelfand AE, Smith AFM (1990) Sampling-based approaches to calculating marginal densities. *J Amer Statist Assoc* 85:398–409

[56] Gelman A, Carlin JB, Stern HS, Rubin DB (1995) *Bayesian Data Analysis.* Chapman & Hall, London

[57] Geman S, Geman D (1984) Stochastic relaxation, Gibbs distributions and the Bayesian restoration of images. *IEEE Trans Pattern Anal Machine Intell* 6:721–741

[58] Geman S, McClure D (1985) Bayesian image analysis: An application to single photon emission tomography. *Proceedings of the Statistical Computing Section,* American Statistical Association, Washington, DC, pp 12–18

[59] Gilks WR, Richardson S, Spiegelhalter DJ (editors) (1996) *Markov Chain Monte Carlo in Practice.* Chapman & Hall, London

[60] Goradia TM (1991) *Stochastic Models for Human Gene Mapping.* Ph.D. Thesis, Division of Applied Sciences, Harvard University

[61] Goradia TM, Lange K (1988) Applications of coding theory to the design of somatic cell hybrid panels. *Math Biosci* 91:201–219

[62] Graham RL, Knuth DE, Patashnik O (1988) *Concrete Mathematics: A Foundation for Computer Science.* Addison-Wesley, Reading, MA

[63] Green P (1990) Bayesian reconstruction for emission tomography data using a modified EM algorithm. *IEEE Trans Med Imaging* 9:84–94

[64] Grimmett GR, Stirzaker DR (2001) *Probability and Random Processes,* 3rd ed. Oxford University Press, Oxford

[65] Guttorp P (1991) *Statistical Inference for Branching Processes.* Wiley, New York

[66] Gzyl H, Palacios JL (1997) The Weierstrass approximation theorem and large deviations. *Amer Math Monthly* 104:650–653

[67] Hardy GH, Wright EM (1960) *An Introduction to the Theory of Numbers,* 4th ed. Clarendon Press, Oxford

[68] Harris TE (1989) *The Theory of Branching Processes.* Dover, New York

[69] Hastings WK (1970) Monte Carlo sampling methods using Markov chains and their applications. *Biometrika* 57:97–109

[70] Henrici P (1982) *Essentials of Numerical Analysis with Pocket Calculator Demonstrations.* Wiley, New York

[71] Herman GT (1980) *Image Reconstruction from Projections: The Fundamentals of Computerized Tomography.* Springer-Verlag, New York

[72] Hestenes MR (1981) *Optimization Theory: The Finite Dimensional Case.* Robert E Krieger Publishing, Huntington, NY

[73] Hille E (1959) *Analytic Function Theory, Vol 1.* Blaisdell, New York

[74] Hirsch MW, Smale S (1974) *Differential Equations, Dynamical Systems, and Linear Algebra.* Academic Press, New York

[75] Hochstadt H (1986) *The Functions of Mathematical Physics.* Dover, New York

[76] Hoel PG, Port SC, Stone CJ (1971) *Introduction to Probability Theory.* Houghton Mifflin, Boston

[77] Hoffman K (1975) *Analysis in Euclidean Space*. Prentice-Hall, Englewood Cliffs, NJ

[78] Hwang JT (1982) Improving on standard estimators in discrete exponential families with applications to the Poisson and negative binomial cases. *Ann Statist* 10:857–867

[79] Jagers P (1975) *Branching Processes with Biological Applications*. Wiley, New York

[80] Jones GA, Jones JM (1998) *Elementary Number Theory*. Springer-Verlag, New York

[81] Kac M (1959) *Statistical Independence in Probability, Analysis and Number Theory*. Mathematical Association of America, Washington, DC

[82] Karlin S, Taylor HM (1975) *A First Course in Stochastic Processes*, 2nd ed. Academic Press, New York

[83] Karlin S, Taylor HM (1981) *A Second Course in Stochastic Processes*. Academic Press, New York

[84] Katz B (1966) *Nerve, Muscle, and Synapse*. McGraw-Hill, New York

[85] Keener JP (1993), The Perron-Frobenius theorem and the ranking of football teams, *SIAM Rev*, 35:80–93

[86] Kelly FP (1979) *Reversibility and Stochastic Networks*. Wiley, New York

[87] Kennedy WJ Jr, Gentle JE (1980) *Statistical Computing*. Marcel Dekker, New York

[88] Kimura M (1980) A simple method for estimating evolutionary rates of base substitutions through comparative studies of nucleotide sequences. *J Mol Evol* 16:111–120

[89] Kingman JFC (1993) *Poisson Processes*. Oxford University Press, Oxford

[90] Kirkpatrick S, Gelatt CD, Vecchi MP (1983) Optimization by simulated annealing. *Science* 220:671–680

[91] Klain DA, Rota G-C (1997) *Introduction to Geometric Probability*. Cambridge University Press, Cambridge

[92] Körner TW (1988) *Fourier Analysis*. Cambridge University Press, Cambridge

[93] Lamperti J (1977) *Stochastic Processes. A Survey of the Mathematical Theory.* Springer-Verlag, New York

[94] Lange K (1982) Calculation of the equilibrium distribution for a deleterious gene by the finite Fourier transform. *Biometrics* 38:79-86

[95] Lange K (1999) *Numerical Analysis for Statisticians.* Springer-Verlag, New York

[96] Lange K (2002) *Mathematical and Statistical Methods for Genetic Analysis,* 2nd ed. Springer-Verlag, New York

[97] Lange K, Boehnke M (1982) How many polymorphic genes will it take to span the human genome? *Amer J Hum Genet* 34:842–845

[98] Lange K, Carson R (1984) EM reconstruction algorithms for emission and transmission tomography. *J Computer Assist Tomography* 8:306–316

[99] Lange K, Fessler JA (1995) Globally convergent algorithms for maximum a posteriori transmission tomography. *IEEE Trans Image Processing* 4:1430–1438

[100] Lange K, Gladstien K (1980) Further characterization of the long-run population distribution of a deleterious gene. *Theor Pop Bio* 18:31-43

[101] Lange K, Hunter DR, Yang I (2000) Optimization transfer using surrogate objective functions (with discussion). *J Comput Graph Statist* 9:1–59

[102] Lawler GF (1995) *Introduction to Stochastic Processes.* Chapman & Hall, London

[103] Lawler GF, Coyle LN (1999) *Lectures on Contemporary Probability.* American Mathematical Society, Providence, RI

[104] Lazzeroni LC, Lange K (1997) Markov chains for Monte Carlo tests of genetic equilibrium in multidimensional contingency tables. *Ann Statist* 25:138–168

[105] Li W-H, Graur D (1991) *Fundamentals of Molecular Evolution.* Sinauer Associates, Sunderland, MA

[106] Liggett TM (1985) *Interacting Particle Systems.* Springer-Verlag, New York

[107] Lindvall T (1992) *Lectures on the Coupling Method.* Wiley, New York

[108] Liu JS (1996) Metropolized independent sampling with comparisons to rejection sampling and importance sampling. *Statist Comput* 6:113–119

[109] Lotka AJ (1931) Population analysis—the extinction of families I. *J Wash Acad Sci* 21:377–380

[110] Lozansky E, Rousseau C (1996) *Winning Solutions.* Springer-Verlag, New York

[111] Luenberger DG (1984) *Linear and Nonlinear Programming,* 2nd ed. Addison-Wesley, Reading, MA

[112] McLeod RM (1980) *The Generalized Riemann Integral.* Mathematical Association of America, Washington, DC

[113] Metropolis N, Rosenbluth A, Rosenbluth M, Teller A, Teller E (1953) Equations of state calculations by fast computing machines. *J Chem Physics* 21:1087–1092

[114] Moler C, Van Loan C (1978) Nineteen dubious ways to compute the exponential of a matrix. *SIAM Rev* 20:801–836

[115] Newman DJ (1980) Simple analytic proof of the prime number theorem. *Amer Math Monthly* 87:693–696

[116] Niven I, Zuckerman HS, Montgomery HL (1991) *An Introduction to the Theory of Numbers.* Wiley, New York

[117] Norris JR (1997) *Markov Chains.* Cambridge University Press, Cambridge

[118] Perelson AS, Macken CA (1985) *Branching Processes Applied to Cell Surface Aggregation Phenomena.* Springer-Verlag, Berlin

[119] Peressini AL, Sullivan FE, Uhl JJ Jr (1988) *The Mathematics of Nonlinear Programming.* Springer-Verlag, New York

[120] Phillips AN (1996) Reduction of HIV concentration during acute infection: Independence from a specific immune response. *Science* 271:497–499

[121] Pollard JH (1973) *Mathematical Models for the Growth of Human Populations.* Cambridge University Press, Cambridge

[122] Powell MJD (1981) *Approximation Theory and Methods.* Cambridge University Press, Cambridge

[123] Press WH, Teukolsky SA, Vetterling WT, Flannery BP (1992) *Numerical Recipes in Fortran: The Art of Scientific Computing,* 2nd ed. Cambridge University Press, Cambridge

[124] Rao CR (1973) *Linear Statistical Inference and its Applications,* 2nd ed. Wiley, New York

[125] Reed TE, Neil JV (1959) Huntington's chorea in Michigan. *Amer J Hum Genet* 11:107–136

[126] Rényi A (1970) *Probability Theory*. North-Holland, Amsterdam

[127] Roman S (1984) *The Umbral Calculus*. Academic Press, New York

[128] Roman S (1997) *Introduction to Coding and Information Theory*. Academic Press, New York

[129] Rosenthal JS (1995) Convergence rates for Markov chains. *SIAM Rev* 37:387–405

[130] Ross SM (1996) *Stochastic Processes*, 2nd ed. Wiley, New York

[131] Royden HL (1968) *Real Analysis*, 2nd ed. Macmillan, London

[132] Rubinow SI (1975) *Introduction to Mathematical Biology*. Wiley, New York

[133] Rudin W (1964) *Principles of Mathematical Analysis,* 2nd ed. McGraw-Hill, New York

[134] Rushton AR (1976) Quantitative analysis of human chromosome segregation in man-mouse somatic cell hybrids. *Cytogenetics Cell Genet* 17:243–253

[135] Silver EA, Costa D (1997) A property of symmetric distributions and a related order statistic result. *Amer Statistician* 51:32–33

[136] Sedgewick R (1988) *Algorithms*, 2nd ed. Addison-Wesley, Reading, MA

[137] Seneta E (1973) *Non-negative Matrices: An Introduction to Theory and Applications*. Wiley, New York

[138] Solomon H (1978) *Geometric Probability*. SIAM, Philadelphia

[139] Spivak M (1965) *Calculus on Manifolds*. Benjamin/Cummings Publishing Co., Menlo Park, CA

[140] Steele JM (1997) *Probability Theory and Combinatorial Optimization*. SIAM, Philadelphia

[141] Stein C (1986) *Approximate Computation of Expectations*. Institute of Mathematical Statistics, Hayward, CA

[142] Tanner MA (1993) *Tools for Statistical Inference: Methods for Exploration of Posterior Distributions and Likelihood Functions*, 2nd ed. Springer-Verlag, New York

294   References

[143] Taylor HM, Karlin S (1984) *An Introduction to Stochastic Modeling.* Academic Press, Orlando, FL

[144] Tenenbaum G, France MM (2000) *The Prime Numbers and Their Distribution.* translated by Spain PG, American Mathematical Society, Providence, RI

[145] The Huntington's Disease Collaborative Research Group (1993). A novel gene containing a trinucleotide repeat that is expanded and unstable on Huntington's disease chromosomes. *Cell* 72:971–983

[146] Tuckwell HC (1989) *Stochastic Processes in the Neurosciences.* SIAM, Philadelphia

[147] Waterman MS (1995) *Introduction to Computational Biology: Maps, Sequences, and Genomes.* Chapman & Hall, London

[148] Weiss M, Green H (1967) Human-mouse hybrid cell lines containing partial complements of human chromosomes and functioning human genes. *Proc Natl Acad Sci USA* 58:1104–1111

[149] Wilf HS (1978) *Mathematics for the Physical Sciences.* Dover, New York

[150] Wilf HS (1986) *Algorithms and Complexity.* Prentice-Hall, New York

[151] Wilf HS (1990) *generatingfunctionology.* Academic Press, San Diego

[152] Williams D (1991) *Probability with Martingales.* Cambridge University Press, Cambridge

[153] Williams D (2001) *Weighing the Odds: A Course in Probability and Statistics.* Cambridge University Press, Cambridge

[154] Yee PL, Vyborný R (2000) *The Integral: An Easy Approach after Kurzweil and Henstock.* Cambridge University Press, Cambridge

[155] Zagier D (1997) Newman's short proof of the prime number theorem. *Amer Math Monthly* 104:705–708

Index

Springer Texts in Statistics (continued from page ii)

Madansky: Prescriptions for Working Statisticians

McPherson: Applying and Interpreting Statistics: A Comprehensive Guide, Second Edition

Mueller: Basic Principles of Structural Equation Modeling: An Introduction to LISREL and EQS

Nguyen and Rogers: Fundamentals of Mathematical Statistics: Volume I: Probability for Statistics

Nguyen and Rogers: Fundamentals of Mathematical Statistics: Volume II: Statistical Inference

Noether: Introduction to Statistics: The Nonparametric Way

Nolan and Speed: Stat Labs: Mathematical Statistics Through Applications

Peters: Counting for Something: Statistical Principles and Personalities

Pfeiffer: Probability for Applications

Pitman: Probability

Rawlings, Pantula and Dickey: Applied Regression Analysis

Robert: The Bayesian Choice: From Decision-Theoretic Foundations to Computational Implementation, Second Edition

Robert and Casella: Monte Carlo Statistical Methods

Rose and Smith: Mathematical Statistics with *Mathematica*

Santner and Duffy: The Statistical Analysis of Discrete Data

Saville and Wood: Statistical Methods: The Geometric Approach

Sen and Srivastava: Regression Analysis: Theory, Methods, and Applications

Shao: Mathematical Statistics

Shorack: Probability for Statisticians

Shumway and Stoffer: Time Series Analysis and Its Applications

Terrell: Mathematical Statistics: A Unified Introduction

Timm: Applied Multivariate Analysis

Toutenburg: Statistical Analysis of Designed Experiments, Second Edition

Whittle: Probability via Expectation, Fourth Edition

Zacks: Introduction to Reliability Analysis: Probability Models and Statistical Methods

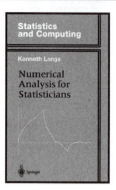
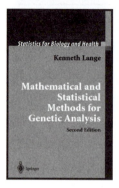
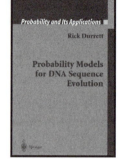